Sustainable Agri-Productivity

Sustainable agriculture is at a critical crossroads as global food demand rises alongside challenges such as soil degradation, climate change, and declining resource efficiency. *Sustainable Agri-Productivity: A Roadmap to Hunger Alleviation using Beneficial Microbes* addresses these pressing concerns by presenting microbes as key drivers of resilient and productive agroecosystems. This book aims to bridge the gap between fundamental microbial science and its practical applications in agriculture.

This volume stands out by offering a holistic and multidisciplinary approach, integrating crop productivity, soil health management, and emerging biotechnological innovations. It provides insights into how beneficial microbes can enhance nutrient availability, improve stress tolerance, and restore ecological balance across diverse farming systems, including stress-prone environments such as saline and arid regions.

Key Features:

- Role of beneficial microbes in major crops including cereals, legumes, and horticultural systems
- Microbial approaches for improving soil fertility and nutrient cycling
- Strategies for managing saline, arid, and water-limited agricultural conditions
- Advances in plant microbiome research and microbial biotechnology
- Application of precision agriculture and genome-based microbial interventions
- Development and use of biofertilizers and sustainable input systems

Overall, this book serves as a comprehensive resource for researchers, academicians, policymakers, and students in agriculture and allied sciences. It is also valuable for extension professionals and practitioners seeking sustainable, microbe-based solutions for enhancing productivity and ensuring global food security.

Sustainable Agri-Productivity

A Roadmap to Hunger Alleviation using Beneficial Microbes

Edited by

Monika Singh, Sandeep Kumar Singh, Ajay Kumar, and Manoj Kumar Solanki

CRC Press
Taylor & Francis Group
Boca Raton London New York

CRC Press is an imprint of the
Taylor & Francis Group, an **informa** business

Designed cover image: Getty Images

First edition published 2027
by CRC Press
2385 NW Executive Center Drive, Suite 320, Boca Raton FL 33431

and by CRC Press
4 Park Square, Milton Park, Abingdon, Oxon, OX14 4RN

CRC Press is an imprint of Taylor & Francis Group, LLC

Library of Congress CataloginginPublication Data
Names: Singh, Monika, (Professor of biotechnology) editor | Singh, Sandeep Kumar, (Microbiologist) editor | Kumar, Ajay, PhD editor | Solanki, Manoj Kumar editor
Title: Sustainable agri-productivity : a roadmap to hunger alleviation using beneficial microbes / edited by Monika Singh, Sandeep Kumar Singh, Ajay Kumar, Manoj Kumar Solanki.
Description: First edition. | Boca Raton, FL : CRC Press, 2026. | Includes bibliographical references and index.
Identifiers: LCCN 2026024778 (print) | LCCN 2026024779 (ebook) | ISBN 9781041123798 hardback | ISBN 9781041123804 paperback | ISBN 9781003664451 ebook
Subjects: LCSH: Sustainable agriculture | Agricultural productivity | Soil fertility | Plant-microbe relationships | Plant growth-promoting rhizobacteria | Drought-tolerant plants | Food security
Classification: LCC S494.5.S86 S8725 2026 (print) | LCC S494.5.S86 (ebook)
LC record available at https://lccn.loc.gov/2026024778
LC ebook record available at https://lccn.loc.gov/2026024779

ISBN: 9781041123798 (hbk)
ISBN: 9781041123804 (pbk)
ISBN: 9781003664451 (ebk)

DOI: 10.1201/9781003664451

Typeset in Times
by codeMantra

Contents

Preface vii
About Authors viii
List of Contributors x

Chapter 1 Endophytes Stimulate the Synthesis of Micronutrients and Function as Scavengers in the Soil 1

Migdalia Sarahy Navidad-Murrieta, Alicia Paulina Cárdenas-Castro, Saúl González-Moya, Edson Rayón-Díaz, Porfirio Gutiérrez-Martínez, Sonia Guadalupe Sáyago-Ayerdi, and Victor Manuel Zamora-Gasga

Chapter 2 Edible Plant Microbiome of Fruits and Vegetables: Research, Development, and Challenges 24

Ritu Singh and Prem Pratap Singh

Chapter 3 Molecular Aspects of Protecting and Promoting Plant Growth by Microbial Consortia 34

Alejandra Chacón-López, Adriana Aranda-Vázquez, José Luis Hernández-Flores, Ulises Miguel López-García, Efigenia Montalvo González, Rosa Isela Ortiz Basurto, María de Lourdes García-Magaña, and Selene Aguilera-Aguirre

Chapter 4 Elucidating Possibilities of Grasses and Their Allied Beneficial Microorganisms for Sustainable Agricultural Production 72

Shriniketan Puranik, Meghalamane Krishnamurthy Murali, Murtheppa Manoj, Priyanka Mulage, Revanna Mythra, Sandeep Kumar Singh, and Livleen Shukla

Chapter 5 Endophytic Bacteria as Natural Allies in Mitigating Abiotic Stress: Mechanisms, Applications, and Prospects for Sustainable Agriculture 103

Rachana Singh, Aditya Sinha Roy, Bhumika Singh, and Shivanshi Tyagi

Chapter 6 Harnessing Ethnobotany and Traditional Knowledge for Sustainable Agri-Productivity: The Role of Beneficial Microbes in Hunger Alleviation 124

Japneet Kaur Gandhi and Anuradha Kumari

Chapter 7 Battling Salinity: Microbial Soil Fertility Programs for Arid Lands 143

Neha Singh, Nisha Yadav, and Shaili Srivastava

Chapter 8 Enriching Tomorrow through Microbes: Unveiling Future Research Demands 174

Yashpal Singh, Vikram Singh, Monika Singh, Sanjana Bhat, Anushka Das, Arun Bhatt, and Rishendra Kumar

Chapter 9 Nurturing the Soil: Microbial Strategies for Soil and Plant Health 185

Anisha Rupashree, Nitin Swamy, and Punam Kumari

Chapter 10 Mycorrhizal Associations: Enhancing Nutrient Acquisition 210

Rajendra Prasad, Kunal Madhav, Shivani Mehra, Shweta Pandey, Neha Pundir, Palak Chauhan, and Shikha Thakur

Chapter 11 Plant Microbiome for Sustainable Agriculture 246

Charu Saklani, Aabir Chakraborty, and Anushka Pragati Lakra

Chapter 12 Microbial Solutions to Drought Agriculture 277

Nisha Yadav, Neha Singh, and Shaili Srivastava

Index 295

Preface

As the global human population is projected to exceed 9.7 billion by 2050, the challenge of ensuring food security has transitioned from a looming concern to an urgent international crisis. Currently, approximately 700 million people lack access to sufficient food, while nearly 2 billion face chronic nutritional deficiencies. Traditionally, agricultural productivity has relied heavily on the intensification of synthetic fertilizers and pesticides; however, while these practices once boosted yields, they have led to widespread environmental degradation, soil biodiversity loss, and significant health risks.

This book, *Sustainable Agri-Productivity: A Roadmap to Hunger Alleviation using Beneficial Microbes*, addresses this critical juncture by exploring the transformative potential of microbial technology. This book provides a comprehensive roadmap for utilizing beneficial microorganisms such as plant growth-promoting rhizobacteria to build a more resilient and eco-friendly food system. By harnessing natural processes such as atmospheric nitrogen fixation, mineral solubilization (including phosphorus, potassium, and zinc), and the modulation of growth-regulating phytohormones, these microbes offer a viable alternative to conventional agrochemicals.

Beyond nutrient acquisition, beneficial microbes serve as a vital defense system, enhancing plant resilience against climate change-induced stresses and suppressing phytopathogens through biocontrol mechanisms. This work emphasizes that integrating these biological solutions into agroecosystems is not merely an option but also a necessity to achieve United Nations Sustainable Development Goal 2: Zero Hunger.

This volume bridges the gap between laboratory-based microbial research and practical field applications, addressing the technical, regulatory, and scalability challenges that remain. It is intended for scientists, policymakers, and agricultural practitioners who seek to transition toward a future where productivity and sustainability coexist, ensuring that the world's most vulnerable populations are nourished through the power of the microbial world.

About Authors

Monika Singh is currently working as an assistant professor in the Department of Biotechnology, School of Applied and Life Sciences, Uttaranchal University, Dehradun, Uttarakhand, India. She completed her doctoral research in the Department of Botany, Institute of Science, Banaras Hindu University, Varanasi, India, on the theme *Plant–Microbe Interactions.* During her doctoral research, she was awarded the prestigious DST-INSPIRE Research Fellowship by the Department of Science and Technology (India).

She has published more than 45 scientific contributions in the form of research articles, review papers, patents, and book chapters in leading international journals and publishers. Her research interests primarily focus on environmental microbiology and microbial nanotechnology, with particular emphasis on endophytes associated with horticultural crops.

Sandeep Kumar Singh is currently working as a senior research fellow in the Division of Microbiology at the Indian Agricultural Research Institute, New Delhi, India. He obtained both his Bachelor's and Master's degrees in Botany from Banaras Hindu University, Varanasi, India, in 2016 and 2018, respectively.

He has been involved in multiple research projects at the Indian Agricultural Research Institute and has worked extensively on plant–microbe interactions, waste recycling, pesticide remediation, plant growth-promoting bacteria, plant disease management, and plant taxonomy. He has published more than 110 scientific contributions, including research articles, review papers, and book chapters in reputed international journals and publishers.

Ajay Kumar is currently working as an assistant professor at the Amity Institute of Biotechnology, Amity University, Noida, India. He completed his doctoral research in the Department of Botany, Institute of Science, Banaras Hindu University, Varanasi, India, in 2015, focusing on *Plant–Microbe Interactions.*

He recently completed his tenure as a visiting scientist at the Agricultural Research Organization, Volcani Center, Rishon LeZion, Israel. Over the course of his research career, he has published more than 300 scientific contributions, including research articles, review papers, books, and book chapters in leading international journals and publishers. His research interests include plant–microbe interactions, microbial biocontrol, postharvest management, endophytes associated with medicinal plants, and cyanobacteria–pesticide interactions.

Manoj Kumar Solanki, Ph.D. in Microbiology, currently serves as an associate professor at IES University, Bhopal, Madhya Pradesh, India. Previously, he worked as a scientist at the University of Silesia, Poland, under the NAWA program (2021–2023). He has also served as a visiting scientist at the Agricultural Research Organization, Volcani Center, Rishon LeZion, Israel (2016–2020), and at the Guangxi Academy of Agricultural Sciences, China (2013–2015). Earlier in his career, he worked as a senior research fellow at the Indian Council of Agricultural Research, India.

He obtained his M.Sc. from Barkatullah University in 2006 and Ph.D. from Rani Durgawati University in 2013. His research focuses on plant–microbe interactions, soil microbiology, and plant disease management, with particular emphasis on microbiomes, nutrient management, and microbial bioformulations. He has published over 80 high-impact research papers and 30 book chapters and has edited more than 10 books and special issues in journals such as *Frontiers in Microbiology*, *Microbiological Research*, and *BMC Plant Biology*.

Contributors

Selene Aguilera-Aguirre
Laboratorio Integral de Investigación en Alimentos/Laboratorio de Microbiología, Departamento de Ingeniería Química y Bioquímica
Instituto Tecnológico de Tepic
Tepic, Nayarit, México

Adriana Aranda-Vázquez
Laboratorio Integral de Investigación en Alimentos/Laboratorio de Microbiología, Departamento de Ingeniería Química y Bioquímica
Instituto Tecnológico de Tepic
Tepic, Nayarit, México

Rosa Isela Ortiz Basurto
Laboratorio Integral de Investigación en Alimentos/Laboratorio de Microbiología, Departamento de Ingeniería Química y Bioquímica
Instituto Tecnológico de Tepic
Tepic, Nayarit, México

Sanjana Bhat
School of Medical and Allied Sciences (Emversity)
K.R. Mangalam University
Gurugram, Haryana, India

Arun Bhatt
Department of Biotechnology
G.B. Pant Institute of Engineering and Technology
Pauri Garhwal, Uttarakhand, India

Alicia Paulina Cárdenas-Castro
Laboratorio Nacional Conahcyt de Apoyo a la Evaluación de Productos Bióticos (LaNAEPBi)
Unidad de Servicio Tecnológico Nacional de México/I.T. de Durango
Durango, Dgo., México

Alejandra Chacón-López
Laboratorio Integral de Investigación en Alimentos/Laboratorio de Microbiología, Departamento de Ingeniería Química y Bioquímica
Instituto Tecnológico de Tepic
Tepic, Nayarit, México

Aabir Chakraborty
Department of Forensic Science
School of Allied Sciences, Devbhoomi Uttarakhand University
Manduwala, Naugaon, Uttarakhand, India

Palak Chauhan
Department of Biotechnology
School of Applied and Life Sciences, Uttaranchal University
Dehradun, Uttarakhand, India

Anushka Das
ICAR-Central Inland Fisheries Research Institute
Barrackpore, West Bengal, India

Japneet Kaur Gandhi
Institute of Science
Nirma University
Gota, Ahmedabad, Gujarat, India

María de Lourdes García-Magaña
Laboratorio Integral de Investigación en Alimentos/Laboratorio de Microbiología, Departamento de Ingeniería Química y Bioquímica
Instituto Tecnológico de Tepic
Tepic, Nayarit, México

Efigenia Montalvo González
Laboratorio Integral de Investigación en Alimentos/Laboratorio de Microbiología, Departamento de Ingeniería Química y Bioquímica
Instituto Tecnológico de Tepic
Tepic, Nayarit, México

Saúl González-Moya
Laboratorio Nacional Conahcyt de Apoyo a la Evaluación de Productos Bióticos (LaNAEPBi)
Unidad de Servicio Tecnológico Nacional de México/I.T. de Tepic
Tepic, Nayarit, México

Porfirio Gutiérrez-Martínez
Laboratorio Nacional Conahcyt de Apoyo a la Evaluación de Productos Bióticos (LaNAEPBi)
Unidad de Servicio Tecnológico Nacional de México/I.T. de Tepic
Tepic, Nayarit, México

José Luis Hernández-Flores
Departamento de Ingeniería Genética
Centro de Investigación y de Estudios Avanzados del IPN, Unidad Irapuato
Irapuato, Guanajuato, México

Rishendra Kumar
Department of Biotechnology
Sir J.C. Bose Campus, Bhimtal, Kumaun University
Nainital, Uttarakhand, India

Anuradha Kumari
School of Environmental Sciences
Jawaharlal Nehru University
New Delhi, India

Punam Kumari
Centre of Advanced Study in Botany
Institute of Science, Banaras Hindu University
Varanasi, Uttar Pradesh, India

Anushka Pragati Lakra
Department of Forensic Science
School of Allied Sciences, Devbhoomi Uttarakhand University
Manduwala, Naugaon, Uttarakhand, India

Ulises Miguel López-García
Laboratorio Integral de Investigación en Alimentos/Laboratorio de Microbiología, Departamento de Ingeniería Química y Bioquímica
Instituto Tecnológico de Tepic
Tepic, Nayarit, México

Kunal Madhav
Department of Microbiology
I.T.S. Dental College, Hospital and Research Centre
Greater Noida, Uttar Pradesh, India

Murtheppa Manoj
Department of Agricultural Microbiology
University of Agricultural Sciences, GKVK
Bengaluru, Karnataka, India

Shivani Mehra
Department of Biotechnology
School of Applied and Life Sciences, Uttaranchal University
Dehradun, Uttarakhand, India

Priyanka Mulage
Department of Agricultural Microbiology, College of Horticulture
University of Horticultural Sciences
Bagalkot, Bidar, India

Meghalamane Krishnamurthy Murali
Department of Agricultural Microbiology
University of Agricultural Sciences, GKVK
Bengaluru, Karnataka, India

Revanna Mythra
Division of Natural Resource Management
ICAR-Indian Agricultural Research Institute
Assam, India

Migdalia Sarahy Navidad-Murrieta
Laboratorio Nacional Conahcyt de Apoyo a la Evaluación de Productos Bióticos (LaNAEPBi)
Unidad de Servicio Tecnológico Nacional de México/I.T. de Tepic
Tepic, Nayarit, México

Neha Pundir
Department of Biotechnology
School of Applied and Life Sciences, Uttaranchal University
Dehradun, Uttarakhand, India

Shweta Pandey
Department of Pharmacy
Chandigarh College of Pharmacy (CCP), Landran
Mohali, Punjab, India

Rajendra Prasad
Department of Agriculture
School of Agriculture, Uttaranchal University
Dehradun, Uttarakhand, India

Shriniketan Puranik
Division of Basic Sciences
ICAR-Indian Institute of Pulses Research
Kanpur, Uttar Pradesh, India

Edson Rayón-Díaz
Laboratorio Nacional Conahcyt de Apoyo a la Evaluación de Productos Bióticos (LaNAEPBi)
Unidad de Servicio Tecnológico Nacional de México/I.T. de Tepic
Tepic, Nayarit, México

Aditya Sinha Roy
Amity Institute of Biotechnology
Amity University
Noida, Uttar Pradesh, India

Anisha Rupashree
Department of Biotechnology
St. Aloysius College
Jabalpur, Madhya Pradesh, India

Charu Saklani
Department of Forensic Science
School of Allied Sciences, Devbhoomi Uttarakhand University
Manduwala, Naugaon, Uttarakhand, India

Sonia Guadalupe Sáyago-Ayerdi
Laboratorio Nacional Conahcyt de Apoyo a la Evaluación de Productos Bióticos (LaNAEPBi)
Unidad de Servicio Tecnológico Nacional de México/I.T. de Tepic
Tepic, Nayarit, México

Livleen Shukla
Division of Microbiology
ICAR-Indian Agricultural Research Institute
New Delhi, India

Bhumika Singh
Amity Institute of Biotechnology
Amity University
Noida, Uttar Pradesh, India

Monika Singh
Department of Biotechnology
School of Applied and Life Sciences, Uttaranchal University
Dehradun, Uttarakhand, India

Neha Singh
Amity School of Earth and Environmental Sciences
Amity University
Gurugram, Haryana, India

Prem Pratap Singh
Department of Viticulture & Enology
University of California
Davis, California

Rachana Singh
Amity Institute of Biotechnology
Amity University
Noida, Uttar Pradesh, India

Ritu Singh
Department of Plant Science
University of California
Davis, California

Sandeep Kumar Singh
Division of Microbiology
ICAR-Indian Agricultural Research Institute
New Delhi, India

Vikram Singh
Department of Biotechnology
Amrapali University
Haldwani, Uttarakhand, India

Yashpal Singh
Department of Biotechnology
Amrapali University
Haldwani, Uttarakhand, India

Shaili Srivastava
Amity School of Earth and Environmental Sciences
Amity University
Gurugram, Haryana, India

Nitin Swamy
Department of Biotechnology
St. Aloysius College
Jabalpur, Madhya Pradesh, India

Shikha Thakur
Department of Biotechnology
School of Applied and Life Sciences, Uttaranchal University
Dehradun, Uttarakhand, India

Shivanshi Tyagi
Amity Institute of Biotechnology
Amity University
Noida, Uttar Pradesh, India

Nisha Yadav
Amity School of Earth and Environmental Sciences
Amity University
Gurugram, Haryana, India

Victor Manuel Zamora-Gasga
Laboratorio Nacional Conahcyt de Apoyo a la Evaluación de Productos Bióticos (LaNAEPBi)
Unidad de Servicio Tecnológico Nacional de México/I.T. de Tepic
Tepic, Nayarit, México

1 Endophytes Stimulate the Synthesis of Micronutrients and Function as Scavengers in the Soil

Migdalia Sarahy Navidad-Murrieta, Alicia Paulina Cárdenas-Castro, Saúl González-Moya, Edson Rayón-Díaz, Porfirio Gutiérrez-Martínez, Sonia Guadalupe Sáyago-Ayerdi, and Victor Manuel Zamora-Gasga

1.1 INTRODUCTION

Endophytic microorganisms inhabit the internal tissues of plants without causing visible symptoms of disease, establishing relationships that range from commensalism to mutualistic symbiosis (Strobel et al., 2004; Kashyap et al., 2023). Unlike pathogens, these microorganisms do not harm their hosts and, on the contrary, often provide multiple benefits such as enhanced plant growth, improved tolerance to abiotic stress, and the production of bioactive compounds with antimicrobial, antifungal, or insecticidal properties (Caruso et al., 2022; Strobel et al., 2004). While traditionally associated with fungi, the term "endophyte" now broadly includes both bacteria and fungi capable of forming endophytic associations in nearly all known plant species (Burragoni & Jeon, 2021; Lodewyckx et al., 2002). Colonization typically occurs through natural openings, such as root cracks or wounds, via horizontal transmission from the surrounding soil or, less commonly, vertically through seeds (Caruso et al., 2022; Kumari et al., 2022).

Endophytes differ from other plant-associated microorganisms, such as epiphytes or rhizospheric microbes, in their localization and type of interaction. As noted by Fadiji and Babalola (2020b), epiphytes colonize plant surfaces, while rhizospheric microbes inhabit the soil adjacent to roots. In contrast, endophytes colonize the interior of living plant tissues without visible damage. This intercellular or intracellular presence grants them access to host nutrients and signaling molecules, enabling

DOI: 10.1201/9781003664451-1

highly specialized metabolic exchanges (Govindasamy et al., 2018). Ecologically, endophytes are key players in the plant holobiont—the integrated unit of plant and microbiota that collectively responds to environmental stress (Eid et al., 2019). Taxonomically, endophytes encompass diverse genera: *Pseudomonas*, *Bacillus*, *Burkholderia*, and *Microbacterium*, and many *Ascomycota* and *Basidiomycota* fungi, often adapted to specific plant organs (Burragoni & Jeon, 2021; Lodewyckx et al., 2002; Pathak et al., 2022).

Within this context, endophytes have gained recognition as agents of sustainability and resilience in agriculture. Their interaction with host plants involves numerous functional mechanisms. One major pathway is plant growth promotion (PGP), achieved through phytohormone production (e.g., indole-3-acetic acid [IAA], gibberellins, cytokinins), solubilization of essential minerals such as phosphorus and potassium, iron acquisition via siderophores, and biological nitrogen fixation (Devi et al., 2020). These processes collectively enhance nutrient assimilation and stimulate root and shoot development, particularly under nutrient-poor conditions. In addition to supporting plant growth, endophytes influence soil health by contributing to its physical and microbial structure. Through the secretion of exopolysaccharides and other organic compounds, some endophytes promote the aggregation of soil particles, improving aeration, moisture retention, and cation exchange capacity (Omomowo & Babalola, 2019). These improvements reduce compaction and create more favorable conditions for plant roots and beneficial microbial interactions.

Endophytic microorganisms not only directly support plant vitality but also contribute to improved soil conditions and overall ecosystem balance. Their integration into agricultural systems represents a significant advance toward more sustainable, productive, and climate-resilient farming practices.

1.2 ENDOPHYTES AND MICRONUTRIENT MOBILIZATION

Endophytes comprise a diverse group of cryptic, often ubiquitous microorganisms that complete at least part of their life cycle inside plant tissues without eliciting visible signs of disease (Domka et al., 2019; Rashmi et al., 2019; Singh et al., 2022). Some are capable of facultative symbiosis, with life cycles that do not require strict synchronization with the host. This flexibility enables them to facilitate the mobilization and uptake of both macro- and micronutrients, offering a valuable advantage in the acquisition of otherwise inaccessible or immobile elements (Vishwakarma et al., 2021). Endophytes have been categorized into four groups based on host specificity, colonization patterns, modes of transmission, and biodiversity. These categories influence how endophytes promote plant health, either indirectly by enhancing soil formation or directly through increased tolerance to metal toxicity and improved biomass productivity (Domka et al., 2019; Rozpądek et al., 2018).

Micronutrients such as boron, copper, iron, manganese, molybdenum, nickel, and zinc are vital for plant development, participating in essential processes including photosynthesis, respiration, and protein synthesis (Kumar et al., 2021). Endophytes support micronutrient uptake through mechanisms such as phosphate solubilization, nitrogen fixation, production of IAA, and iron mobilization via siderophores (Villanueva et al., 2022). A notable biochemical process is the activity

TABLE 1.1
Enzymatic Activities and Metabolites Produced by Endophytic Microorganisms that Enhance Micronutrient Availability in the Soil

Enzyme or Metabolite	Mechanism	Impact on the Plant and Soil	References
Phosphatases	Hydrolyzes organic and inorganic phosphorus compounds, releasing soluble phosphate.	Enhances root development, flowering, and seed production by increasing phosphorus availability.	Bogas et al. (2022)
Phytases	Break down phytate, the main phosphorus storage form in seeds and grains.	Increase phosphorus availability to the plant from internal reserves.	Pires et al. (2019)
Siderophores	Molecules with high affinity for iron (Fe^{3+}) and other micronutrients.	Improve uptake of iron, cobalt, nickel, manganese, and molybdenum.	Adedayo and Babalola (2023)
Organic acids	Release of acids such as malic, succinic, oxalic, citric, and lactic acids by endophytes.	Solubilize zinc and manganese, enhance their absorption, and promote the release of glomalin, which stabilizes organic matter and serves as a bioindicator of soil quality.	Barajas González et al. (2024); Son et al. (2024)

of 1-aminocyclopropane-1-carboxylate (ACC) deaminase, an enzyme that converts ACC into α-ketobutyrate and ammonia, thereby reducing ethylene levels in host plants. This reduction supports root elongation and overall plant development under stress conditions (Chaudhary et al., 2022).

Endophytes also secrete enzymes that aid in the breakdown and solubilization of nutrients within the soil or plant tissues, enhancing nutrient availability and uptake (Ameen et al., 2024; Mehta et al., 2019; Waqar et al., 2024). Table 1.1 summarizes the major enzymatic and metabolic pathways by which endophytes facilitate micronutrient mobilization.

Together, these findings demonstrate the versatile and essential role of endophytes in improving plant nutrition and productivity, highlighting their growing importance in sustainable agricultural practices. Building upon these insights into nutrient acquisition, the next section explores the specific mechanisms by which endophytes mobilize or enhance micronutrient availability, emphasizing their biochemical versatility and adaptive strategies in the plant–soil interface.

1.3 MECHANISMS OF MICRONUTRIENT SYNTHESIS INDUCED BY ENDOPHYTES

Endophytes inhabit a wide range of plant tissues—including leaves, stems, roots, flowers, and fruits—while maintaining an asymptomatic presence (Aktar et al., 2023; Hurek & Reinhold-Hurek, 2003). Their ability to improve plant nutrition has been the subject of numerous recent studies. This section focuses on the mechanisms by which these microorganisms enhance nutrient availability and facilitate plant defense. Specifically, we explore how endophytes fix nitrogen, solubilize phosphorus and potassium, enhance iron uptake, and produce secondary metabolites like organic acids and volatile compounds with pest-repelling properties (I. Afzal et al., 2019; Najjar, 2025).

These plant–endophyte interactions, particularly during the early stages of development, are critical for sustainable agriculture and crop biofortification (Hurek & Reinhold-Hurek, 2003). Advances in plant microbiology have revealed that these relationships are dynamic and adaptive, supporting plants under nutritional stress. Endophytes are thus recognized not only for their ubiquity and functional diversity but also for their considerable biotechnological potential.

Recent studies have identified the genera *Bacillus*, *Pseudomonas*, *Enterobacter*, *Streptomyces*, and *Pantoea* as dominant in plant tissues during vegetative stages, with *Bacillus* being particularly prevalent. These endophytes demonstrate a wide array of PGP traits, including the production of phytohormones, antioxidants, and stress-responsive enzymes. Their metabolic outputs support adaptation to both biotic stress (e.g., diseases, phytoparasites) and abiotic stress (e.g., ultraviolet light, temperature extremes, water stress, salinity) (Figure 1.1), with potential applications extending into the food and pharmaceutical sectors (Aktar et al., 2023; Cui et al., 2024).

In addition, endophytes are increasingly viewed as a promising source of bioactive compounds. They produce antimicrobials and antioxidants that not only benefit

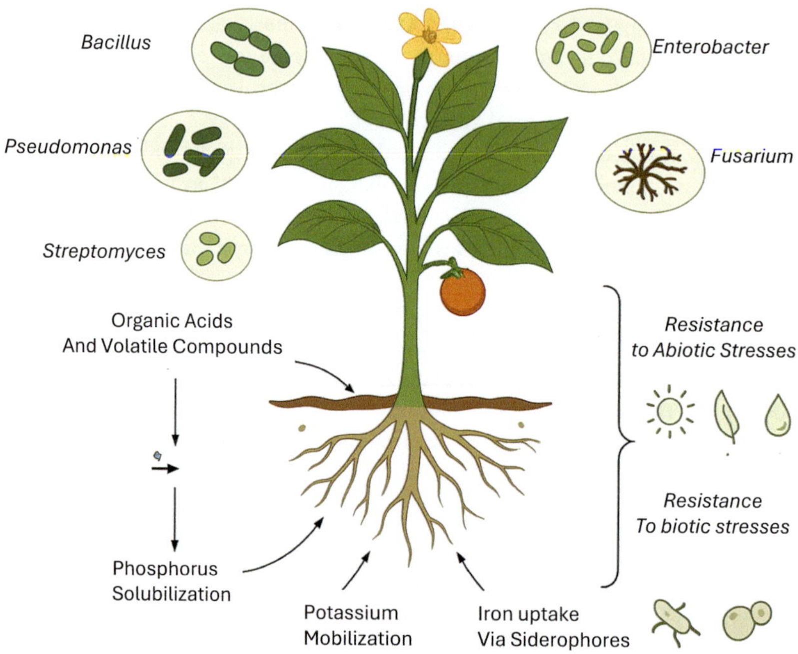

FIGURE 1.1 Mechanisms of micronutrient synthesis induced by endophytes.

plants but also reduce reliance on synthetic agrochemicals, contributing to safer and more sustainable food systems (Dias et al., 2009; Fadiji & Babalola, 2020a).

1.3.1 Biological Nitrogen Fixation: Key Biochemical Processes and Their Impact on Soil Fertility

The *nifH* gene encodes the dinitrogenase reductase subunit, a key component of the nitrogenase enzyme complex responsible for converting atmospheric nitrogen (N_2) into ammonium (NH_4^+), a bioavailable form for plant uptake. This process is crucial for sustainable agriculture as it reduces the dependence on synthetic nitrogen fertilizers. The expression of *nifH* is influenced by environmental factors such as oxygen levels, nitrogen availability, and signaling compounds from the host plant (Figure 1.2) (Gaby & Buckley, 2012).

Biological nitrogen fixation is among the most extensively studied functions of endophytic bacteria due to its potential to enhance plant growth in nitrogen-poor soils (Hurek & Reinhold-Hurek, 2003). The nitrogenase enzyme operates under anaerobic conditions and demands substantial ATP input. Endophytic and rhizospheric bacteria such as *Azospirillum*, *Herbaspirillum*, and some *Pseudomonas* species possess the genetic and physiological tools to carry out efficient nitrogen fixation (I. Afzal et al., 2019).

Cueva-Yesquén et al. (2020) demonstrated the ability of *Pseudomonas* strains to grow in nitrogen-free media, confirming nitrogen fixation potential through polymerase chain reaction (PCR) detection of the *nifH* gene in *Pseudomonas cichorii* EP220. This experiment, conducted in LGI medium under nitrogen-free and nitrogen supplemented conditions, highlighted the strain's capacity to function as a biofertilizer.

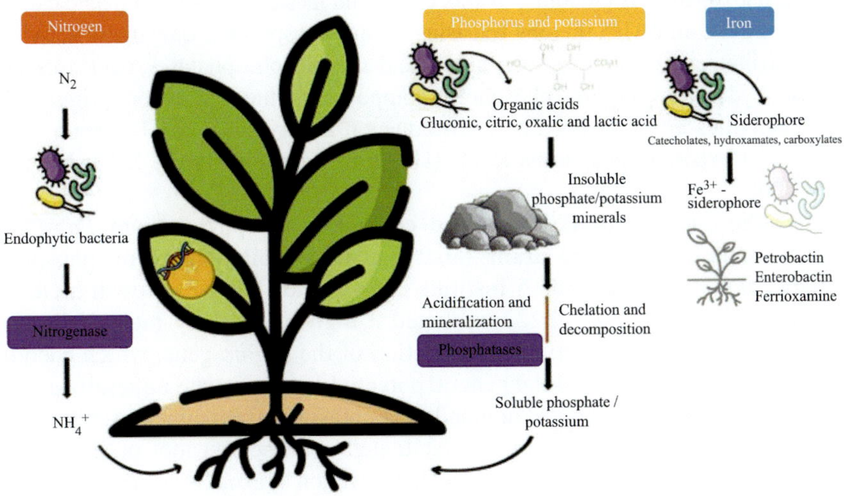

FIGURE 1.2 Key biochemical mechanisms of nutrient mobilization by endophytic microorganisms.

Evidence of nitrogen transfer from endophytes to plants has been provided using ^{15}N isotopic labeling, which confirmed nitrogen incorporation into plant tissues. Fadiji and Babalola (2020a) also observed that nitrogen-fixing endophytes positively influence root morphology, increasing root hair density and enhancing nutrient absorption. Similarly, Gupta et al. (2013) reported enhanced nitrogen accumulation in plants colonized by endophytic bacteria in nitrogen-limited soils. However, it is acknowledged that endophytes are generally less efficient nitrogen fixers than *Rhizobium* species associated with root nodules.

1.3.2 Phosphorus and Potassium Solubilization: Production of Enzymes and Organic Acids that Facilitate the Mobilization of These Nutrients

Endophytes can solubilize phosphates and other nutrients by producing organic acids (such as gluconic acid) and phosphatase enzymes that convert mineral and organic phosphorus into bioavailable forms. Additionally, bacterial exudates acidify the surrounding environment and release potassium from minerals, thereby increasing the availability of these nutrients to plants (I. Afzal et al., 2019; Hurek & Reinhold-Hurek, 2003).

The availability of phosphorus (P) and potassium (K) in agricultural soils is one of the most critical limiting factors in crop productivity. Although these elements are often abundant in total form, they frequently exist in chemically insoluble complexes that are inaccessible to plants. Both bacterial and fungal endophytes employ specialized biochemical mechanisms to solubilize these nutrients. Notably, nearly 75% of phosphorus applied as fertilizer forms stable complexes in the soil and becomes unavailable to crops (I. Afzal et al., 2019).

One of the main mechanisms involves the production of low-molecular-weight organic acids—such as gluconic, oxalic, citric, and lactic acids—that chelate cations like Ca^{2+}, Fe^{3+}, and Al^{3+}, thereby releasing bound phosphorus. Enzymatic activities also contribute, with enzymes such as acid and alkaline phosphatases, nucleases, and phytases hydrolyzing organic phosphorus compounds in the rhizosphere (Figure 1.2). These activities are influenced by environmental factors such as pH, carbon availability, and existing phosphorus levels (Hurek & Reinhold-Hurek, 2003; Nautiyal et al., 2000).

Cueva-Yesquén et al. (2020) tested cultivable endophytes from *Passiflora incarnata* and found that 33 strains (56.9%) could solubilize inorganic phosphate in National Botanical Research Institute's phosphate (NBRIP) growth medium. The strain *Pseudomonas* EP222 exhibited a high halo/colony ratio, indicating strong solubilization capacity. PCR detection of the *AcPho* gene, which encodes acid phosphatase, confirmed the genetic potential for phosphate mineralization.

Endophytes are also involved in mobilizing potassium from mineral sources such as feldspars and mica. This is primarily facilitated by the acidification of the environment through organic acid production and the secretion of exopolysaccharides, which disaggregate mineral structures. Some endophytes may also induce the expression of transporter proteins and ion channels in the host plant, aiding in the assimilation of mobilized potassium (I. Afzal et al., 2019).

Studies by Fadiji and Babalola (2020a) and Najjar (2025) have shown that *Bacillus* and *Enterobacter* strains isolated from wheat and tomato enhanced phosphorus and potassium availability under both field and greenhouse conditions. These increases were associated with higher biomass and yield, measured using spectrophotometry and atomic absorption spectrometry in sterilized and natural soil samples. The ability of endophytes to solubilize phosphorus and mobilize potassium provides a sustainable alternative to chemical fertilizers, enhancing crop nutrition while supporting environmentally friendly agricultural practices.

1.3.3 Siderophore Production: Release of Compounds that Enhance Iron Uptake in the Rhizosphere

Iron is an essential micronutrient involved in vital plant metabolic functions, including chlorophyll synthesis, electron transport, and redox enzyme activity. However, its bioavailability in soils is often limited due to its low solubility—particularly under alkaline or well-aerated conditions where ferric iron (Fe^{3+}) predominates in insoluble forms. To overcome this constraint, many endophytes have developed the capacity to produce siderophores—low-molecular-weight compounds that chelate Fe^{3+} and facilitate its transport toward plant roots as Fe^{3+}–siderophore complexes (Figure 1.2) (I. Afzal et al., 2019; Cueva-Yesquén et al., 2020; Cui et al., 2024; Ma et al., 2016).

Once secreted into the soil, siderophores bind ferric iron and transport it to the root surface, where it is subsequently taken up by the plant and utilized in physiological processes such as photosynthesis and respiration (Ma et al., 2016).

Siderophores are broadly categorized into three chemical classes (Ahmed & Holmström, 2014; Miethke & Marahiel, 2007; Renshaw et al., 2002):

- Catecholates, which utilize catechol or dihydroxybenzoic acid derivatives—such as enterobactin produced by *Escherichia coli*—are particularly effective under neutral to slightly acidic conditions.
- Hydroxamates, the most common among endophytic bacteria and fungi, include desferrioxamine and ferricrocin, synthesized by species such as *Streptomyces* and *Fusarium.*
- Carboxylates, which rely on carboxyl groups for iron coordination, are less efficient but particularly relevant in alkaline soils; examples include rhodotorulic acid and aerobactin-type siderophores.

Specific examples of siderophores include petrobactin (produced by some *Bacillus* species), enterobactin (produced by enteric bacteria and utilized by some *Pseudomonas* species), and ferrioxamines (by *Streptomyces*) (Miethke & Marahiel, 2007). These compounds form stable Fe^{3+}–siderophore complexes that are recognized and transported through specialized uptake systems present in both bacterial and plant cells (Ahmed & Holmström, 2014).

Once inside the plant—whether in the apoplast or cytoplasm—iron is released from the complex and becomes available for metabolic use (I. Afzal et al., 2019). Calvente et al. (2001) also demonstrated that media containing bacterial siderophores could inhibit the growth of phytopathogenic fungi, with antifungal activity correlating with siderophore concentration.

In a study using blue chrome azurol S (CAS) medium, Cueva-Yesquén et al. (2020) found that 77.6% of endophytic strains from *P. incarnata* produced detectable siderophores, as indicated by halos of orange discoloration. This method, based on competition for iron between the siderophores and the chromogenic dye chrome azurol S, enabled quantification of siderophore production. PCR detection of the *asb* gene confirmed the genetic basis for petrobactin biosynthesis in these strains.

Beyond nutrition, siderophores contribute to plant health by limiting pathogen access to iron—effectively acting as a biocontrol mechanism. Fadiji and Babalola (2020a) and Najjar (2025) observed that *Pseudomonas fluorescens* and *Bacillus subtilis* strains not only increased foliar Fe content in tomato and wheat but also reduced the incidence of diseases such as bacterial wilt and leaf blight.

Controlled inoculation studies have further demonstrated that plants associated with siderophore-producing endophytes exhibit enhanced iron accumulation, greater biomass, increased chlorophyll content, and improved photosynthetic efficiency (Cueva-Yesquén et al., 2020).

1.3.4 Secondary Metabolites and Absorption of Micronutrients: Interaction with Plant and Other Microorganisms

One of the most extensively studied mechanisms in endophytes is the biosynthesis of phytohormones such as IAA, gibberellins, and cytokinins. IAA is typically synthesized from tryptophan through pathways such as indole-3-pyruvate, tryptamine, and other microbial routes (I. Afzal et al., 2019; Glick, 2012). González Ortega-Villaizán et al. (2024) demonstrated that the endophytic fungus *Serendipita indica* not only produces IAA but also reprograms gene expression in *Arabidopsis thaliana* roots, particularly affecting *GH3.5* and *GH3.17* genes, which regulate IAA homeostasis. Furthermore, it inhibits the PIN2 transporter, altering the auxin gradient and promoting localized auxin accumulation. This, in turn, stimulates vigorous root development and enhances nutrient uptake by increasing root length, the number of lateral roots, and overall water and nutrient absorption efficiency (Dias et al., 2009).

Beyond phytohormones, endophytes produce a variety of bioactive compounds, including osmoprotectants, antioxidant enzymes such as superoxide dismutase, catalase, and peroxidases, as well as organic acids. These molecules help plants withstand abiotic stresses such as salinity, drought, extreme temperatures, and the presence of heavy metals. Among the five major plant hormones—abscisic acid, cytokinins, ethylene, gibberellins, and IAA—IAA and ethylene play central roles in plant–bacteria interactions (I. Afzal et al., 2019; Phetcharat & Duangpaeng, 2012).

Endophytes also secrete cell wall–degrading enzymes, including cellulases and pectinases, which facilitate internal colonization. For example, polygalacturonidase and endoglucanase from *Xylella fastidiosa* enlarge the pore size of pit membranes, suggesting that effective endophytes likely possess a tailored set of localized and confined cell wall–degrading enzymes (Aktar et al., 2023).

Aktar et al. (2023) identified several non-rhizobial bacterial species from the roots of *Corchorus capsularis*, with most isolates belonging to the genera *Bacillus*,

Staphylococcus, and *Ensifer*. In particular, *Bacillus cereus* and *Staphylococcus hominis* consortia enhanced growth phenotypes—such as root length, shoot length, germination rate, and biomass accumulation—in jute cultivars CVL-1, O-9897, and Robi-1. These effects were associated with increased expression of gibberellin-related genes, GA-20 and GA-03, which encode gibberellin oxidases.

Cui et al. (2024) compiled studies demonstrating that endophytic fungi like *Exophiala*, *Curvularia protuberata*, and strain FZT214 improved plant tolerance to cadmium, heat, and salinity in tomato and *Panicum* species. These benefits correlated with enhanced biomass production, reduced oxidative stress, and modulation of gene expression linked to cellular defense and ionic homeostasis. In addition to promoting growth and stress tolerance, endophytes are notable producers of metabolites with antimicrobial, antifungal, insecticidal, and even therapeutic properties for human health. These include alkaloids, flavonoids, terpenes, polyacetylenes, and phenolic compounds. Fadiji and Babalola (2020a) documented multiple cases of *Bacillus*, *Streptomyces*, and *Aspergillus* strains producing bioactive compounds that effectively inhibit pathogens such as *Fusarium* and *Cladosporium*. These substances were identified using high-performance liquid chromatography coupled with mass spectrometry, *in vitro* bioassays, and activity assessments on model plant leaves. Some of these compounds also exhibited antioxidant, anticancer, and anti-inflammatory activities, suggesting novel applications in pharmaceutical biotechnology.

The secondary metabolites produced by endophytes—ranging from phytohormones and antioxidants to alkaloids and terpenes—constitute a biochemical arsenal with diverse roles in plant development, defense, and resilience. These compounds, often synthesized through specific metabolic pathways such as the tryptophan route for auxins, play a key role in enhancing nutrient uptake and conferring protection against environmental stressors (Najjar, 2025).

1.4 SUBSTANCES FOR PROTECTION AGAINST PHYTOPATHOGENS

The presence of endophytic bacteria enhances host plant growth not only through nutrient-related mechanisms but also indirectly by suppressing plant pathogens and pests. These microorganisms produce a variety of bioactive compounds with antagonistic effects against pathogens, including antibiotics, toxins, siderophores, hydrolytic enzymes, and antimicrobial volatile organic compounds. Both bacterial and fungal phytopathogens can be effectively targeted by endophytic bacteria (I. Afzal et al., 2019; Sheoran et al., 2015). In particular, antifungal activity may result from the secretion of enzymes such as chitinases, proteases, and glucanases, which degrade key components of fungal cell walls (Zhang et al., 2012).

Additionally, endophytic bacteria can confer protection against fungal, bacterial, and viral pathogens through a process known as induced systemic resistance (ISR) (I. Afzal et al., 2019; Alvin et al., 2014). ISR functions by priming the plant's defense systems, enabling more rapid and effective responses to subsequent pathogenic or herbivorous attacks, even in tissues that were not directly exposed. This systemic protection is typically mediated by complex signaling networks involving salicylic acid (SA), jasmonic acid (JA), and ethylene (ET) pathways, which are central to the activation of ISR (Pieterse et al., 2012). Several genera, including *Bacillus*,

Pseudomonas, and *Serratia*, have been shown to trigger ISR and thereby strengthen the host plant's resistance to disease (I. Afzal et al., 2019; Pieterse et al., 2012).

As demonstrated throughout this chapter, understanding the mechanisms employed by endophytes has led to the development of innovative biofertilizers and integrated agricultural practices. These strategies leverage the natural benefits of endophytes to promote crop productivity while reducing reliance on chemical inputs, offering a more sustainable path for modern agriculture.

1.5 ENDOPHYTIC MICROORGANISMS AS SOIL SCAVENGERS

Endophytic microorganisms, long recognized for their roles in promoting plant growth and stress resilience, are now emerging as key agents in soil restoration through their scavenging capabilities. These microorganisms colonize plant tissues asymptomatically and interact with the rhizosphere and surrounding soil environment in multiple beneficial ways (Figure 1.3). Their activity contributes not only to plant health but also to the ecological balance and biogeochemical cycling within soils.

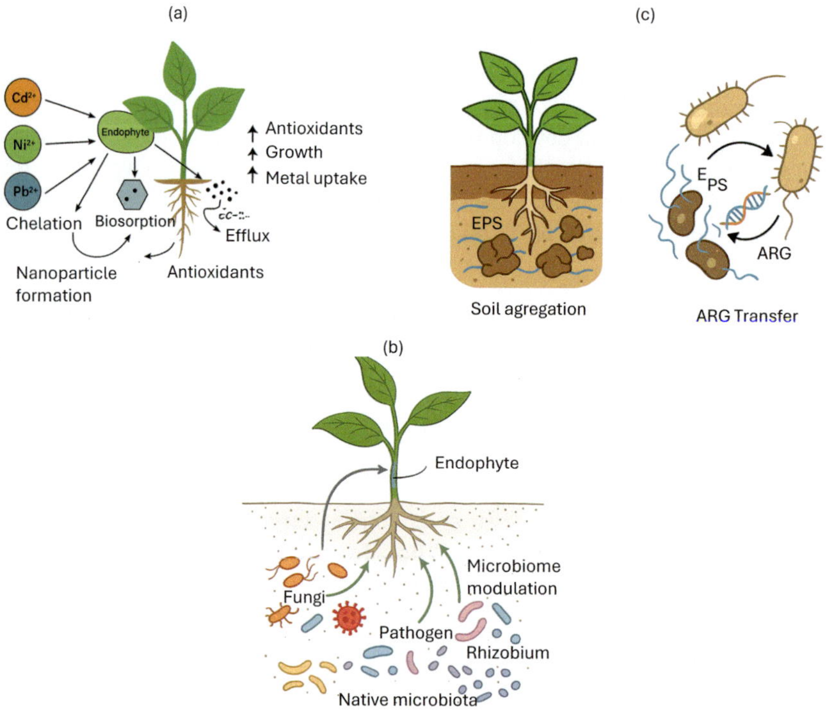

FIGURE 1.3 Functional mechanisms of endophytic microorganisms enhancing soil health and resilience: (a) heavy metal detoxification by endophytes via chelation, enzymatic transformation, and oxidative stress mitigation; (b) modulation of the rhizosphere microbiome through plant-mediated signaling and microbial cooperation; (c) secretion of extracellular polymeric substances (EPS) improving soil structure and influencing gene transfer dynamics.

1.5.1 Heavy Metal Detoxification: Biochemical Mechanisms and Their Application in Bioremediation

One of the most studied roles of endophytes is their ability to detoxify heavy metals, such as cadmium (Cd), nickel (Ni), and lead (Pb), through biochemical mechanisms involving chelation, enzymatic transformation, and oxidative stress mitigation. These mechanisms make endophytes valuable allies in microbial-assisted bioremediation, offering a sustainable alternative to traditional chemical treatments in contaminated soils. Recent studies have shown that *Enterobacter ludwigii* SAK5 and *Exiguobacterium indicum* SA22 can detoxify heavy metals through physiological and molecular mechanisms. These strains exhibited strong resistance to cadmium (Cd) and nickel (Ni), with high uptake capacities and improved plant growth under metal stress. Their action involved glutathione-mediated reactive oxygen species (ROS) reduction and downregulation of stress-related genes (*OsGST, OsMTP1, OsPCS1*). Inoculated plants also showed higher levels of proline and sugars, reduced abscisic acid, and increased SA, indicating a more balanced stress response (Jan et al., 2019). Similarly, endophytic fungi such as *Fusarium proliferatum* SL3 and *Aspergillus terreus* MGRF2 have demonstrated the ability to mitigate cadmium and nickel toxicity in *Solanum lycopersicum* by enhancing metal immobilization and protecting root cell architecture. Treated plants exhibited improved biomass, restored chlorophyll and carotenoid levels, and increased antioxidant activity, including flavonoid accumulation and regulated glutathione levels. Moreover, fungal inoculation modulated stress-related hormone levels, reducing abscisic acid while enhancing salicylic and gibberellic acid pathways, contributing to improved physiological performance and metal stress tolerance (Khan et al., 2025). According to Y. Zhang et al. (2024), *Cupriavidus metallidurans* CML2 is a newly identified strain with exceptional cadmium (Cd) resistance and PGP traits. Capable of tolerating up to 13.1 mM Cd, CML2 detoxifies the metal through biosorption, nanoparticle formation, periplasmic immobilization, and efflux systems involving genes such as *copA*, *czcE*, and *czcCBA*. It also enhances root and shoot development in rice by producing IAA, solubilizing phosphorus, and releasing siderophores. Shahzad et al. (2025) demonstrated that endophytic *Bacillus thuringiensis*, especially when combined with *Salix alba* root powder, significantly enhances wheat tolerance to cadmium stress by improving growth, nutrient assimilation, and antioxidant defense. The combined treatment improved protein, proline, and sugar contents, while reducing Cd accumulation by up to 80% compared to non-inoculated controls. Additionally, key antioxidant enzymes—including superoxide dismutase (SOD), peroxidase (POD), catalase (CAT), and ascorbate peroxidase (APX)—were upregulated, indicating enhanced oxidative stress mitigation.

Together, these findings highlight the crucial role of endophytic microorganisms in mitigating heavy metal toxicity through a combination of biochemical and physiological strategies. By reducing metal uptake, enhancing antioxidant defenses, and promoting plant growth, these microbes offer a practical and eco-friendly alternative to conventional remediation approaches. Their multifaceted actions position them as valuable allies in the development of sustainable solutions for agriculture in metal-contaminated environments.

1.5.2 Endophyte–Soil Microbiome Interactions: Beyond Internal Colonization

The role of endophytes in plant health extends well beyond their internal colonization. Recent studies have highlighted their capacity to shape the surrounding microbial environment, particularly in the rhizosphere, where complex microbial communities interact with both plant roots and external inoculants. These interactions are not limited to direct antagonism against pathogens but involve a broader modulation of microbial dynamics, often enhancing the resilience of plants and contributing to soil ecosystem stability. One compelling example is *Metarhizium brunneum*, an endophytic fungus known for its biocontrol potential. In tomato plants, *M. brunneum* significantly reduced spider mite populations. However, this protective effect diminished in the presence of native soil microbiota, suggesting that plant-mediated signaling or microbial interference may influence the outcome of such inoculations (Rasool et al., 2025). Interestingly, the fungus also reshaped the rhizosphere microbiota—suppressing dominant fungal genera while promoting beneficial bacteria like *Pseudomonas* and *Rhizobium*. These community shifts occurred even in the absence of direct contact, likely triggered by changes in root exudation patterns induced by the endophyte. Such findings underscore the importance of indirect plant–microbe–microbe signaling in mediating rhizosphere outcomes.

Similarly, Tang et al. (2024) demonstrated that in *Berchemia polyphylla* var. *leioclada*, bacterial and fungal genera were frequently shared between endophytic and rhizospheric compartments. This overlap implies a dynamic exchange of microbes and signaling cues between internal plant tissues and the external soil environment. Network analysis further revealed that, while the soil microbiome exhibited higher structural complexity and resilience, the presence of endophytes subtly influenced its organization—suggesting that endophytic colonization can modulate microbial interactions even without dominating the microbial landscape.

The introduction of non-native endophytes can also reshape native microbial communities in meaningful ways. A notable case is strain RF67, a member of the *Rhizobium* group, which was initially isolated from *Vaccinium angustifolium* roots. Although RF67 showed limited persistence in tomato roots, its inoculation induced clear shifts in microbial structure and diversity within both root and soil environments (Yurgel et al., 2022). The presence of RF67 correlated with increased bacterial richness and the enrichment of families like Enterobacteriaceae and Rhizobiaceae, both of which are associated with PGP activities. What stands out from co-occurrence network analyses is that RF67 inoculation fostered stronger microbial cooperation in roots, with Proteobacteria, Actinobacteriota, and Bacteroidota forming tightly connected networks. These interactive networks persisted even after the initial establishment phase, suggesting that transient colonizers can initiate longer-lasting structural changes in the microbiome. In some cases, these shifts promoted the dominance of beneficial taxa, even if the inoculant itself was no longer present in significant numbers (Yurgel et al., 2022).

Altogether, these studies reinforce a more nuanced view of endophyte–soil microbe interactions. Rather than merely competing for niche space, endophytes may coordinate with native microbes—modulating signaling, promoting beneficial assemblages, and enhancing the functional capacity of the microbial ecosystem. This cooperative dynamic plays a pivotal role in helping plants respond to environmental pressures, underlining the potential of endophytic inoculants as tools for sustainable agriculture.

1.5.3 The Dual Role of Microbial EPS in Soil Health and Gene Transfer

Endophytic microorganisms also play a pivotal role in improving soil structure by releasing extracellular polymeric substances (EPS)—complex mixtures of polysaccharides, glycoproteins, and other macromolecules. These secretions promote the aggregation of soil particles, enhancing key physical properties such as water retention, aeration, and root penetration, all of which contribute to a more stable and fertile soil environment. Exopolysaccharides secreted by soil bacteria have emerged as critical agents in the immobilization of heavy metals and the enhancement of soil structure. These complex polymers, rich in functional groups such as carboxyl, phosphate, and hydroxyl, bind positively charged metal ions like Cd^{2+} and Pb^{2+}, facilitating their precipitation and reducing their mobility and bioavailability. In a study using lettuce (*Lactuca sativa*) as the model plant, EPS-producing *Bacillus* strains Z23 and Z39 achieved over 90% removal efficiency of Cd and Pb through biosorption and EPS-mediated mineralization. Inoculation with these strains reduced heavy metal accumulation in lettuce roots and leaves, while simultaneously increasing soil polysaccharide content and promoting the formation of stable microaggregates in the rhizosphere (H. Zhang et al., 2024). In wheat grown under lead (Pb) stress, the application of *Aspergillus flavus* AUMC 16068 and its EPS led to marked improvements in plant performance. These fungal EPS—rich in functional groups such as hydroxyl, carboxyl, phosphate, and amine—acted as natural chelators, binding Pb^{2+} ions and reducing their bioavailability. As a result, plants treated with EPS accumulated less Pb in shoots and grains, with higher retention in roots, minimizing systemic toxicity. Physiologically, EPS-treated plants showed enhanced biomass, elevated carbohydrate and proline levels, and increased activity of antioxidant enzymes like CAT and POD, which are vital for mitigating oxidative stress (El-khawaga et al., 2024). In a study focused on enhancing drought resilience in maize, two exopolysaccharide-producing bacterial strains—*Pseudomonas aeruginosa* EPB9 and *B. cereus* EPB17—were isolated from arid soils in Pakistan and evaluated for their biochemical properties and agronomic benefits (Naseem et al., 2024). EPB9, a Gram-negative and catalase-positive strain, produced a higher EPS yield (1.62 g/L) than the Gram-positive EPB17 (1.20 g/L). Both strains exhibited phosphate solubilizing abilities and formed mucoid colonies rich in functional groups such as hydroxyl, carboxyl, and sulfate esters. When applied to maize seeds, particularly as 10-day-old inocula, these bacteria improved soil moisture retention (up to 24.9% with EPB9), increased relative water content in leaves, enhanced shoot and root biomass, and boosted leaf protein and sugar levels under drought conditions. Notably, EPS treatments also mitigated oxidative stress by reducing proline accumulation and the activity of antioxidant enzymes (SOD, POD, CAT), suggesting a stabilizing effect on plant metabolism during water deficit (Naseem et al., 2024).

However, beyond their beneficial role in soil remediation, EPS may also influence microbial dynamics in less favorable ways. Certain EPS fractions—specifically tightly bound (TBEPS), loosely bound, and soluble EPS—have been shown to significantly enhance the horizontal transfer of antibiotic resistance genes through bacterial conjugation. TBEPS, in particular, increased transfer frequency by over 15-fold, a phenomenon linked to the redox activity and molecular properties of the polymers. These characteristics modulate microbial communication, oxidative stress regulation, and energy production, facilitating gene exchange in dense microbial environments such as activated sludge (Wang et al., 2024).

1.6 APPLICATIONS IN SUSTAINABLE AGRICULTURE: RECENT RESEARCH, USE OF BIOFERTILIZERS, AND REMEDIATION OF DEGRADED SOILS

Innovative strategies to enhance crop productivity and sustainability are needed due to the increasing global population (Albahri et al., 2023). In general, the endophytes have been shown to increase biomass and root length. Their critical role in nutrient acquisition includes solubilizing phosphate and fixing atmospheric nitrogen, which makes these essential nutrients more bioavailable to plants and improves crop yields (Mishra et al., 2020).

Endophytes also contribute to plant stress tolerance. Endophytes mitigate the effects of abiotic stresses (salinity, drought, and heavy metal contamination) by producing stress-related enzymes and metabolites (Sena et al., 2024).

The utilization of endophytes is an important sustainable alternative to chemical pesticides and fertilizers since studies have reported their natural ability to suppress plant pests and pathogens (M. Afzal et al., 2019). In general, the microbial strains that comprise endophytic communities can be used as biofertilizers helping promote plant growth and the recovery of diseased plants, protecting plants from stress, or acting as biocontrol agents (Yadav et al., 2021).

However, challenges remain in understanding the long-term impacts on agricultural ecosystems and in maintaining specific endophytic communities (Sena et al., 2024). Also, Ramírez-Cariño et al. (2024) reviewed *in vivo* studies evaluating rhizobacterial biofertilization in plants, seedlings, and final crop products, concluding that this approach represents a sustainable alternative for addressing current global agricultural challenges. Table 1.2 summarizes recent research about the mechanisms by which endophytes protect plants from pathogens or pests and promote healthy crop growth. For instance, certain bacterial endophytes can exploit the hyphae of fungal pathogens as conduits to access plant roots from the soil, thereby serving as protective agents against infection (Palmieri et al., 2020). Besides, the root endophytic fungus *S. indica* secretes a histidine-rich protein, which not only improves its ability to access micronutrients but also modulates oxidative stress and reactive oxygen homeostasis to enhance host colonization (Nostadt et al., 2020). Additionally, the endophytic rhizobacterium *Paraburkholderia phytofirmans* PsJN has been reported to induce the upregulation of cold stress-related genes in grapevine (Chen et al., 2022). Similarly, the inoculation of rice plants with the endophytic bacterium

TABLE 1.2
Endophyte-Mediated Mechanisms of Plant Protection

Endophyte Species	Host Plant	Target	Effects	Mechanisms	References
Bacillus velezensis, *Bacillus megaterium*, and *Herbaspirillum huttiense*	Tomato (*Solanum lycopersicum*)	*Rhizoctonia solani*	Reduced root-rot disease and improved plant protection	Antifungal activity, production of indole acetic acid, gibberellic acid, siderophores, and phosphate solubilization	Abbas et al. (2024)
Bacillus subtilis A9	Morel mushroom (*Morchella esculenta*)	*Lecanicillium aphanocladii*	Reduced rot disease development	Antimicrobial activity, inhibition of pathogen growth, and activation of defense-related responses	Chen et al. (2024)
Aspergillus flavus	Potato (*Solanum tuberosum*)	Root-knot nematodes (*Meloidogyne* spp.)	Reduced nematode survival, root penetration, and disease progression	Nematicidal activity of fungal culture filtrates and induction of systemic resistance	Ghareeb et al. (2024)
Endophytic *Bacillus* strains	Tomato (*Solanum lycopersicum*)	*Fusarium oxysporum* f. sp. *lycopersici*	Reduced Fusarium wilt severity and supported plant growth	Direct and indirect antagonism, pathogen suppression, and plant growth-promoting activity	Viana et al. (2024)

Brevibacterium linens RS16 reduces ethylene emissions, a common stress hormone, due to its ACC deaminase activity. This inoculation also enhances the expression of small heat shock proteins, which contribute to improved stress resilience (Choi et al., 2022). Moreover, *P. phytofirmans* PsJN has been demonstrated to positively influence root system architecture in *A. thaliana* under both normal and high-temperature conditions, promoting root expansion and thereby improving access to nutrients and water while mitigating abiotic stresses such as heat and drought (Macabuhay et al., 2022). The co-existence of heat and salinity stress can significantly impair crop productivity, but the inoculation of soybean (*Glycine max* L.) plants with the endophytic fungus *Trichoderma virens* SB10, in conjunction with glycine betaine (GB) treatments, has been shown to confer substantial tolerance to these stresses. In the presence of GB, *T. virens* SB10 enhances the synthesis of key hormones such as gibberellins, IAA, and SA, while reducing proline accumulation and Na^+ uptake. Furthermore, co-treatment with the fungus and GB improves the uptake of essential macronutrients, including nitrogen (N), calcium (Ca), and potassium (K). Modulation of *GmHKT1* and *GmSOS1* gene expression, both of which are critical for salt tolerance (Singh & Roychoudhury, 2021), results in the maintenance of a high K^+/Na^+ ratio, ultimately leading to enhanced growth rates and antioxidant defense mechanisms. These responses include the enhancement of antioxidant defense components such as ascorbate peroxidases (APX), superoxide dismutases (SOD), peroxidases (POD), and reduced glutathione (GSH) (Bilal et al., 2023).

By 2029, a 50% increase in food production will be required to meet the rising food demand driven by population growth (Chittora et al., 2020). Thus, the intensive use of chemical fertilizers may further damage agricultural land and reduce soil nutrient availability (Ghorai & Ghosh, 2022). Recently, Bibi et al. (2024) reported that cyanobacteria produce metabolites that improve nutrient bioavailability and soil status, while also helping plants defend against abiotic and biotic stress. In addition, cyanobacteria are among the first microorganisms to form and colonize biological soil crusts, which are involved in ecological succession. Therefore, addressing the knowledge gap related to climate change and its impact on terrestrial cyanobacteria is essential, particularly as increasing attention is being directed toward their use as biofertilizers in sustainable agriculture.

1.7 PROSPECTS AND PROJECTIONS

Endophytic microorganisms represent a key strategy in promoting plant growth and improving crop nutrition. Through mechanisms such as biological nitrogen fixation, phosphate and potassium solubilization, iron acquisition through siderophores, and the production of secondary metabolites, these symbionts not only optimize the use of soil resources but also strengthen plant resistance to stress conditions.

The scientific evidence gathered shows that endophytes possess a remarkable functional versatility, supported by specialized metabolic pathways and molecular adaptations that can be exploited in agricultural biotechnology.

Plant-beneficial endophytic bacteria have great potential for agricultural applications. Selected microbiomes can be used as bioinoculants to reduce agrochemical use, control pathogens, produce bioactive compounds, remediate contaminated soils, and improve micronutrient biofortification. In addition, the development of carrier materials for the preservation and application of plant growth-promoting bacteria, as well as "plant probiotics" technologies, may support their use in nurseries and agroindustry as biofertilizers and/or biopesticides (Aktar et al., 2023).

In addition, emerging developments in omics and gene editing open new opportunities for the rational design of microbial consortia aimed at improving the productivity, sustainability, and resilience of agroecosystems. The potential of endophytes is not limited to agriculture; their bioactive metabolites have pharmacological uses that could revolutionize sectors such as medicine and the functional food industry.

Although many endophytic microorganisms have been identified through metagenomics, which provides access to the genetic diversity of microbial communities without the need for cultivation, transcriptomics, and proteomics are also required to identify gene expression and activity patterns under specific plant–microbe interaction conditions. Together, these tools enable the rational design of microbial consortia with complementary functions, optimized for particular agronomic needs. However, further development is still required to obtain products with consistent and reproducible performance under field conditions.

This is especially useful in the context of climate change, where crops need to be more resilient to water, thermal, and nutritional variability. Current efforts are aimed at scaling up the production of bioinoculants based on multiresistant and multifunctional endophytes, validated under actual field conditions and adjusted to different agroecological zones.

Finally, the exploration of the therapeutic potential of metabolites produced by endophytes is of increasing interest, and compounds derived from plant endophytes, particularly medicinal plants, are being evaluated for pharmaceutical applications. This positions endophytes not only as an allies in agriculture but also as a valuable source for medical and pharmaceutical biotechnology.

1.8 CONCLUSION

Endophytic microorganisms have proven to be essential allies in sustainable agriculture, not only by promoting plant growth and nutrient uptake but also by enhancing soil structure and ecosystem stability. Their ability to fix nitrogen, solubilize phosphorus and potassium, produce siderophores, and secrete beneficial metabolites positions them as multifunctional agents that reduce dependence on chemical inputs. Moreover, their interactions with native soil microbes—whether through modulation of microbial communities or secretion of compounds like EPS—contribute to improved plant resilience under stress conditions, including heavy metal contamination and drought.

Importantly, their role goes beyond nutrient cycling. Endophytes also act as biocontrol agents and soil scavengers, influencing microbial networks in ways that benefit the entire plant–soil system. These qualities make them valuable tools for ecological restoration and climate-resilient farming practices.

Looking ahead, future research should focus on the rational design of endophytic consortia tailored to specific crops and environments. Advances in omics, bioinformatics, and gene editing will be key to unlocking their full potential—allowing us to engineer microbiomes that improve nutrient efficiency, boost crop yields, and reduce environmental impact. At the same time, field validation and regulatory frameworks will be essential to bring lab discoveries into real-world application. Endophytes are not just a promising alternative; they may well become a cornerstone of the next generation of regenerative agriculture.

REFERENCES

Abbas, M. M., Ismael, W. H., Mahfouz, A. Y., Daigham, G. E., & Attia, M. S. (2024). Efficacy of endophytic bacteria as promising inducers for enhancing the immune responses in tomato plants and managing Rhizoctonia root-rot disease. *Scientific Reports*, *14*(1), 1331. https://doi.org/10.1038/s41598-023-51000-8

Adedayo, A. A., & Babalola, O. O. (2023). Fungi that promote plant growth in the Rhizosphere boost crop growth. *Journal of Fungi*, *9*(2), 239. https://doi.org/10.3390/jof9020239

Afzal, I., Shinwari, Z. K., Sikandar, S., & Shahzad, S. (2019). Plant beneficial endophytic bacteria: Mechanisms, diversity, host range and genetic determinants. *Microbiological Research*, *221*, 36–49. https://doi.org/10.1016/j.micres.2019.02.001

Afzal, M., Rehman, K., Shabir, G., Tahseen, R., Ijaz, A., Hashmat, A. J., & Brix, H. (2019). Large-scale remediation of oil-contaminated water using floating treatment wetlands. *NPJ Clean Water*, *2*(1), 3.

Ahmed, E., & Holmström, S. J. M. (2014). Siderophores in environmental research: Roles and applications. *Microbial Biotechnology*, *7*(3), 196–208. https://doi.org/10.1111/1751-7915.12117

Aktar, N., Hossain, M. S., Amin, M. R., Ahmed, R., Ahmed, B., Ullah, M. W., Hossain, Q. M., & Islam, M. S. (2023). Endophytic bacteria improve jute growth through induction of phytohormones-related gene expression. *Physiological and Molecular Plant Pathology*, *127*, 102115. https://doi.org/10.1016/j.pmpp.2023.102115

Albahri, G., Alyamani, A. A., Badran, A., Hijazi, A., Nasser, M., Maresca, M., & Baydoun, E. (2023). Enhancing essential grains yield for sustainable food security and bio-safe agriculture through latest innovative approaches. *Agronomy*, *13*(7), 1709.

Alvin, A., Miller, K. I., & Neilan, B. A. (2014). Exploring the potential of endophytes from medicinal plants as sources of antimycobacterial compounds. *Microbiological Research*, *169*(7), 483–495. https://doi.org/10.1016/j.micres.2013.12.009

Ameen, M., Mahmood, A., Sahkoor, A., Zia, M. A., & Ullah, M. S. (2024). The role of endophytes to combat abiotic stress in plants. *Plant Stress*, *12*, 100435. https://doi.org/10.1016/j.stress.2024.100435

Barajas González, J. A., de la Rosa, Y. E. K., Carrillo-González, R., González-Chávez, M. d. C. Á., Hidalgo Lara, M. E., Soto Hernández, R. M., & Herrera Cabrera, B. E. (2024). NaCl modifies biochemical traits in bacterial endophytes isolated from halophytes: Towards salinity stress mitigation using consortia. *Plants*, *13*(12), 1626. https://doi.org/10.3390/plants13121626

Bibi, S., Saadaoui, I., Bibi, A., Al-Ghouti, M., & Abu-Dieyeh, M. H. (2024). Applications, advancements, and challenges of cyanobacteria-based biofertilizers for sustainable agro and ecosystems in arid climates. *Bioresource Technology Reports*, *25*, 101789.

Bilal, S., Shahzad, R., Asaf, S., Imran, M., Al-Harrasi, A., & Lee, I.-J. (2023). Efficacy of endophytic SB10 and glycine betaine duo in alleviating phytotoxic impact of combined heat and salinity in Glycine max L. via regulation of redox homeostasis and physiological and molecular responses. *Environmental Pollution*, *316*, 120658.

Bogas, A. C., Cruz, F. P. N., Lacava, P. T., & Sousa, C. P. (2022). Endophytic fungi: An overview on biotechnological and agronomic potential. *Brazilian Journal of Biology*, *84*, e258557. https://doi.org/10.1590/1519-6984.258557

Burragoni, S. G., & Jeon, J. (2021). Applications of endophytic microbes in agriculture, biotechnology, medicine, and beyond. *Microbiological Research*, *245*, 126691. https://doi.org/10.1016/j.micres.2020.126691

Calvente, V., de Orellano, M. E., Sansone, G., Benuzzi, D., & Sanz de Tosetti, M. I. (2001). Effect of nitrogen source and pH on siderophore production by *Rhodotorula* strains and their application to biocontrol of phytopathogenic moulds. *Journal of Industrial Microbiology and Biotechnology*, *26*(4), 226–229. https://doi.org/10.1038/sj.jim.7000117

Caruso, D. J., Palombo, E. A., Moulton, S. E., & Zaferanloo, B. (2022). Exploring the promise of endophytic fungi: A review of novel antimicrobial compounds. *Microorganisms*, *10*(10), 1990. https://www.mdpi.com/2076-2607/10/10/1990

Chaudhary, P., Agri, U., Chaudhary, A., Kumar, A., & Kumar, G. (2022). Endophytes and their potential in biotic stress management and crop production. *Frontiers in Microbiology*, *13*, 933017. https://doi.org/10.3389/fmicb.2022.933017

Chen, X., Zhang, Y., Chao, S., Song, L., Wu, G., Sun, Y., Chen, Y., & Lv, B. (2024). Biocontrol potential of endophytic *Bacillus subtilis* A9 against rot disease of *Morchella esculenta*. *Frontiers in Microbiology*, *15*. https://doi.org/10.3389/fmicb.2024.1388669

Chen, X.-l., Sun, M.-c., Chong, S.-l., Si, J.-p., & Wu, L.-s. (2022). Transcriptomic and metabolomic approaches deepen our knowledge of plant–endophyte interactions. *Frontiers in Plant Science*, *12*, 700200.

Chittora, D., Meena, M., Barupal, T., Swapnil, P., & Sharma, K. (2020). Cyanobacteria as a source of biofertilizers for sustainable agriculture. *Biochemistry and Biophysics Reports*, *22*, 100737.

Choi, J., Roy Choudhury, A., Walitang, D. I., Lee, Y., & Sa, T. (2022). ACC deaminase-producing *Brevibacterium linens* RS16 enhances heat-stress tolerance of rice (*Oryza sativa* L.). *Physiologia Plantarum*, *174*(1), e13584.

Cueva-Yesquén, L. G., Goulart, M. C., Attili de Angelis, D., Nopper Alves, M., & Fantinatti-Garboggini, F. (2020). Multiple plant growth-promotion traits in endophytic bacteria retrieved in the vegetative stage from passionflower. *Frontiers in Plant Science*, *11*, 621740. https://doi.org/10.3389/fpls.2020.621740

Cui, J., Fanxuan, N., Yuquan, Z., Dawei, Z., Dinggang, Z., Jinfeng, W., Liang, Q., Lu, X., & and Liu, L. (2024). A review on plant endophytes in response to abiotic stress. *Environmental Pollutants and Bioavailability*, *36*(1), 2323123. https://doi.org/10.1080/26395940.2024.2323123

Devi, R., Kaur, T., Kour, D., Rana, K. L., Yadav, A., & Yadav, A. N. (2020). Beneficial fungal communities from different habitats and their roles in plant growth promotion and soil health. *Microbial Biosystems*, *5*(1), 21–47. https://doi.org/10.21608/mb.2020.32802.1016

Dias, A. C. F., Costa, F. E. C., Andreote, F. D., Lacava, P. T., Teixeira, M. A., Assumpção, L. C., Araújo, W. L., Azevedo, J. L., & Melo, I. S. (2009). Isolation of micropropagated strawberry endophytic bacteria and assessment of their potential for plant growth promotion. *World Journal of Microbiology and Biotechnology*, *25*(2), 189–195. https://doi.org/10.1007/s11274-008-9878-0

Domka, A. M., Rozpaądek, P., & Turnau, K. (2019). Are fungal endophytes merely mycorrhizal copycats? The role of fungal endophytes in the adaptation of plants to metal toxicity. *Frontiers in Microbiology*, *10*, 371. https://doi.org/10.3389/fmicb.2019.00371

Eid, A. M., Salim, S. S., Hassan, S. E.-D., Ismail, M. A., & Fouda, A. (2019). Role of endophytes in plant health and abiotic stress management. In V. Kumar, R. Prasad, M. Kumar, & D. K. Choudhary (Eds.), *Microbiome in Plant Health and Disease: Challenges and Opportunities* (pp. 119–144). Springer Singapore. https://doi.org/10.1007/978-981-13-8495-0_6

El-khawaga, H. A., Mustafa, A. E., El khawaga, M. A., Mahfouz, A. Y., & Daigham, G. E. (2024). Bio-stimulating effect of endophytic *Aspergillus flavus* AUMC 16068 and its respective ex-polysaccharides in lead stress tolerance of Triticum aestivum plant. *Scientific Reports*, *14*(1), 11952. https://doi.org/10.1038/s41598-024-61936-0

Fadiji, A. E., & Babalola, O. O. (2020a). Elucidating mechanisms of endophytes used in plant protection and other bioactivities with multifunctional prospects. *Frontiers in Bioengineering and Biotechnology*, *8*, 467. https://doi.org/10.3389/fbioe.2020.00467

Fadiji, A. E., & Babalola, O. O. (2020b). Metagenomics methods for the study of plant-associated microbial communities: A review. *Journal of Microbiological Methods*, *170*, 105860. https://doi.org/10.1016/j.mimet.2020.105860

Gaby, J. C., & Buckley, D. H. (2012). A comprehensive evaluation of PCR primers to amplify the nifH gene of nitrogenase. *PLoS One*, *7*(7), e42149. https://doi.org/10.1371/journal.pone.0042149

Ghareeb, R. Y., Jaremko, M., Abdelsalam, N. R., Abdelhamid, M. M. A., El-Argawy, E., & Ghozlan, M. H. (2024). Biocontrol potential of endophytic fungi against phytopathogenic nematodes on potato (*Solanum tuberosum* L.). *Scientific Reports*, *14*(1), 15547. https://doi.org/10.1038/s41598-024-64056-x

Ghorai, P., & Ghosh, D. (2022). Ameliorating the performance of NPK biofertilizers to attain sustainable agriculture with special emphasis on bioengineering. *Bioresource Technology Reports*, *19*, 101117.

Glick, B. R. (2012). Plant growth-promoting bacteria: Mechanisms and applications. *Scientifica*, *2012*(1), 963401. https://doi.org/10.6064/2012/963401

González Ortega-Villaizán, A., King, E., Patel, M. K., Pérez-Alonso, M. M., Scholz, S. S., Sakakibara, H., Kiba, T., Kojima, M., Takebayashi, Y., Ramos, P., Morales-Quintana, L., Breitenbach, S., Smolko, A., Salopek-Sondi, B., Bauer, N., Ludwig-Müller, J., Krapp, A., Oelmüller, R., Vicente-Carbajosa, J., & Pollmann, S. (2024). The endophytic fungus *Serendipita indica* affects auxin distribution in *Arabidopsis thaliana* roots through alteration of auxin transport and conjugation to promote plant growth. *Plant, Cell & Environment*, *47*(10), 3899–3919. https://doi.org/10.1111/pce.14989

Govindasamy, V., George, P., Raina, S. K., Kumar, M., Rane, J., & Annapurna, K. (2018). Plant-associated microbial interactions in the soil environment: Role of endophytes in imparting abiotic stress tolerance to crops. In S. K. Bal, J. Mukherjee, B. U. Choudhury, & A. K. Dhawan (Eds.), *Advances in Crop Environment Interaction* (pp. 245–284). Springer Singapore. https://doi.org/10.1007/978-981-13-1861-0_10

Gupta, G., Panwar, J., & Jha, P. N. (2013). Natural occurrence of Pseudomonas aeruginosa, a dominant cultivable diazotrophic endophytic bacterium colonizing Pennisetum glaucum (L.) R. Br. *Applied Soil Ecology*, *64*, 252–261. https://doi.org/10.1016/j.apsoil.2012.12.016

Hurek, T., & Reinhold-Hurek, B. (2003). *Azoarcus* sp. strain BH72 as a model for nitrogen-fixing grass endophytes. *Journal of Biotechnology*, *106*(2), 169–178. https://doi.org/10.1016/j.jbiotec.2003.07.010

Jan, R., Khan, M. A., Asaf, S., Lubna, Lee, I. J., & Kim, K. M. (2019). Metal resistant endophytic bacteria reduces cadmium, nickel toxicity, and enhances expression of metal stress related genes with improved growth of *Oryza sativa*, via regulating its antioxidant machinery and endogenous hormones. *Plants (*Basel, Switzerland), *8*(10), 363. https://doi.org/10.3390/plants8100363

Kashyap, N., Singh, S.K., Yadav, N., Singh, V.K., Kumari, M., Kumar, D., Shukla, L., Kaushalendra, Bhardwaj, N. and Kumar, A., 2023. Biocontrol screening of endophytes: Applications and limitations. *Plants*, *12*(13), 2480. https://doi.org/10.3390/plants12132480

Khan, I., Asaf, S., Kang, S.-M., & Lee, I.-J. (2025). Physiological mechanisms of heavy metal detoxification in tomato plants mediated by endophytic fungi under nickel and cadmium stress. *Plant Physiology and Biochemistry*, *221*, 109589. https://doi.org/10.1016/j.plaphy.2025.109589

Kumar, S., Kumar, S., & Mohapatra, T. (2021). Interaction between macro- and micronutrients in plants. *Frontiers in Plant Science*, *12*, 665583. https://doi.org/10.3389/fpls.2021.665583

Kumari, M., Qureshi, K.A., Jaremko, M., White, J., Singh, S.K., Sharma, V.K., Singh, K.K., Santoyo, G., Puopolo, G. and Kumar, A., 2022. Deciphering the role of endophytic microbiome in postharvest diseases management of fruits: Opportunity areas in commercial up-scale production. *Frontiers in Plant Science*, *13*, 4390. https://doi.org/10.3389/fpls.2022.1026575

Lodewyckx, C., Jaco, V., Fiona, P., B., M. E. R., Safieh, T., Max, M., & and der Lelie, D. v. (2002). Endophytic Bacteria and Their Potential Applications. *Critical Reviews in Plant Sciences*, *21*(6), 583–606. https://doi.org/10.1080/0735-260291044377

Ma, Y., Rajkumar, M., Zhang, C., & Freitas, H. (2016). Beneficial role of bacterial endophytes in heavy metal phytoremediation. *Journal of Environmental Management*, *174*, 14–25. https://doi.org/10.1016/j.jenvman.2016.02.047

Macabuhay, A., Arsova, B., Watt, M., Nagel, K. A., Lenz, H., Putz, A., Adels, S., Müller-Linow, M., Kelm, J., & Johnson, A. A. (2022). Plant growth promotion and heat stress amelioration in *Arabidopsis* inoculated with *Paraburkholderia phytofirmans* PsJN rhizobacteria quantified with the GrowScreen-Agar II phenotyping platform. *Plants, 11*(21), 2927.

Mehta, P., Sharma, R., Putatunda, C., & Walia, A. (2019). Endophytic fungi: Role in phosphate solubilization. In B. P. Singh (Ed.), *Advances in Endophytic Fungal Research: Present Status and Future Challenges* (pp. 183–209). Springer International Publishing. https://doi.org/10.1007/978-3-030-03589-1_9

Miethke, M., & Marahiel, M. A. (2007). Siderophore-based iron acquisition and pathogen control. *Microbiology and Molecular Biology Reviews, 71*(3), 413–451. https://doi.org/10.1128/mmbr.00012-07

Mishra, R., Joshi, R. K., & Zhao, K. (2020). Base editing in crops: Current advances, limitations and future implications. *Plant Biotechnology Journal, 18*(1), 20–31.

Najjar, A. A. (2025). Therapeutic potential of endophytic microbes: Emphasizing both fungal and bacterial endophytes. *Applied Microbiology, 5*(1), 5. https://www.mdpi.com/2673-8007/5/1/5

Naseem, M., Chaudhry, A. N., Jilani, G., Alam, T., Naz, F., Ullah, R., Zahoor, M., Zaman, S., & Sohail. (2024). Exopolysaccharide-producing bacterial cultures of *Bacillus cereus* and *Pseudomonas aeruginosa* in soil augment water retention and maize growth. *Heliyon, 10*(4). https://doi.org/10.1016/j.heliyon.2024.e26104

Nautiyal, C. S., Bhadauria, S., Kumar, P., Lal, H., Mondal, R., & Verma, D. (2000). Stress induced phosphate solubilization in bacteria isolated from alkaline soils. *FEMS Microbiology Letters, 182*(2), 291–296. https://doi.org/10.1016/S0378-1097(99)00605-9

Nostadt, R., Hilbert, M., Nizam, S., Rovenich, H., Wawra, S., Martin, J., Küpper, H., Mijovilovich, A., Ursinus, A., & Langen, G. (2020). A secreted fungal histidine-and alanine-rich protein regulates metal ion homeostasis and oxidative stress. *New Phytologist, 227*(4), 1174–1188.

Omomowo, O. I., & Babalola, O. O. (2019). Bacterial and fungal endophytes: Tiny giants with immense beneficial potential for plant growth and sustainable agricultural productivity. *Microorganisms, 7*(11), 481. https://www.mdpi.com/2076-2607/7/11/481

Palmieri, D., Vitale, S., Lima, G., Di Pietro, A., & Turrà, D. (2020). A bacterial endophyte exploits chemotropism of a fungal pathogen for plant colonization. *Nature Communications, 11*(1), 5264.

Pathak, P., Rai, V. K., Can, H., Singh, S. K., Kumar, D., Bhardwaj, N., Roychowdhury, R., de Azevedo, L. C. B., Kaushalendra, Verma, H., & Kumar, A. (2022). Plant-endophyte interaction during biotic stress management. *Plants, 11*(17), 2203. https://doi.org/10.3390/plants11172203

Phetcharat, P., & Duangpaeng, A. (2012). Screening of endophytic bacteria from organic rice tissue for indole acetic acid production. *Procedia Engineering, 32*, 177–183. https://doi.org/10.1016/j.proeng.2012.01.1254

Pieterse, C. M. J., Van der Does, D., Zamioudis, C., Leon-Reyes, A., & Van Wees, S. C. M. (2012). Hormonal modulation of plant immunity. *Annual Review of Cell and Developmental Biology, 28*, 489–521. https://doi.org/10.1146/annurev-cellbio-092910-154055

Pires, E. B. E., de Freitas, A. J., Souza, F. F. e., Salgado, R. L., Guimarães, V. M., Pereira, F. A., & Eller, M. R. (2019). Production of Fungal Phytases from agroindustrial byproducts for pig diets. *Scientific Reports, 9*(1), 9256. https://doi.org/10.1038/s41598-019-45720-z

Ramírez-Cariño, H. F., Morales, I., Guadarrama-Mendoza, P. C., Tlilayatzi-Bello, A. A., & Valadez-Blanco, R. (2024). Beneficial traits of crop-associated rhizobacteria toward plant growth promotion, biofertilization, and nutrient enhancement of harvested agro-products. In K. A. Abd-Elsalam & H. I. Mohamed (Eds.), *Bacterial Secondary Metabolites* (Chapter 17, pp. 299–334). Elsevier. https://doi.org/https://doi.org/10.1016/B978-0-323-95251-4.00005-3

Rashmi, M., Kushveer, J. S., & Sarma, V. V. (2019). A worldwide list of endophytic fungi with notes on ecology and diversity. *Mycosphere*, *10*(1), 798–1079. https://doi.org/10.5943/mycosphere/10/1/19

Rasool, S., Groos, M., Hannula, S. E., & Biere, A. (2025). Bioinoculant-induced plant resistance is modulated by interactions with resident soil microbes. *Environmental Microbiome*, *20*(1), 7. https://doi.org/10.1186/s40793-025-00667-9

Renshaw, J. C., Robson, G. D., Trinci, A. P. J., Wiebe, M. G., Livens, F. R., Collison, D., & Taylor, R. J. (2002). Fungal siderophores: Structures, functions and applications. *Mycological Research*, *106*(10), 1123–1142. https://doi.org/10.1017/S0953756202006548

Rozpądek, P., Domka, A., Ważny, R., Nosek, M., Jędrzejczyk, R., Tokarz, K., & Turnau, K. (2018). How does the endophytic fungus *Mucor* sp. improve *Arabidopsis arenosa* vegetation in the degraded environment of a mine dump? *Environmental and Experimental Botany*, *147*, 31–42. https://doi.org/10.1016/j.envexpbot.2017.11.009

Sena, L., Mica, E., Valè, G., Vaccino, P., & Pecchioni, N. (2024). Exploring the potential of endophyte-plant interactions for improving crop sustainable yields in a changing climate. *Frontiers in Plant Science*, *15*, 1349401.

Shahzad, A., Hameed, S., Qin, M., Li, H., Zafar, S., Siddiqui, S., Sattar, S., Mahmood, Z., & Mehwish, S. (2025). Cadmium (Cd) detoxification and activation of plant defense enzymes in wheat (*Triticum aestivum*) through the use of endophytic *Bacillus thuringiensis* and *Salix alba* root powder. *Environmental Pollution*, *364*, 125147. https://doi.org/10.1016/j.envpol.2024.125147

Sheoran, N., Valiya Nadakkakath, A., Munjal, V., Kundu, A., Subaharan, K., Venugopal, V., Rajamma, S., Eapen, S. J., & Kumar, A. (2015). Genetic analysis of plant endophytic *Pseudomonas putida* BP25 and chemo-profiling of its antimicrobial volatile organic compounds. *Microbiological Research*, *173*, 66–78. https://doi.org/10.1016/j.micres.2015.02.001

Singh, A., & Roychoudhury, A. (2021). Gene regulation at transcriptional and post-transcriptional levels to combat salt stress in plants. *Physiologia Plantarum*, *173*(4), 1556–1572.

Singh, R., Pandey, K.D., Singh, M., Singh, S.K., Hashem, A., Al-Arjani, A.B.F., Abd_Allah, E.F., Singh, P.K. and Kumar, A., 2022. Isolation and characterization of endophytes bacterial strains of *Momordica charantia* L. and their possible approach in stress management. *Microorganisms*, 10(2), p.290. https://doi.org/10.3390/microorganisms10020290

Son, Y., Martinez, C. E., & Kao-Kniffin, J. (2024). Three important roles and chemical properties of glomalin-related soil protein. *Frontiers in Soil Science*, *4*. https://doi.org/10.3389/fsoil.2024.1418072

Strobel, G., Daisy, B., Castillo, U., & Harper, J. (2004). Natural products from endophytic microorganisms. *Journal of Natural Products*, *67*(2), 257–268. https://doi.org/10.1021/np030397v

Tang, Y., Zhou, S., Xiao, Y., Zhang, T., Tao, X., Shi, K., Lu, Y., Yang, Y., Zhao, Y., & Zhao, T. (2024). Exploring the microbial ecosystem of *Berchemia polyphylla* var. *leioclada*: A comprehensive analysis of endophytes and rhizospheric soil microorganisms. *Frontiers in Microbiology*, *15*, 1338956.

Viana, G. C., Médici, L. O., Vidal, M. S., & Baldani, J. I. (2024). Bacillus endophytic strains control Fusarium wilt caused by *Fusarium oxysporum* f. sp. *lycopersici* in tomato cv. Perinha. *Brazilian Journal of Microbiology*, *55*(4), 4019–4034. https://doi.org/10.1007/s42770-024-01539-z

Villanueva, N. S., Bermudez, A. T., Mamani, G. M., Chavez, M. A., Guzman, W. G., & Espinoza-Montes, F. (2022). Contribution of native endophytic bacteria in nutrients accumulation in high Andean Poaceae [Contribución de bacterias endófitas nativas en la acumulación de nutrientes en poaceas altoandinas.]. *Revista de Investigaciones Veterinarias del Perú (RIVEP)*, *33*(5). https://doi.org/10.15381/rivep.v33i5.23790

Vishwakarma, K., Kumar, N., Shandilya, C., & Varma, A. (2021). Unravelling the role of endophytes in micronutrient uptake and enhanced crop productivity. In N. Shrivastava, S. Mahajan, & A. Varma (Eds.), *Symbiotic Soil Microorganisms: Biology and Applications* (pp. 63–85). Springer International Publishing. https://doi.org/10.1007/978-3-030-51916-2_4

Wang, Q., Li, X., Zhou, K., Li, Y., Wang, Y., Zhang, G., Guo, H., Zhou, J., & Wang, T. (2024). Mechanisms of conjugative transfer of antibiotic resistance genes induced by extracellular polymeric substances: Insights into molecular diversities and electron transfer properties. *Journal of Hazardous Materials*, *476*, 135181. https://doi.org/10.1016/j.jhazmat.2024.135181

Waqar, S., Bhat, A. A., & Khan, A. A. (2024). Endophytic fungi: Unravelling plant-endophyte interaction and the multifaceted role of fungal endophytes in stress amelioration. *Plant Physiology and Biochemistry*, *206*, 108174. https://doi.org/10.1016/j.plaphy.2023.108174

Yadav, G., Srivastva, R., & Gupta, P. (2021). Endophytes and their applications as biofertilizers. In P. Bhatt, S. Gangola, D. Udayanga, & G. Kumar (Eds.), *Microbial Technology for Sustainable Environment* (pp. 95–123). Springer Singapore. https://doi.org/10.1007/978-981-16-3840-4_7

Yurgel, S. N., Ajeethan, N., & Smertenko, A. (2022). Response of plant-associated microbiome to plant root colonization by exogenous bacterial endophyte in perennial crops. *Frontiers in Microbiology*, *13*, 863946. https://doi.org/10.3389/fmicb.2022.863946

Zhang, D., Spadaro, D., Valente, S., Garibaldi, A., & Gullino, M. L. (2012). Cloning, characterization, expression and antifungal activity of an alkaline serine protease of *Aureobasidium pullulans* PL5 involved in the biological control of postharvest pathogens. *International Journal of Food Microbiology*, *153*(3), 453–464. https://doi.org/10.1016/j.ijfoodmicro.2011.12.016

Zhang, H., Wang, K., Liu, X., Yao, L., Chen, Z., & Han, H. (2024). Exopolysaccharide-producing bacteria regulate soil aggregates and bacterial communities to inhibit the uptake of cadmium and lead by lettuce. *Microorganisms*, *12*(11), 2112. https://www.mdpi.com/2076-2607/12/11/2112

Zhang, Y., Zhou, Q., Gao, C., Lu, Y., Sheng, Y., Xiao, M., Yun, Y., Selvaraj, J. N., Zhang, X., Li, Y., & Yu, X. (2024). Endophytic bacteria for Cd remediation in rice: Unraveling the Cd tolerance mechanisms of *Cupriavidus metallidurans* CML2. *Journal of Hazardous Materials*, *469*, 133846. https://doi.org/10.1016/j.jhazmat.2024.133846

2 Edible Plant Microbiome of Fruits and Vegetables
Research, Development, and Challenges

Ritu Singh and Prem Pratap Singh

2.1 INTRODUCTION

The modern understanding of human nutrition is undergoing a paradigm shift. For decades, fruits and vegetables have been promoted for their rich content of vitamins, minerals, and dietary fiber. While this perspective remains fundamentally correct, it has been incomplete in its understanding. Recent research revealed that these sources are not merely source of nutrition but are, in fact, complex biological ecosystems that contain thousands to billions of microorganisms per serving (Berg et al., 2025). Therefore, the consumption of fruits and vegetables translates to a significant and direct interaction between human microbiome and the diverse environmental microbiome.

With this context, the term 'edible plant microbiome' was introduced in 2014 to describe the diverse microbial communities associated with raw-eaten as well as fermented plant products (Berg et al., 2025). The purpose of this concept is to redescribe food-associated microbes, considering them not as contaminants, but as a functional dietary component that contribute to human health. This concept extends our growing understanding of the complex microbiota present on the surface and within the tissues of fruits and vegetables, which harbor diverse bacteria and fungi, including Proteobacteria, Bacteroidetes, Firmicutes, Actinobacteria, Ascomycota, and Basidiomycota, comprising both beneficial and pathogenic members.

Further, the edible plant microbiome also integrates with the broader public health agendas. One of the major goals of edible plant microbiome is the 'One Health' model which describes how microbes interchange between humans, animals, and the environment, building an interconnected continuum in which the food chain serves as a key route of microbial transfer (Berg et al., 2025). Similarly, another agenda is the 'Exposome', which identifies the full spectrum of environmental exposures across a person's lifetime. Within this model, food-associated microbes represent a critical yet historically overlooked factor influencing human physiology and disease risk (Berg et al., 2025).

The present chapter focuses on the role of raw fruits and vegetables as a major source of environmental microbes that reach the human gut. It discusses the journey

DOI: 10.1201/9781003664451-2

of food-associated microbes from field to gut and how this community is modified along the journey. Further, the chapter explores the dynamic interactions between plant microbiome, plant-derived chemicals, and the gut microbiota, and how these interactions influence human physiology. The chapter also highlights the translation of current understating of dynamic microbial world into microbial inoculants, synthetic communities, and functional foods. Finally, it outlines key challenges, including food safety, knowledge gaps, and the need for integrated 'soil–plant–gut' frameworks to guide research, regulation, and practice.

2.2 THE DIVERSE MICROBIAL WORLD OF FRUITS AND VEGETABLES

Fresh fruits and vegetables are not merely sources of vitamins and nutrients; they are vibrant ecosystems teeming with a diverse community of microorganisms, collectively known as the fruit and vegetable microbiome. This microbiome includes a wide array of harmful and beneficial microbes that reside both on the surface (epiphytes) and within the tissues (endophytes) of the produce. These microbial communities play a crucial role in maintaining the health and quality of the fruit, managing postharvest diseases, and significantly influencing human health by interacting with the gut microbiota. The composition of this microbiome is vast, with certain microbial groups consistently found to be dominant and functionally important. For instance, the most prevalent bacterial phyla associated with fruits and vegetables are Proteobacteria, Bacteroidetes, Firmicutes, and Actinobacteria (He et al., 2024). Genera such as *Bifidobacterium*, *Lactobacillus*, *Prevotella*, and *Bacteroides* are frequently highlighted for their beneficial roles both in plant health and as probiotics for human consumption. In the fungal realm, the phyla Ascomycota and Basidiomycota are prominent. This includes many yeasts, such as *Aureobasidium pullulans*, *Pichia membranifaciens*, and *Candida oleophila*, which act as powerful antagonists against postharvest pathogens (Zhang et al., 2020). However, this also includes fungi like *Aspergillus*, *Alternaria*, *Penicillium*, and *Botrytis cinerea*, which are common postharvest pathogens that cause significant crop loss (Chen et al., 2022).

The microbial composition of produce is not static; it is shaped by a multitude of factors from the farm to the consumer. Preharvest factors, such as geographical location, farming practices (organic vs. conventional), soil type, and climate, are primary determinants of which microbes initially colonize the plant. The plant itself contributes to this ecosystem by creating a nutrient-rich niche on its surface through glandular trichomes and the leaching of compounds, which supports dense microbial communities. After harvesting, the microbiome continues to evolve based on handling, storage conditions, and treatments. Temperature and relative humidity in cold storage are critical factors; for example, a relative humidity of 90% is often ideal for storing fruit, as it prevents water loss while inhibiting the growth of most fungi. Additionally, postharvest treatments like ozone application can also significantly alter the microbial community on fruits like grapes (Table 2.1).

TABLE 2.1
Key Factors Influencing the Edible Plant Microbiome

Stage	Factor	Mechanism of Influence	Impact on Microbiome	Reference
Preharvest	Plant genotype	Host genetics and co-evolution select for specific microbial partners.	Determines the baseline community structure and composition.	Berg et al. (2025)
	Agricultural practice	Application of pesticides/fertilizers (conventional) vs. fostering soil biodiversity (organic).	Conventional practices may reduce diversity; organic may increase it.	Droby and Wisniewski (2018)
	Soil microbiome	Acts as the primary 'seed bank' for root and endophytic microbes.	The diversity of the soil directly influences the diversity of the plant's internal microbiome.	Santoyo (2022)
	Climate	Abiotic stressors like UV, temperature, and rainfall select for adapted microbes.	Shapes community composition, e.g., selects for UV-resistant taxa on surfaces.	Leveau (2019)
Postharvest	Handling and transport	Physical damage releases nutrients; cross-contamination from surfaces.	Can introduce spoilage organisms and create niches for rapid growth.	De Wit et al. (2024)
	Storage temperature	Cold storage selects for psychrotolerant (cold-adapted) microbes.	Reduces overall diversity and shifts community toward cold-tolerant species.	Palumbo et al. (2022)
	Packaging	Modified atmosphere packaging alters oxygen and carbon dioxide levels.	Selects for anaerobic or microaerophilic microbes and inhibits aerobes.	Palumbo et al. (2022)
	Sanitization	Chemical agents (e.g., chlorine) cause widespread microbial death.	Causes quantitative and qualitative shifts; reduces diversity and selects for resistant strains.	Gu et al. (2019)

2.3 THE ROLE OF THE MICROBIOME IN POSTHARVEST HEALTH

The interaction between the host fruit, its native microbiome, and environmental factors determines its susceptibility to disease, a concept illustrated by the 'disease triangle'. A healthy, diverse microbiome can protect fruit from decay through several mechanisms:

- *Competition*: Beneficial microbes compete with pathogens for essential nutrients and space on the fruit's surface, effectively crowding them out.
- *Antagonism and Antibiosis*: Many beneficial microbes produce antimicrobial compounds that directly inhibit or kill pathogens.

- *Bacillus* species, for example, produce cyclic lipopeptides like fengycin and iturin that have potent antifungal activity. Similarly, yeasts can secrete enzymes like β-(1,3) glucanase that break down fungal cell walls.
- *Induced Resistance*: Beneficial microbes can trigger the fruit's own defense mechanisms. This 'crying for help strategy' allows the plant to recruit beneficial microbes from the environment to fight off pathogens. The production of reactive oxygen species by yeasts can create an oxidative burst on the fruit's surface, signaling the activation of host defense genes.

2.4 IMPACT ON HUMAN GUT MICROBIOTA

- The consumption of fruits and vegetables serves as a direct link between the environmental microbiome and the human gut microbiome. The rich array of phytochemicals in produce, such as polyphenols, flavonoids, and dietary fibers (polysaccharides), are not easily digested by humans but act as substrates for gut microbes.
- *Metabolism of Phytochemicals*: When these compounds reach the large intestine, they are fermented by the gut microbiota into beneficial metabolites, most notably short-chain fatty acids like acetate, propionate, and butyrate.
- *Modulation of Gut Bacteria*: This process selectively promotes the growth of beneficial bacteria. Polyphenols from berries, for instance, encourage the proliferation of *Lactobacillus*, *Bifidobacterium*, and *Akkermansia*. Dietary fibers from sources like citrus pectin and dragon fruit oligosaccharides also boost *Bifidobacteria* and *Lactobacilli*.
- *Health-Promoting Effects*: This modulation of the gut microbiota leads to numerous health benefits, including anti-obesity, anti-inflammatory, and antidiabetic effects. By altering the Firmicutes/Bacteroidetes ratio and increasing the abundance of beneficial microbes, the phytochemicals help rescue gut dysbiosis and maintain a balanced intestinal ecosystem.

2.5 JOURNEY OF RHIZOSPHERE-DERIVED BENEFICIALS FROM SOIL TO PLATE

The rhizosphere, a narrow yet bustling region of soil directly influenced by root secretions, is a foundational hub of microbial life critical to plant health. Within this dynamic environment, a specialized group of bacteria known as plant growth–promoting rhizobacteria (PGPR) thrives, establishing a symbiotic relationship with the plant that profoundly enhances its vitality, while their story begins in the soil. In the last decade, it has been repeatedly demonstrated that the plant interior is colonized by a range of endophytes mostly deriving from the rhizosphere, and many of them have been reported to improve plant growth or health (Sturz and Nowak, 2000; Hardoim et al., 2008; Pathak et al., 2022; Singh et al., 2022). It does not end there; many of these beneficial microbes undertake a remarkable journey, migrating from the roots into the very tissues we consume, thereby linking soil health directly to human nutrition.

At the heart of this symbiotic exchange are well-studied PGPR genera such as *Pseudomonas*, *Bacillus*, *Azospirillum*, and *Rhizobium*. Their primary contributions to the plant are twofold: enhancing nutrient uptake and producing a host of bioactive compounds. For instance, *Rhizobium* and *Azospirillum* are famed for their ability to fix atmospheric nitrogen, converting it into a form the plant can readily absorb, thereby reducing the need for synthetic fertilizers (Compant et al., 2010). Meanwhile, species of *Pseudomonas* and *Bacillus* excel at solubilizing phosphorus, unlocking this essential but often inaccessible nutrient from the soil matrix. Beyond direct nutrient provision, these bacteria synthesize critical bioactive compounds. They produce phytohormones like auxins and cytokinins that regulate plant growth and development, as well as siderophores, molecules that scavenge iron from the environment, making it available to the plant while simultaneously depriving pathogenic microbes of this vital element.

While many of these bacteria are root-associated, their influence is not confined to belowground. A significant subset of PGPR can transition from being rhizosphere inhabitants to becoming endophytes, microbes that live within plant tissues without causing disease. This migration from the soil to the plant's aerial parts is a crucial step in their journey (Frank et al., 2017). These bacteria can penetrate the root epidermis and cortex, traveling through the plant's vascular system (the xylem and phloem) to colonize stems, leaves, and ultimately, the fruits and vegetables themselves.

Once established within these edible tissues, the bacteria continue to confer benefits. They can bolster the plant's systemic resistance against pathogens, making the fruit or vegetable more resilient both in the field and after harvest. This internal microbial community becomes an integral part of the fruit and vegetable microbiome. When we consume these raw or minimally processed foods, we also ingest this rich consortium of beneficial, plant-derived microbes. This process serves as a natural and direct pathway for introducing environmentally sourced beneficial bacteria into the human gut, where they can interact with and enrich our native microbiota.

In essence, the journey of PGPR from the rhizosphere to our plate represents a profound connection between agriculture and human health. The life-giving interactions that begin in the soil culminate in the nutritional and microbial quality of the food we eat, highlighting that a healthy plant, nurtured by a vibrant community of beneficial bacteria, is not just a source of vitamins and fiber, but also a living vector for microbes that can positively shape our own internal ecosystems.

2.6 RESEARCH AND DEVELOPMENT: ENGINEERING THE EDIBLE MICROBIOME

The growing understanding of the edible plant microbiome is propelling a paradigm shift toward the engineering-based advancements in agriculture and food sector. Researchers started designing approaches to harness full potential of plant microbiomes to improve sustainability and creating functional foods with healthcare advantages.

2.6.1 Enhancing Crops with 'Plant Probiotics' and Biostimulants

One of the key applications of plant microbiome is the development of plant probiotics, i.e., plant growth–promoting microorganisms (PGPMs). PGPMs are the useful microbes that offer a sustainable alternative to chemical fertilizers and pesticides (Woo and Pepe, 2018). They are being used as biofertilizers to improve nutrient uptake, as well as biostimulants and biocontrol agents to enhance growth and stress tolerance and suppress pathogens, respectively (Compant et al., 2019; Anand et al., 2023; Kesawat et al., 2023). Besides, PGPMs also enhance the nutritional value of the crops. For instance, inoculating strawberries or tomatoes with specific bacterial strains increases vitamin C and antioxidant values in the crops (Jiménez-Gómez et al., 2017; Pathak et al., 2023).

2.6.2 Synthetic Microbial Communities (SynComs)

While research has shown the potential of single-strain PGPMs, it is also apparent that complex natural environments and microbial functions are often carried out by microbial communities instead of single strain. This understanding, along with the need to move beyond the vagaries of single-strain PGPMs, has led to the development of stable and synergistic microbial communities known as synthetic microbial communities, or SynComs. SynComs are the engineering innovations in agro-microbiology that are designed by multiple but well-defined microbial species, that are selected for specific functions such as nitrogen fixation, pathogen suppression, drought tolerance, and many more (Tariq et al., 2025). SynComs are currently being under development to deliver more beneficial functions to crop plants, viz. improving crop resilience to drought, improving the uptake of nutrients from the soil, and additionally conferring predictable protection against a range of diseases (Martins et al., 2023). The challenge associated with SynComs remains in ensuring the stability and efficacy in variable natural conditions of fields (Martins et al., 2023).

2.6.3 Novel Functional Foods

The knowledge of the edible plant microbiome has unveiled new avenues for developing the next generation of functional foods. The most direct approach involves the addition of well-characterized probiotic strains, such as *Lactobacillus* and *Bifidobacterium* species, to the plant-based ingredients, viz. fruit juices, yogurts, and fermented vegetables (Grujović et al., 2025). However, it presents key challenges, including preserving the sensory properties of the food after the probiotic addition and ensuring the viability of the probiotic microbes throughout the product's shelf life, even though the approach combines the health benefits of both the plant material and the added probiotics (Grujović et al., 2025). Building on this concept, an alternative strategy involves using fruit and vegetable extracts (rich in specific phytochemicals) as functional ingredients to selectively modulate the existing gut microbiota. The approach tends to rebalance the dysbiotic gut ecosystem by promoting the growth of beneficial resident bacteria (Jia et al., 2025). Further, SynCom design framework has also been applied to food fermentation processes, where researchers use defined microbial communities instead of spontaneous fermentation to improve reproducibility, flavor, and nutritional value associated with the fermented foods like sauerkraut or kimchi (Nikoloudaki et al., 2024).

Together, these approaches represent an evolution, from identifying natural plant microbiomes to utilizing them as beneficial microbes, and ultimately to constructing synthetic ecosystems with multiple microbial species. Notably, there is a clear shift toward deliberately designing plant–microbe and food–microbe interactions, for sustainable agriculture, improved food quality, and enhanced human health.

2.7 CHALLENGES, RISKS, AND FUTURE OUTLOOK

Although the benefits of edible plant microbiomes are tremendous, there are many significant challenges and risks that need careful management. The advancement requires maintaining food safety while ensuring microbial diversity by addressing major knowledge gaps and integrating ecological and health perspectives into agricultural and food systems.

2.7.1 The Food Safety Imperative: A Double-Edged Sword

The foremost challenge associated with fresh produce is ensuring its safety from human pathogen. Fresh fruits and vegetables are a potential source of foodborne diseases caused by pathogens such as *Escherichia coli* O157:H7, *Salmonella* spp., and *Listeria monocytogenes* (León et al., 2009). To control these contaminations (from cultivation to processing), food industries rely on microbial risk assessment (MRA), a systemic process that involves contaminant identification, exposure assessment, and risk characterization to ensure food safety (Institute of Food Science and Technology, 2025). However, in fresh produce, contamination is often sporadic and occurs at low levels, posing a major challenge for MRA (León et al., 2009). As a result, the 'sanitization paradox' comes into focus, referring to a long-term unintended consequence of microbial eradication through washing and chemical sanitizers. Although it is necessary to control acute pathogenic risks, aggressive sanitization reduces the edible microbiome diversity and significantly reduces the 'colonization resistance' that protect against pathogens. This provides a clear rationale for developing more nuanced, ecologically informed strategies that can efficiently target pathogens while maintaining the beneficial microbial communities (Wei et al., 2017).

2.7.2 Knowledge Gaps and Key Research Questions

Despite various advancements, many fundamental questions remain and continue to guide future research. Recent findings are largely correlational, such as association between dietary plant diversity and gut microbial richness, leaving the causal relationship between specific plant-associated microbes and its measurable health outcomes fundamentally unexplored (Ma et al., 2025). Further, it has become increasingly apparent that microbial function varies at the strain level, with different strains of the same bacterial species exerting markedly different metabolic and immunological effects on their hosts. A major challenge is the development of new tools that can identify and track beneficial microbes at strain level along the entire soil-plant-gut axis, rather than limiting identification to the species or genus level (Ma et al., 2025).

Because microbiome function develops from a dynamic network of microbe–microbe and host–microbe interactions, a deeper understanding of microbial interactions within the soil-plant-gut axis is needed to determine its stability, resilience, and response to disturbances (Ma et al., 2025). There is also urgent need for the systemic analysis of how plant domestication and modern agricultural practices, which have prioritized yield and appearance over other traits, have altered the microbial quality and diversity of our crops (Ma et al., 2025).

2.7.3 Future Outlook: A 'Farm-to-Gut' Integrated Vision

Building on the challenges, the future direction of edible microbiome research is moving toward the technology-driven approaches. Through the integrative application of 'omics' tools, viz. metagenomics, metatranscriptomics, and metabolomics, researchers can develop a comprehensive understanding of the edible plant microbiome and its impact on human health (Droby and Wisniewski, 2018). Multi-omics strategies will enable the mapping of an individual's unique gut microbiome and support the development of personalized dietary recommendations. Further, an emerging frontier would be extending the plant-gut microbiome axis up to the neurological and mental health. It is apparent that gut microbiome communicates with the brain via the 'microbiota-gut-brain axis'; therefore, any dietary modulation of the gut might serve as a viable strategy for supporting brain health (Yassin et al., 2025). The ultimate target is an agricultural system that prioritizes not only yield and quality but also cultivating healthy and diverse microbiome, producing food that is safe, nutritious, microbially rich, and functionally optimized to promote human health.

2.8 CONCLUSION

The relationship between fruits, vegetables, their associated microbial communities, and human health is far more complex and systemic than traditionally understood. Each serving of raw produce acts as a microbial inoculum that connects environmental and human associated microbiomes. This intake shapes gut microbiota composition, delivers fermentable substrates and microbes, and influences immune maturation and disease susceptibility across the life span. The main goal is to balance two essential requirements: the need for microbial food safety, especially against pathogens such as *E. coli*, *Salmonella*, and *Listeria*, and the recognition that plant-derived microbial communities support gut ecosystem stability and colonization resistance.

The sanitation paradigm based on complete inhibition can not only decrease acute risk but also destroy beneficial microbes and weaken ecological infection barriers. Future progress will depend on a fully integrated 'soil–plant–human–gut' perspective which connects plant growth–promoting microbes, fruit and vegetable microbiomes, and the gut microbiota into a single ecological axis. This effort will enable the rational design of plant-based edible microbiome and application of multi-omics tools to track strains and functions across this axis. The future agricultural systems will need to consider yield, quality, and safety together with the microbial signatures of crops. The ultimate goal should be to cultivate and deliver food that is nutritionally dense, microbiologically safe, and microbially configured to promote human health.

REFERENCES

Anand, U., T. Pal, N. Yadav, V. K. Singh, V. Tripathi, K. K. Choudhary, A. K. Shukla, K. Sunita, A. Kumar, E. Bontempi, and Y. Ma. 2023. Current scenario and future prospects of endophytic microbes: Promising candidates for abiotic and biotic stress management for agricultural and environmental sustainability. *Microbial Ecology* 86: 1455–1486. https://doi.org/10.1007/s00248-023-02190-1

Berg, G., G. V. Toledo, J. Schierstaedt, H. Hyöty, and W. A. Wicaksono. 2025. Linking the edible plant microbiome and human gut microbiome. *Gut Microbes* 17(1): 2551113. https://doi.org/10.1080/19490976.2025.2551113

Chen, Y., Z. Zhang, S. Tian, and B. Li. 2022. Application of omic technologies in postharvest pathology: Recent advances and perspectives. *Current Opinion in Food Science* 45: 100820. https://doi.org/10.1016/j.cofs.2022.100820

Compant, S., C. Clément, and A. Sessitsch. 2010. Plant growth-promoting bacteria in the rhizo- and endosphere of plants: Their role, colonization, mechanisms involved and prospects for utilization. *Soil Biology and Biochemistry* 42(5): 669–678.

De Wit, M., G. Liguori, and V. Vaštakaitė-Kairienė. 2024. Preharvest and postharvest factors affecting fruit and vegetables quality, physiology, and shelf-life. *Frontiers in Plant Science* 15: 1522530.

Droby, S., and M. Wisniewski. 2018. The fruit microbiome: A new frontier for postharvest biocontrol and postharvest biology. *Postharvest Biology and Technology* 140: 107–112.

Frank, A., J. Saldierna Guzmán, and J. Shay. 2017. Transmission of bacterial endophytes. *Microorganisms* 5(4): 70.

Grujović, M. Ž., T. Semedo-Lemsaddek, and K. G. Marković. 2025. Application of probiotics in foods: A comprehensive review of benefits, challenges, and future perspectives. *Foods* 14(17): 3088.

Gu, G., A. Ottesen, S. Bolten, et al. 2019. Impact of routine sanitation on the microbiomes in a fresh produce processing facility. *International Journal of Food Microbiology* 294: 31–41.

Hardoim, P. R., L. S. van Overbeek, and J. D. van Elsas. 2008. Properties of bacterial endophytes and their proposed role in plant growth. *Trends in Microbiology* 16(10): 463–471.

He, S., L. Gao, Z. Zhang, et al. 2024. Diversity analysis of microorganisms on the surface of four summer fruit varieties in Baotou, Inner Mongolia, China. *PeerJ* 12: e18752. https://doi.org/10.7717/peerj.18752

Institute of Food Science and Technology. n.d. Fresh produce safety. Institute of Food Science and Technology. Accessed September 16, 2025. https://www.ifst.org/resources/information-statements/fresh-produce-safety

Jia, Y., Y. Liu, Y. Wu, F. Ren, and H. Liu. 2025. From diet to microbiota: How fruits and vegetables influence gut microbiota. *International Journal of Food Science and Technology* 60(1): vvae008.

Jiménez-Gómez, A., L. Celador-Lera, M. Fradejas-Bayón, and R. Rivas. 2017. Plant probiotic bacteria enhance the quality of fruit and horticultural crops. *AIMS Microbiology* 3(3): 483–502.

Kesawat, M. S., N. Satheesh, B. S. Kherawat, A. Kumar, H. U. Kim, S. M. Chung, and M. Kumar. 2023. Regulation of reactive oxygen species during salt stress in plants and their crosstalk with other signaling molecules—Current perspectives and future directions. *Plants* 12(4): 864. https://doi.org/10.3390/plants12040864

León, J. S., L. A. Jaykus, and C. L. Moe. 2009. Food safety issues and the microbiology of fruits and vegetables. In *Microbiologically Safe Foods*, eds. N. Heredia, I. Wesley, and S. García, pp. 255–290. Hoboken, NJ: John Wiley & Sons.

Leveau, J. H. 2019. A brief from the leaf: Latest research to inform our understanding of the phyllosphere microbiome. *Current Opinion in Microbiology* 49: 41–49.

Ma, H., D. Cornadó, and J. M. Raaijmakers. 2025. The soil-plant-human gut microbiome axis into perspective. *Nature Communications* 16(1): 7748.

Martins, S. J., J. Pasche, H. A. O. Silva, et al. 2023. The use of synthetic microbial communities to improve plant health. *Phytopathology* 113(8): 1369–1379.

Nikoloudaki, O., F. Aheto, R. Di Cagno, and M. Gobbetti. 2024. Synthetic microbial communities: A gateway to understanding resistance, resilience, and functionality in spontaneously fermented food microbiomes. *Food Research International* 192: 114780.

Palumbo, M., G. Attolico, V. Capozzi, et al. 2022. Emerging postharvest technologies to enhance the shelf-life of fruit and vegetables: An overview. *Foods* 11(23): 3925.

Pathak, P., V. K. Rai, H. Can, S. K. Singh, D. Kumar, N. Bhardwaj, R. Roychowdhury, L. C. B. de Azevedo, Kaushalendra, H. Verma, and A. Kumar. 2022. Plant-endophyte interaction during biotic stress management. *Plants* 11(17): 2203. https://doi.org/10.3390/plants11172203

Pathak, P., M. Singh, A. Naskar, S. K. Singh, N. Bhardwaj, and A. Kumar. 2023. Isolation and phenotypic microarray profiling of different pseudomonas strains isolated from the rhizosphere of Curcuma longa L. *Stresses* 3(4): 749–761.

Santoyo, G. 2022. How plants recruit their microbiome? New insights into beneficial interactions. *Journal of Advanced Research* 40: 45–58.

Singh, R., K. D. Pandey, M. Singh, S. K. Singh, A. Hashem, A. B. F. Al-Arjani, E. F. Abd_Allah, P. K. Singh, and A. Kumar. 2022. Isolation and characterization of endophytes bacterial strains of *Momordica charantia* L. and their possible approach in stress management. *Microorganisms* 10(2): 290. https://doi.org/10.3390/microorganisms10020290

Sturz, A. V., and J. Nowak. 2000. Endophytic communities of rhizobacteria and the strategies required to create yield enhancing associations with crops. *Applied Soil Ecology* 15(2): 183–190.

Tariq, A., S. Guo, F. Farhat, and X. Shen. 2025. Engineering synthetic microbial communities: Diversity and applications in soil for plant resilience. *Agronomy* 15(3): 513. https://doi.org/10.3390/agronomy15030513

Wei, W., X. Wang, Z. Xie, et al. 2017. Evaluation of sanitizing methods for reducing microbial contamination on fresh strawberry, cherry tomato, and red bayberry. *Frontiers in Microbiology* 8: 2397.

Woo, S. L., and O. Pepe. 2018. Microbial consortia: Promising probiotics as plant biostimulants for sustainable agriculture. *Frontiers in Plant Science* 9: 1801.

Yassin, L. K., M. M. Nakhal, A. Alderei, et al. 2025. Exploring the microbiota-gut-brain axis: Impact on brain structure and function. *Frontiers in Neuroanatomy* 19: 1504065.

Zhang, X., B. Li, Z. Zhang, Y. Chen, and S. Tian. 2020. Antagonistic yeasts: A promising alternative to chemical fungicides for controlling postharvest decay of fruit. *Journal of Fungi* 6(3): 158. https://doi.org/10.3390/jof603015

3 Molecular Aspects of Protecting and Promoting Plant Growth by Microbial Consortia

Alejandra Chacón-López,
Adriana Aranda-Vázquez,
José Luis Hernández-Flores,
Ulises Miguel López-García,
Efigenia Montalvo González,
Rosa Isela Ortiz Basurto,
María de Lourdes García-Magaña,
and Selene Aguilera-Aguirre

3.1 INTRODUCTION

Plant growth promotion by microorganisms characteristically involves symbiotic associations, where both the plant and microorganisms are benefited. These interactions can occur within the soil surrounding plant roots (rhizosphere), inside plant tissues, or leaf surfaces. The microbial species can be single organisms or microbial consortia that work synergistically to promote plant growth and health (Hasan et al., 2024; Puranik et al., 2023; Sruthy et al., 2023). Microbial consortia, formed by complex communities of bacteria, fungi, and other microorganisms, play a key role in plant growth promotion, nutrition improvement, abiotic stress tolerance, and protection against phytopathogens (Ram et al., 2022; Singh et al., 2025; Aravindharajan et al., 2025). These consortia facilitate the accessibility of vital nutrients such as nitrogen, potassium, and phosphorus and participate in the decomposition of organic matter, improving the soil structure. Concerning plant growth promotion, they can affect several metabolic processes at the molecular level, such as the production of phytohormones, the cell growth rate, and differentiation processes in plants (Santoyo et al., 2021). Beneficial microorganisms play a critical role in defense against diseases and pests by producing antimicrobial metabolites that inhibit phytopathogen growth or induce mechanistic defenses, improving plant abilities to resist attacks. Among these molecular defense mechanisms are the activation of plant resistance, the production of volatile compounds, and the alteration of the balance of the microbiome in

DOI: 10.1201/9781003664451-3

favor of beneficial organisms against pathogens (Minchev et al., 2021). Thus, microbial consortia can act as biological regulators, promoting greater resilience of plants in the face of stressful conditions, such as extreme temperatures, droughts, or contaminated soils (Petrillo et al., 2022; Sruthy et al., 2025). Integrating diverse microorganisms with complementary capabilities leads to sustainable agriculture and less dependence on pesticides and fertilizers, which can negatively affect human health and the environment (Soltero-Rios et al., 2024). Therefore, studying the molecular basis of microbiome–plant interaction is critical to safeguard and enhance plant growth. The knowledge on the molecular level can revolutionize various aspects of agriculture, delivering better and more sustainable solutions.

3.2 MICROBIAL CONSORTIA

Microbial consortia are dynamic and diverse communities of microorganisms that interact with each other and plants in a symbiotic or mutualistic relationship. The different types of microbial consortia in plant growth promotion and protection can be recognized according to their specific roles in supporting plant health through direct and indirect mechanisms (Tabacchioni et al., 2021).

3.2.1 Direct Mechanisms

Among the direct mechanisms are nitrogen fixation, iron acquisition, and phosphorus and potassium solubilization (Figure 3.1). Bacteria that fix atmospheric nitrogen and make it available to plants for growth could form a consortium, which helps plants grow in nitrogen-poor soils.

The genera *Rhizobium*, *Bradyrhizobium*, *Azospirillum*, and *Azotobacter* are recognized as the principal bacterial genera involved mainly in nitrogen fixation (Barua et al., 2023; Novello et al., 2024). Also, other microorganisms, such as the Actinomycetes *Parasponia* spp. and *Frankia* sp., are well documented in nitrogen fixation (Santi et al., 2013; Soumare et al., 2020).

Solubilization of phosphate and potassium is another microbial strategy that includes microorganisms that can turn insoluble phosphorus or potassium into soluble forms, increasing their availability and promoting healthy root growth and better plant development (Suleman et al., 2018). *Bacillus* spp., *Enterobacter* spp., *Burkholderia* spp., *Pseudomonas* spp., *Rhizobium* spp., *Mesorhizobium* spp., and *Serratia* spp. are some of the genera that form consortia and can solubilize insoluble phosphorus molecules in the soil, making them more accessible to plants (Hasan et al., 2024). Numerous microorganisms can solubilize potassium in the soil, particularly bacterial and fungal species such as *Bacillus edaphicus*, *Bacillus mucilaginosus*, *Ferrooxidans* sp., *Acidothiobacillus* sp., *Pseudomonas* sp., and *Burkholderia arearesome* (Prajapati et al., 2013).

Regarding environments where iron is limited, siderophore-producing consortia can play a crucial role in acquiring iron, which is necessary for survival and growth. The siderophore-producing consortia are composed of microorganisms that produce siderophores that bind and sequester iron, essential for many physiological

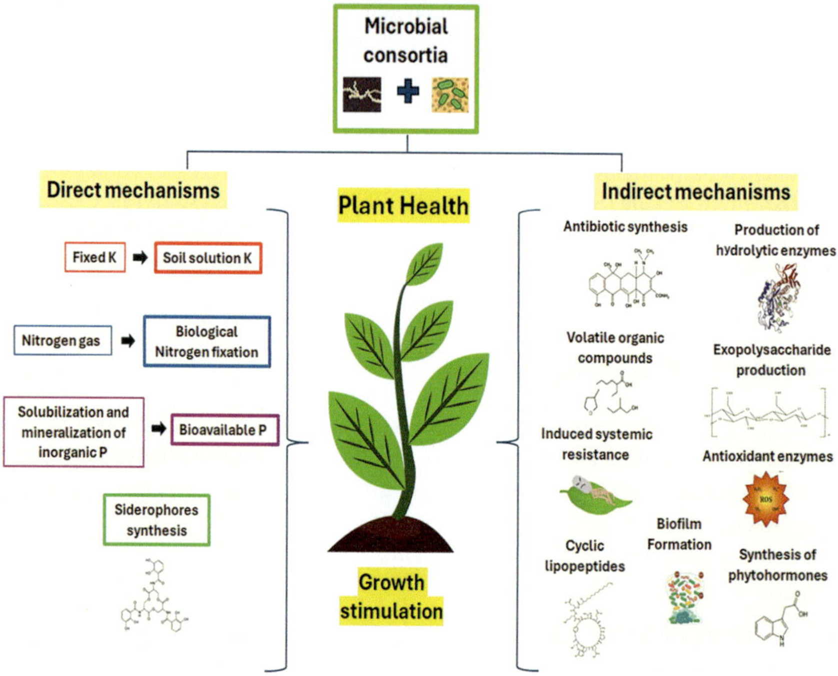

FIGURE 3.1 Direct and indirect mechanisms mediated by microbial consortia with beneficial effects on host plants.

processes in plants, especially in iron-deficient soils. The main Gram-negative bacteria that synthesize siderophores are *Enterobacter* and *Pseudomonas*. On the other hand, some Gram-positive species, like *Bacillus* and *Rhodococcus*, can also produce these metabolites (Goswami et al., 2016; Xie et al., 2024a). Siderophore-producing consortia underline the importance of microbial collaboration in sustaining life and highlight their potential in agriculture.

Mycorrhizal fungi form symbiotic relationships with plant roots, improving water and nutrient uptake, such as phosphorus, and providing plants with resistance to soil-borne pathogens. Several genera, such as *Glomus* spp. and *Trichoderma* spp., can decompose organic matter, including cellulose and lignin, converting them into nutrients that plants can use. Upamanya et al. (2020) reported a fungi consortium composed of *Trichoderma harzianum*, *Metarhizium anisopliae*, and *Beauveria bassiana*, which promote eggplant growth through the production of growth hormones (Upamanya et al., 2020). It is interesting how bacteria and fungi join forces and encourage plant growth. Recent studies have evaluated the combination of fungi-bacteria consortia, which form a network that helps them absorb nutrients such as phosphorus and minerals (Nunes et al., 2024) (Table 3.1). For example, *Trichoderma* and *Bacillus* species effectively promote plant growth and health, revealing their potential as a consortium and improving the efficacy of increasing agro-productivity in multiple crops (Santoyo et al., 2024).

TABLE 3.1
Examples of Microbial Consortia and Their Beneficial Effects on Crop Plants

Microbial Consortia	Plant Growth Promoting and Protection Activities	Crop	Reference
Bacillus atrophaeus *Pseudomonas* sp. *Trichoderma harzianum*	Promotion of growth. Attenuation of drought stress.	Maize	Yim et al. (2025)
Bacillus subtilis *Bradyrhizobium japonicum* *T. harzianum*	Plant growth promotion. Enhanced yield.	Soybean	Travençoli-Rossetim et al. (2025)
Pseudomonas sp. *Azotobacter* sp.	Promotion of growth.	Wheat	Dobrzyński et al. (2025)
Cellulomonas iranensis *Pseudomonas* sp.	Phosphorus solubilization.	Rice	Che et al. (2024)
Rhizophagus irregularis *Bacillus amyloliquefaciens* *Pseudomonas brassicacearum*	Increasing shoot concentrations of ABA, JA, and IAA (Indole-3-acetic acid) Drought protection.	Potato	Mamun et al. (2024)
Chryseobacterium indologenes *Stenotrophomonas maltophilia*	Promotion of growth. Salinity tolerance.	Chickpea	Sathya et al. (2024)
Bacillus sp. *Chryseobacterium* sp. *Enterobacter* sp. *Pseudomonas* sp. *Stenotrophomonas* sp.	Promotion of growth. Tolerance to salinity stress.	Rice	Wang et al. (2024a)
Claroideoglomus claroideum *Naganishia albida* *Burkholderia caledonica*	Increment in biomass. Tolerance to drought stress.	Strawberry	Perez-Moncada et al. (2024)
Trichoderma afroharzianum *Pseudomonas fluorescens* *Bacillus licheniformis* *B. subtilis*	Solubilization of zinc and phosphorus. Biocontrol against *Rhizoctonia bataticola*	Bean	Singh et al. (2024)
B. subtilis *Bacillus aryabhattai*	Tolerance to salinity stress.	Rice	Siddika et al. (2024)
B. subtilis *B. amyloliquefaciens* *P. fluorescens*	Protection and promotion of the growth. Stimulation of indigenous soil microbiome.	Barley	Niewiadomska et al. (2023)
Azotobacter chroococcum *T. afroharzianum*	Plant growth promotion. Drought tolerance.	Tomato	Cirillo et al. (2023)

(Continued)

TABLE 3.1 (*Continued*)
Examples of Microbial Consortia and Their Beneficial Effects on Crop Plants

Microbial Consortia	Plant Growth Promoting and Protection Activities	Crop	Reference
Paenibacillus peoriae *B. subtilis* *Lysinibacillus* sp. *Bacillus simplex* *Pseudomonas chlororaphis* *Beauveria bassiana* *Scopulariopsis brevicaulis* *Trichoderma gamsii* *Trichoderma ghanense* *Trichoderma lignicola*	Biocontrol against *Fusarium oxysporum* f. sp. *physali*.	Gooseberry	Garcia et al. (2023)
Serendipita indica *Rhizophagus intraradices* *A. chroococcum*	Improved properties under both well-watered and drought stress conditions.	Maize	Tyagi et al. (2023)
Burkholderia phytofirmans *Enterobacter aerogenes* *P. fluorescens*	Growth promotion under salinity stress.	Maize	Afzal et al. (2023)
B. amyloliquefaciens *Bacillus* sp.	Plant growth promotion and alleviation of drought stress.	Spinach	Petrillo et al. (2022)
Bacillus cereus *Bacillus thuringiensis* *Bacillus tropicus*	Nitrogen fixation, phosphate solubilization, ammonia production, and tolerance to drought stress.	Maize	Azeem et al. (2022)
Glomus versiforme *Bacillus methylotrophicus*	Increase chlorophyll, carotenoids Drought tolerance.	Tobacco	Begum et al. (2022)
Bacillus sp. Mycorrhizal fungi	Plant growth promotion.	Wheat	Yadav et al. (2022)
Aeromonas encheleia *Pseudomonas azotoformans*	Increase germination rate, root elongation, seedlings fresh weight, dry weight.	Maize	Houida et al. (2022)
Arthrobacter sp. *Enterobacter* sp. *Brevibacterium* sp. *Plantibacter* sp.	Increase seed germination, plant shoot/root, and productivity; improve soil fertility.	Barley	Kaur et al. (2022)
B. bassiana Mycorrhizal fungi	Biocontrol against *Spodoptera littoralis*	Cotton	Metwally et al. (2022)
Piriformospora indica *Azorhizobium caulinodans*	Promotion of growth. Mitigation of saline stress	Tomato	Xu et al. (2022)
Glomus versiforme *Bacillus methylotrophicus*	Increased chlorophyll, carotenoids, photosynthesis. Drought tolerance.	Tobacco	Begum et al. (2022)

(Continued)

TABLE 3.1 (*Continued*)
Examples of Microbial Consortia and Their Beneficial Effects on Crop Plants

Microbial Consortia	Plant Growth Promoting and Protection Activities	Crop	Reference
B. licheniformis *Aspergillus violaceofuscus*	Improve catalase activity and proline content.	Tomato	Muthuraja and Muthukumar (2022)
Erwinia sp. *Chryseobacterium arthrosphaerae* *Pseudomonas gessardii*	Nitrogen fixation and mineral solubilization.	Barley	Kaur et al. (2022)
Glomus spp. *Bacillus* sp.	Reduction of water stress.	Corn	Wilmowicz et al. (2022)
Trichoderma koningiopsis *Trichoderma asperellum*	Osmoprotection and defense system. Improved grain production.	Rice	Anshu et al. (2022)
Paenibacillus sp. *Arthrobacter* sp.	Induction of resistance against *Zymoseptoria tritici.* Tolerance to drought and salinity stresses.	Wheat	Samain et al. (2022)
Enterobacter cloacae Serratia liquefaciens	Increment of proline, glycine betaine, free amino acids, and soluble sugars.	Maize	Ali et al. (2022)
T. gamsii *T. harzianum*	Growth promotion. Biocontrol against *Fusarium pseudograminearum.*	Wheat	Stummer et al. (2022)
Funneliformis mosseae *Glomus fasciculatum* *Bacillus* sp.	Production of lytic enzymes and siderophores. Biocontrol against *Fusarium oxysporum* f. sp. *lycopersici*	Tomato	Devi et al. (2022)
Bacillus nematocida *Pochonia chlamydosporia*	Production of volatile substances, proteases, and chitinases. Biocontrol against *Meloidogyne incognita.*	Tomato	Bo et al. (2022)
Bacillus pumilus *B. subtilis*	Production of surfactin, iturin, and fengycin. Biocontrol against *Botrytis cinerea.*	Tomato	Bouchard-Rochette et al. (2022)
P. fluorescens *Bacillus megaterium*	Promotion of growth. Salinity stress tolerance.	Wheat	Khan et al. (2022)
Pantoea cypripedii *Pseudomonas putida* *Cyanobacteria*	Increment in the aerial growth of tomato seedlings.	Tomato	Toribio et al. (2022)

(Continued)

TABLE 3.1 (*Continued*)
Examples of Microbial Consortia and Their Beneficial Effects on Crop Plants

Microbial Consortia	Plant Growth Promoting and Protection Activities	Crop	Reference
Klebsiella sp. *Kluyvera sp.* *Enterobacter sp.*	*Improve soil nitrogen, phosphorus, and potassium content.* *Promote plant growth.*	Maize	El-Saadony et al. (2022)
Bacillus sp. *Enterobacter sp.* *P. fluorescens*	Regulation of water uptake. Photosynthetic performance. Tolerance to drought stress.	Maize	Saleem et al. (2021)
Bacillus sp. *Delftia sp.* *Enterobacter sp.* *Achromobacter sp.*	Improve the growth under salinity stress.	Tomato	Kapadia et al. (2021)
P. fluorescens *Paenibacillus alvei*	Enhanced resistance against *Phytophthora capsici*.	Tomato	Mhlongo et al. (2021)
Arthrobacter chlorophenolicus *Enterobacter* sp. *Pseudomonas aeruginosa*	Reduce application rates of N-fertilizer.	Wheat	Kumar et al. (2021)
P. fluorescens *Actinomycetes* *Arthrobacter*	Enhance yield and nutrient availability.	Soybean	Yaduwanshi et al. (2021)
Bacillus *Pseudomonas* *Trichoderma*	Defense induction. Biocontrol against *B. cinerea* and *F. oxysporum*.	Tomato	Minchev et al. (2021)
P. putida *P. fluorescens B. thuringiensis* *Pseudomonas synxantha*	Decreased Cd, Pb, and Zn accumulation throughout the plant. Increase plant biomass.	Spinach	Shilev et al. (2020)
B. subtilis *P. fluorescens*	Increment in plant growth and yield.	Bean	Kumar et al. (2020)
T. harzianum *B. bassiana* *Metarhizium anisopliae*	Growth promotion. Resistance induction against *Phomopsis*, v*Alternaria*, and *Fusarium*.	Eggplant	Upamanya et al. (2020)
B. megaterium *B. subtilis*	Increase dry weight.	Alfalfa	Zhu et al. (2020)
B. licheniformis, *T. afroharzianum* *Trichoderma yunnanense*	Enhance net photosynthesis and growth.	Wheat	Oljira et al. (2020)
T. harzianum *Bacillus* spp.	Mitigation of cold stress.	Corn	Moradtalab et al. (2020)
T. harzianum *M. anisopliae* *B. bassiana*	Growth promotion.	Eggplant	Upamanya et al. (2020)

3.2.2 Indirect Mechanisms

Microbial consortia also use indirect mechanisms to protect plants from or reduce the impact of phytopathogens. These strategies include stress management, hydrolytic enzyme production, synthesis of antibiotics in response to plant pathogens or disease resistance, the production of volatile organic compounds (VOCs), exopolysaccharide (EPS) production, and the synthesis of phytohormones (Hasan et al., 2024) (Figure 3.1). Additionally, microbial consortia promote plant growth and play an essential role in improving tolerance to abiotic stress. The ability to promote plant growth and improve stress tolerance simultaneously offers several key advantages in agriculture, especially in adverse soil conditions. For example, salt-tolerant consortia are particularly useful in saline soils, as has been reported for *Bacillus subtilis* and *Bacillus aryabhattai* consortium, which were able to enhance rice growth under saline stress conditions (Siddika et al., 2024) or *Bacillus amyloliquefaciens*, an arbuscular mycorrhizal fungus, and *Pseudomonas brassicacearum* consortium which significantly reduced irreversible drought stress on potato crop (Mamun et al., 2024).

One of the most important ways that beneficial microorganisms help plants defend against pathogens is through a process called induced systemic resistance (ISR), which is a defense mechanism where the plant activates its immune system, called priming, in response to beneficial microorganisms, enabling it to recognize and defend against phytopathogens, improving plant health, growth, and resilience (Srivastava et al., 2024). Abbasi et al. (2022) developed a consortium of *Streptomyces* species that improved the quality of bell pepper fruits and induced plant defense priming (Abbasi et al., 2022). In tomato plants, *Pseudomonas fluorescens* and *Paenibacillus alvei* consortia primed with a metabolic reprogramming, and as a result, an enhanced resistance against *Phytophthora capsici* infections was observed (Mhlongo et al., 2021). Microbial consortia offer significant advantages over individual microbial strains, mainly when applied in areas such as agriculture. In a consortium, microorganisms can have synergistic interactions that enhance the overall effect through direct and indirect mechanisms to promote plant growth, protect plants from pathogens, and improve their resilience to environmental stresses (Negi et al., 2024; Nunes et al., 2024). Certainly, microbial consortia offer advantages regarding synergy, efficiency, and adaptability compared to individual strains with more limited capabilities. From an agricultural standpoint, microbial consortia emerge as a favorable solution due to the sustained advances in crop yields that these systems offer, which provide greater agrarian sustainability.

The Figure 3.1 shows two main sections: the direct and indirect mechanisms of plant growth–promoting rhizobacteria (PGPR). These mechanisms involve a direct interaction between the bacteria and the plant, actively promoting its growth.

3.3 MOLECULAR PATHWAYS OF PLANT GROWTH PROMOTION

Microbial consortia are essential players in plant nutrition. Many vital plant nutrients, such as nitrogen, phosphorus, and iron, occur in the soil in an inaccessible form for plants. Through the ability to fix nitrogen, mobilize phosphorus and iron, and improve soil structure, microbial consortia contribute to sustainable agronomic practices by reducing the requirement for chemical fertilizers (Santoyo et al., 2021). The molecular pathways through which microbial consortia enhance plant growth are complex and can vary depending on the specific types of microorganisms involved in the consortium and the environmental conditions (Arif et al., 2020). For this reason, molecular studies about plant growth promotion by microbial consortia are essential to explore the key molecular interactions, gene expression changes, and signaling pathways that microbes trigger in plants.

3.3.1 Nitrogen Fixation

Along with many reports, nitrogen-fixing soil bacteria source significant nitrogen required in agriculture (Wang et al., 2022). These bacteria contain different genes needed for the metabolism of nitrogen, whose expression depends on the nitrogen availability in the environment. Variants of the soil bacterial consortia have been reported, which can perform nitrogen fixation and contribute to nitrogen availability (Robson & Postgate, 1980; Timofeeva et al., 2023). Thus, the application of bacteria consortia (comprised *Pseudomonas putida* and *P. fluorescens*) reduced application rates of N-fertilizer for rice plants (Etesami & Alikhani, 2016).

Nitrogen-fixing is carried out by diazotrophs using a nitrogenase, an enzyme catalyzing the biological N-reducing reaction. The *nifH* gene, encoding nitrogenase reductase, is widely used to analyze the ability of bacteria to fix nitrogen (Dos Santos et al., 2012). The molecular pathway of nitrogen fixation is a highly regulated and intricate process involving several key molecular steps, enzymes, and regulatory factors (Curatti et al., 2005). The core of nitrogen fixation is the nitrogenase, which catalyzes the reduction process of atmospheric nitrogen to ammonia. This process requires a significant amount of energy and the presence of several cofactors. The nitrogenase complex consists of a dinitrogenase, the central enzyme responsible for the reduction of atmospheric nitrogen, and a dinitrogenase reductase, which provides the necessary electrons to dinitrogenase for nitrogen reduction. It contains an iron-sulfur cluster and is responsible for transferring electrons from reduced ferredoxin or nicotinamide adenine dinucleotide (NADH) to dinitrogenase (Xu & Wang, 2023).

Nitrogen fixation is typically activated when the nitrogen levels are low in the microbial environment. The bacteria sense nitrogen availability through regulatory proteins like NifA, which activates the expression of the *nif* genes when nitrogen is scarce. On the other hand, the ammonia produced through nitrogen fixation can inhibit the expression of *nif* genes when the nitrogen supply becomes sufficient (Zhang et al., 2023). Once bacteria fix nitrogen into ammonia, it must be assimilated by the plant. Plants can incorporate ammonia directly into amino acids and other nitrogen-containing compounds (Zayed et al., 2023). The enzyme glutamine synthetase plays a critical role in the assimilation of ammonia in plants because it catalyzes the conversion of ammonia and glutamate into glutamine, an essential amino acid. This process is the primary mechanism of nitrogen incorporation into plant metabolic pathways (Fortunato et al., 2023).

Rhizobia soil bacteria are the best-known nitrogen-fixing microbes. They form symbiotic relationships with leguminous plants, forming nodules where nitrogen fixation occurs. Rhizobia fix nitrogen using the enzyme nitrogenase, while the plant provides sugars and other organic compounds to the bacteria (Wang et al., 2018). The consortium of Rhizobia and other microvorganisms, like actinobacteria or mycorrhizal fungi, contributes efficiently to nitrogen fixation (Calderon & Dangi, 2024; Singh et al., 2021).

3.3.2 Phosphorous Solubilization

Another essential nutrient for plants is phosphorus; unfortunately, most of the phosphorus in soil is insoluble and not readily available to plants. Phosphorus-solubilizing bacteria use several molecular pathways to solubilize and release phosphorus from insoluble inorganic compounds (such as calcium phosphate, iron phosphate, or aluminum phosphate) or organic phosphorus compounds, making them accessible to plants (Elhaissoufi et al., 2022; Khourchi et al., 2022). Dephosphorylating bacteria-based microbial consortia are promising alternatives to supply phosphorus to plants, as demonstrated by the consortium of *Bacillus licheniformis* and *Pseudomonas aeruginosa*, which enhanced the growth of *Brassica campestris* through phosphorus solubilization (Maheshwari et al., 2015). The primary molecular mechanisms to solubilize phosphorus include the production of organic acids, enzymes, and various metabolic processes, including secretions of enzymes (phytase and phosphatase), acids (organic, inorganic), and chelation (siderophores, extracellular polysaccharides) to produce dissolved phosphate (Liang et al., 2020).

The production and secretion of organic acids are the main mechanisms of phosphorus solubilization by phosphorus-solubilizing bacteria. It has been reported a few genes that regulate phosphate solubilization, such as *gdh* (gluconate dehydrogenase), *gcd* (glucose dehydrogenase), *gadγ* (gluconate 2-dehydrogenase), and *ylil* (aldose sugar dehydrogenase) (Kour et al., 2021; Suleman et al., 2018). Other genes associated with organic acid metabolism, such as *ldh* (L-lactate dehydrogenase), *gyaR* (glyoxylate reductase), and *poxB* (pyruvate dehydrogenase), among others, are also relevant to phosphate solubilization (Garcia-Sanchez et al., 2023). Many Gram-negative bacteria use periplasmic glucose oxidation, mediated by pyrroloquinoline quinone (PQQ) and glucose dehydrogenase (GDH) enzymes, to produce gluconic acid that encodes glucose dehydrogenase (Elhaissoufi et al., 2022; Rawat et al., 2021; Suleman et al., 2018). Most *pqq* genes are present in *Pseudomonas*, *Burkholderia*, *Azotobacter*, and *Mycobacterium* (Shi et al., 2022).

The most studied enzymes involved in phosphate solubilization are the alkaline phosphatases, coded by the *phoD* and *phoA* genes; acid phosphatases, coded by the *phoC* gene; phytase, coded by the *appA* gene; and C-P cleavage enzyme coded by the *phn* gene. The extracellular polyphosphatase and the polyphosphate kinase are coded by the *ppx* and *ppk* genes, respectively (Dai et al., 2020). The phosphorus deprivation response activates numerous genes that can participate in the regulation of this process; among them are *phoU*, *phoR*, and *phoB* genes, which control the expression of genes encoding alkaline phosphatases, especially under low phosphorus conditions (Wan et al., 2020). Phytase enzymes produced by *Aspergillus* and *Bacillus* species catalyze phytate hydrolysis to release phosphate, making it available to plants. The *bpp* gene mainly encodes phytase, and currently, this gene has been isolated and investigated only in a few strains, such as *B. subtilis* and *B. licheniformis* (Suleimanova et al., 2015). Phosphorus-solubilizing bacteria are a key component in sustainable agriculture, as they optimize the use of available nutrients in the soil, improve plant health, and reduce dependence on chemical inputs.

3.3.3 Siderophores for Iron Bioavailability

Iron is indispensable for fundamental cellular processes in microorganisms, yet its bioavailability is inherently limited due to its oxidized state (Fe^{3+}), which is mainly insoluble under natural conditions (Hasnain & Kanwal, 2020). Some bacteria genera, such as *Pseudomonas*, overcome the iron limitation through the synthesis and secretion of siderophores, which are secondary metabolites that form soluble complexes with free Fe^{3+} in the environment (Lozano-Gonzalez et al., 2023). Siderophores are vital in bacterial and plant growth, particularly in iron-limited environments. Siderophore-producing bacteria promote plant health and productivity by facilitating iron acquisition and enhancing plant–microbe interactions (Liu et al., 2022).

Siderophores are synthesized through two principal pathways: one involves the nonribosomal peptide synthetase (NRPS), and the other is the nonribosomal-independent synthesis (NIS). The NRPS pathway involves complex, multi-modular enzymes that perform coordinated steps to synthesize nonribosomal peptides. Each NRPS module typically consists of three key domains: the adenylation (A) domain, the peptidyl carrier protein (PCP) domain, and the condensation (C) domain. NRPS modules assemble the amino acids into a siderophore peptide chain with a defined structure, ensuring their structural and functional integrity (Carroll & Moore, 2018; Xie et al., 2024a). Siderophores can also be synthesized through the NIS pathway, which operates through several enzymes, utilizing dicarboxylic acids and either diamines or amino alcohols as the foundational building blocks. These components are linked via specific chemical bonds to produce hydroxamate or carboxylate-type siderophores (Xie et al., 2024a).

With the expansion of genomic sequencing and the progress in bioinformatics, researchers have identified a broader array of genes that encode NIS synthases. Gene clusters containing these enzymes have been discovered in over 40 different microorganisms, including plant pathogens, saprophytic fungi, and the symbionts of both plants (Blanco Nouche et al., 2023; Carroll & Moore, 2018). In Streptomyces *griseus*, a NIS synthase (DesD), was identified. This enzyme is crucial to producing desferrioxamine, a siderophore involved in the growth and differentiation of this bacterium (Kadi et al., 2007; Yamanaka et al., 2005). The discovery of DesD and similar enzymes underlines the importance of the NIS pathway in siderophore synthesis.

The regulation of siderophore biosynthesis is a highly intricate process that involves several molecular mechanisms, including iron-sensing systems, regulatory proteins, and transcriptional control. The Fur protein is the primary regulator of iron homeostasis in most bacteria, including those that produce siderophores. Fur responds to the intracellular concentration of iron and controls the expression of many genes, including those involved in siderophore biosynthesis (Butcher et al., 2012; Santos et al., 2020). Under sufficient iron, Fur binds to Fe^{2+} (ferrous iron) and forms a Fur-Fe^{2+} complex. This complex binds to specific DNA sequences called Fur boxes, located in the promoters of genes involved in siderophore synthesis. When iron levels are low, the complex dissociates from the Fur boxes, allowing the transcription of the genes involved in the siderophore biosynthesis, thus enabling the bacterium to produce siderophores (Gonzalez et al., 2014; Xie et al., 2024b).

While Fur is the primary regulator, other regulatory systems also control siderophore biosynthesis in response to low iron availability. Global regulators such as DksA and RpoS coordinate the induction of siderophore biosynthesis in stressed environments (Qin et al., 2020; Zhang et al., 2019). In some bacteria, such as *P. aeruginosa, quorum sensing* (*Qs*) regulates the production of siderophores (Wei et al., 2013). *Qs* allows bacteria to sense their population density and adjust gene expression accordingly. This system often involves small molecules like acyl-homoserine lactones (AHLs) and regulatory proteins like LasR and RhlR that modulate siderophore biosynthesis in coordination with other metabolic processes (Wilder et al., 2011; Wu et al., 2013). These regulatory systems ensure that bacteria only produce siderophores when required for iron acquisition. The molecular processes of nitrogen fixation, phosphate solubilization, and siderophore synthesis are interrelated. The synergetic effect of the simultaneous introduction of a consortium of bacteria exhibiting these activities into the soil can stimulate plant growth and development, meeting the goals and objectives of sustainable agriculture.

3.3.4 Bacterial Hormones

The growth hormones produced by microbial consortia are essential to improve plant growth, development, and productivity. These microorganisms synthesize and release various hormones, such as auxins, cytokinins, gibberellins, and abscisic acid (ABA), that interact with the hormonal systems of plants, promoting beneficial effects on multiple levels (Sun et al., 2024). Auxins are essential for cell elongation and root formation. The most common of these bacterial auxins is indoleacetic acid (IAA), although other auxins can also be produced. Microbial biosynthesis of IAA can be classified into tryptophan-dependent and tryptophan-independent pathways. In the tryptophan-dependent pathways, there are different intermediate metabolites (Fiodor et al., 2023). Current research has roughly divided them into five pathways: the indole-3-acetonitrile pathway (IAN), the tryptophan side-chain oxidase pathway (TSO), the tryptamine pathway (TAM), the indole-3-pyruvic acid pathway (IPA/IPyA), and the indole-3-acetamide (IAM) pathway (Tang et al., 2023). The IAA biosynthesis pathway through IAM has been extensively studied. In *Pseudomonas savastanoi*, two key genes, *iaaM* and *iaaH*, encoding tryptophan monooxygenase and indole-3-acetamide hydrolase, respectively, have been identified first (Ahemad & Kibret, 2014). In this pathway, tryptophan is first converted to IAM by tryptophan monooxygenase, followed by hydrolysis of IAM to IAA and ammonia by indole-3-acetamide hydrolase. Current studies have shown that the IAM pathway exists in most species of bacteria and other microorganisms, including fungi, actinomycetes, and cyanobacteria, which comprise microbial consortia (Tsavkelova et al., 2006).

The gibberellins are small molecules of a large group of tetracyclic diterpenoid carboxylic acids based on the ent-gibberellane carbon skeleton. Gibberellin biosynthesis involves the key intermediate, isopentenyl pyrophosphate (IPP) (Hedden, 2020). Three molecules of IPP are joined successively to one dimethylallyl diphosphate (DMAPP) molecule in a condensation reaction by isoprenyl diphosphate synthase (GGPPS), forming GGPP. The enzymes GGPPS, encoded by *blr2148*, *mlr6368*, and *NGR_a02740* in *Bradyrhizobium japonicum*, synthesize GGPP for gibberellin biosynthesis. GGPP serves as a precursor for ent-kaurene in two steps via ent-copalyl diphosphate. These sequential reactions are catalyzed by separate diterpene cyclases, ent-copalyl diphosphate synthase (CPS), and ent-kaurene synthase (ent-KS). Ent-kaurene is converted by a series of cytochrome P450s into gibberellins (Keswani et al., 2022). Bacterial gibberellins stimulate the division and expansion of cells in plants, which promotes vertical growth, plant height, and the size of plant structures.

Cytokinins also play a crucial role in multiple physiological processes, including cell division, organ formation, shoot and root development, chlorophyll biosynthesis, and nutrient transport. Cytokinin biosynthesis is initiated by the enzyme isopentenyltransferase (IPT) encoded by the *ipt* gene. Several bacteria were demonstrated to synthesize cytokinins, such as zeatin, zeatin riboside, and isopentenyladenine. These bacteria include *B. japonicum*, *B. licheniformis*, *P. fluorescens*, and *P. putida* (Timofeeva et al., 2024).

Beneficial bacteria can also produce ABA, which modulates numerous plant physiological processes such as stress tolerance, senescence, and bud rest. Several PGPR, including *Azospirillum lipoferum*, *Achromobacter xylosoxidans*, *B. licheniformis*, *Bacillus pumilus*, and *Brevibacterium halotolerans*, synthesize ABA (Tsukanova et al., 2017). The biosynthesis of hormones by microbial consortia is essential to promote plant growth, improve nutrient absorption, increase resistance to stress, and contribute to more sustainable agriculture.

3.4 MOLECULAR PATHWAYS OF PLANT PROTECTION

Microbial consortia are recognized for their role in plant protection against pathogens. The molecular mechanisms through which these consortia protect plants involve a variety of pathways that enhance plant immunity. Microbial consortia can directly combat phytopathogens through several antagonistic strategies that involve the production of antimicrobial compounds and induction of plant defense responses.

3.4.1 Antimicrobial Compounds

Beneficial microorganisms, such as *Pseudomonas* and *Bacillus* species, produce various antibiotics and secondary metabolites that inhibit plant phytopathogens. These compounds directly affect pathogenic microorganisms and trigger plant responses that strengthen their defense mechanisms. Antibiotics synthesized by microbial consortia play a crucial role in plant protection. *Bacillus* and *Pseudomonas* could produce several kinds of secondary metabolites with antimicrobial activities (Wang et al., 2024b).

Nonribosomal peptide synthetases (NRPS) and polyketide synthases (PKS) synthesize the two most common antimicrobial metabolites produced by *Bacillus*. Lipopeptides disrupt the integrity of the cell membranes, leading to the outflow of cellular components, while polyketides target protein synthesis or disrupt the cell membrane (Rungsirivanich et al., 2021). In addition to lipopeptides and polyketides, *Pseudomonas* also produces other antimicrobial metabolites, such as pyrrolnitrin and phenazines. These antibiotics display effects such as blocking DNA replication, disrupting membrane integrity (Park et al., 2011).

The lipopeptides produced by *Bacillus* are mainly categorized as iturins, fengycins, and surfactins. Iturin and fengycin groups exhibit limited antibacterial but strong antifungal activities, while surfactin exhibits antibacterial and antifungal activities (Xiao et al., 2021). Surfactin disrupts cell membrane integrity, causing an osmolarity imbalance, inhibits pathogen protein synthesis, and interferes with cellular biochemical pathways (Yuan et al., 2018). In *B. subtilis*, surfactin was also crucial in plant rhizosphere colonization, swarming motility, and biofilm formation (Zhen et al., 2023). Fengycin and iturins exert antifungal activity by targeting and disrupting cell membrane integrity (Romero et al., 2007). In *Bacillus*, the regulation of lipopeptide production is mediated by several factors, such as PhrC or ComX with the ComA/ComP two-component system, which plays a key role in regulating lipopeptide biosynthesis (Auchtung et al., 2006). Also, several global transcription regulators, including AbrB, Spo0A, and CodY, participate in regulating lipopeptide biosynthesis (Sun et al., 2021).

Polyketides represent another major group of antimicrobial metabolites in *Bacillus* strains. Among these polyketides are bacillaene, difficidin, macrolactin, and amicoumacin, which exhibit antibacterial activities (Wang et al., 2024b). These polyketides are the major synthesized by *Bacillus velezensis*, *B. amyloliquefaciens*, and *B. subtilis* (Fazle-Rabbee & Baek, 2020; Kaspar et al., 2019). Regarding their regulation, it has been suggested that Spo0A regulates the biosynthesis of *Bacillus* polyketide (Liu et al., 2023).

The *Pseudomonas* genus also produces various types of nonribosomal lipopeptides (NRLPs). The NRLPs produced by *Pseudomonas* can be divided into cyclic lipopeptides (CLPs) and linear lipopeptides (LLPs) (Gotze et al., 2019). LLPs include syringafactin and corrugatin. Among the most studied CLPs are viscosin, amphisin, tolaasin, and syringomycin (Gotze & Stallforth, 2020).

Phenazines are a group of heterocyclic compounds in *Pseudomonas*, considered one of the most versatile secondary metabolites (Chi et al., 2024). These compounds exhibited broad-spectrum antagonistic activities against several phytopathogens, such as *Fusarium* sp., *Colletotrichum* sp., and *Aspergillus* sp. (Biessy & Filion, 2018). Phenazines inhibit microbial growth through multiple mechanisms, including blocking DNA replication, targeting the electron transport chain, disrupting membrane integrity, and generating hydroxyl radicals that destroy macromolecules (Schmitz & Rosenbaum, 2020). Therefore, using these microorganisms and their antimicrobial compounds offers an environmentally friendly strategy to promote plant health, reduce dependence on chemicals, and promote more environmentally friendly agriculture.

3.4.2 Induction of Plant Defense Responses

Beneficial bacteria can trigger plant defense responses through ISR. ISR is typically associated with beneficial microbes and is generally independent of plant salicylic acid (SA) signaling. It often involves jasmonic acid (JA) and ethylene (ET) signaling pathways that activate defense genes and increase phytopathogen resistance. Beneficial microorganisms induce a physiological state of plant host called "priming", in which plants display stronger and faster defense responses against pathogen invasion (Yu et al., 2022).

The activation of JA signaling by applying methyl jasmonate (MeJA) regulates the resistance level and impacts the structure of the rhizosphere microbial community. SA acts as an endogenous signal, triggering the accumulation of pathogenesis-related (PR) proteins. The *EDS1*, *EDS4*, and *PAD4* genes play regulatory functions in SA biosynthesis (Roychowdhury et al., 2024). Although the initial research disregarded the involvement of SA in the ISR, it was demonstrated that beneficial microorganisms, such as *Bacillus* and *Trichoderma*, can control plant disease by activating SA and JA/ET signaling pathways (Dutta et al., 2023; Yang et al., 2023).

In a microbial consortium, beneficial bacteria could produce microbial-associated molecular patterns, which can trigger host immunity (Zamioudis & Pieterse, 2012). The root colonization induces a systemic response in the plants by secondary metabolites named elicitors. These elicitors have multiple roles, for example, the CLPs, which are multifunctional secondary metabolites involved in biofilm formation, motility, and biocontrol due to their antimicrobial activity. These CLPs also have emerged as a significant immunity elicitor of plants (Geudens & Martins, 2018). Surfactin is the best-described CLP among *Bacillus* compounds and is recognized as a systemic resistance trigger. Surfactins are not synthesized ribosomally but rather by NRPSs. In *B. subtilis*, surfactins are formed by a heptamodular NRPS in which the biosynthetic core is coded by the structural genes *srfAA*, *srfAB*, *srfAC*, and *srfAD*. In addition to the structural NRPS genes, several accessory genes encoding transporters and regulatory proteins are coded by *sfp*, *ycxA*, *krsE*, *yerP*, and *comS* genes, which are mainly related to the production yield of surfactin (Theatre et al., 2021). It was reported that surfactin induced the ISR in peanuts, contributing to protecting against *Sclerotium rolfsii*. A similar effect was observed in beans, tomatoes, and tobacco against *Botrytis cinerea* (Cawoy et al., 2014; Ongena et al., 2007) and in maize against *Rhizoctonia solani* (Ali et al., 2022). Induction of ISR by other CLPs produced by *Bacillus* has also been reported. Thus, fengycin produced by *B. subtilis* BBG111 induced a defense state in rice against *R. solani* (Chandler et al., 2015), and in grapevine and tomato triggered ISR against *Plasmopara viticola* and *Sclerotinia sclerotiorum* (Farzand et al., 2019; Li et al., 2019). Other genera, such as *Pseudomonas*, synthesize CLPs involved in plant resistance stimulation, for example, the CLP Massetolide A produced by *P. fluorescens*, which induced ISR in tomato against *Phytophthora infestans* or *Pseudomonas* sp. strain CMR12a, which produces orfamide and sessilin, two other types of CLPs, involved in ISR-triggering in rice and beans against *Magnaporthe oryzae* and *R. solani*, respectively (Ma et al., 2016).

The AHLs, involved in cell-to-cell communication, are another kind of immunity elicitor. These molecules must monitor bacterial behavior according to population density (Hartmann et al., 2021). AHLs of bacteria can elicit different response patterns in the plants depending on the water-soluble AHLs with C4- to C10-residues in the side chain in contrast to hydrophobic ones with C12- and C14-residues (Schikora et al., 2016). An AHL named oxo-C14-HSL, synthesized by several rhizobacteria, can elicit ISR in tomato against *P. infestans*, in wheat against *Puccinia graminis* f. sp. *tritici*, and in barley against *Blumeria graminis* f. sp. *hordei*, *Puccinia hordei*,

Golovinomyces orontii, and *Pseudomonas syringae* DC3000 (Hernandez-Reyes et al., 2014; Schikora et al., 2011; Wehner et al., 2019).

ISR induction can also be achieved by VOCs, which have attracted attention because they induce systemic resistance in the host plant and can act as info chemicals involved in inter-kingdom communication. Based on chemical structure, VOCs can be divided into different classes: benzenoids, alcohols, ketones, alkenes, pyrazines, terpenes, and sulfides. One of the most studied VOC immunity elicitors is 2,3-butanediol, produced by *Bacillus* strains (Grahovac et al., 2023). The 2,3-butanediol bioactivity was demonstrated in the pathosystems maize/*Setosphaeria turcica*, tobacco/*Erwinina carotovora*, and pepper against multiple viruses, where the immunity-eliciting ability was evidenced (Prsic & Ongena, 2020). VOCs produced by *P. fluorescens* WR-1 inhibited the growth of *R*alstonia *solanacearum* in tomatoes. Interestingly, the activity of these volatile compounds extends beyond growth inhibition, as they can modulate *Ralstonia* metabolism to suppress virulence, impacting plant pathogen evolution toward decreased virulence (Raza et al., 2016). The diverse functions of VOCs make these compounds essential players in the interactions between plant pathogens and bacterial consortiums, whose impact must be considered in biological plant protection studies (Schulz-Bohm et al., 2017).

Recently, beneficial bacteria's production of VOCs has been considered a promising biocontrol method that enables short- and long-distance phytopathogen control (Thomas et al., 2020). VOCs' capability to act as suppressive agents on pathogens and signal to plants confirms their potential role as sustainable alternatives to chemical fertilizers and pesticides.

Siderophores can also be a key factor activating ISR in certain beneficial bacteria. These metabolites were primarily known for their ability to bind and transport iron; however, they are also known to act as plant immunity elicitors (Kramer et al., 2020). Pyoverdines, produced by various fluorescent pseudomonads, have been reported to play a role in the induction of systemic resistance in bean, tomato, tobacco, and rice crops against a wide range of microbial pathogens (De Vleesschauwer et al., 2008; Meziane et al., 2005). In addition to pyoverdine, the chelating agent pyochelin produced by *P. aeruginosa* 7NSK2 was also involved in the plant-eliciting activity (Audenaert et al., 2002). The studies on molecular pathways of plant protection by microbial consortia highlight the intricate and multifaceted relationship between plants and their associated microbial consortia, which are crucial in enhancing plant health and resilience to biotic and abiotic stresses. The molecular mechanisms through which these microbial communities exert protective effects are diverse, involving multiple signaling pathways in the plant. For this reason, a deep understanding of the molecular pathways will allow the design of more effective and specific microbial consortia adapted to different environmental conditions and agricultural needs.

3.5 MICROBIAL-MEDIATED PROTECTION AGAINST STRESS

3.5.1 Protection against Abiotic Stress

Crop fields are regulated by two primary components: biotic and abiotic factors. Unfavorable changes in these conditions result in stress, categorized as abiotic or biotic stress, depending on the type of factor involved (Bhat et al., 2022).

Biotic stress arises from diseases caused by pathogens, which are reported to reduce crop yields by approximately 40% annually (Khan et al., 2021). Conversely, abiotic stress is induced by factors such as high temperatures, drought, high salinity, and the presence of heavy metals. The effects of abiotic stresses usually have the most significant impact on crop yields, since losses of up to 80% have been reported (Oshunsanya et al., 2019). Abiotic and biotic stressors can co-occur, leading to even more adverse effects on plant productivity.

Microbial consortia offer a wide range of benefits to plants against abiotic and biotic stresses, improving their tolerance to adverse environmental conditions, protecting them from pathogens and pests, and promoting healthy growth (Figure 3.2). Several investigations are focused on increasing plant stress tolerance through plant–microbe interactions that improve the ability of plants to resist biological and environmental pressure (Negi et al., 2024; Panda & Das, 2024). Numerous microorganisms confer beneficial traits to plants by releasing compounds that regulate various stress factors through multiple metabolites. These compounds can directly protect the plant or indirectly enhance its ability to mitigate stress effects (Kumar & Verma, 2018). PGPRs are a soil bacterial community capable of competitively colonizing plant roots and stimulating growth. These rhizobacteria exhibit beneficial traits that allow them to mitigate the toxic effects caused by high salt concentrations, improving plant growth in a salt environment by synthesizing anti-stress molecules and activating or modulating the response systems of plants during exposure to salt (Fouda et al., 2019). A higher concentration of NaCl stimulates bacteria to show

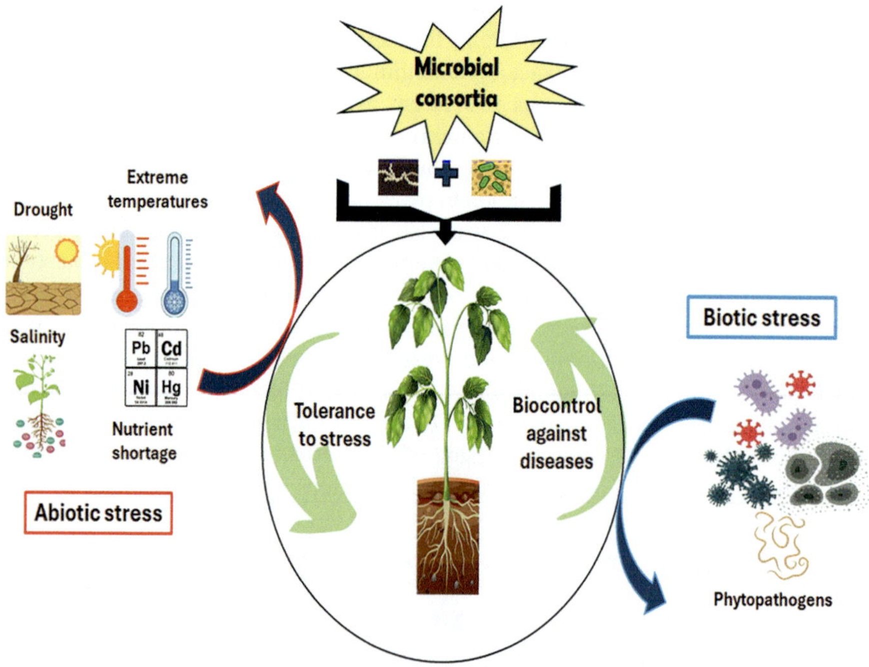

FIGURE 3.2 Effect of microbial consortia in alleviating abiotic and biotic stresses.

an expression of specific genes responsive to salt stress. A transcriptional profile in *B. subtilis* JH642 at 1.2 M NaCl showed the induction of upregulated 123 genes and downregulated 101 genes (Steil et al., 2003). These results suggest that the bacteria activate specific systems to manage the saline stress.

Salts in soil water hinder plant growth in two main ways: they reduce water uptake by creating an osmotic effect, slowing growth, and, when absorbed in excess, damage leaf cells through the transpiration stream, further impairing development. In this scenario, microbial consortia enhance nutrient availability by facilitating nitrogen fixation, phosphorus solubilization, and moderate plant hormone levels through producing 1-aminocyclopropane-1-carboxylate (ACC) deaminase enzyme or indirectly by producing VOCs, EPS and biofilm, antioxidant enzymes, and osmoprotectants (Chompa et al., 2024). Through these mechanisms, microbial consortia increase the resilience of plants to adverse environmental conditions, reduce the dependency on chemical fertilizers, and mitigate environmental stressors. Several microbial consortia have been tested to help plants overcome saline conditions (Table 3.1).

EPS are high-molecular weight compounds produced by microbes that can help plants manage abiotic stresses like drought and salinity. EPS helps in moisture retention, forming a protective layer around the root zone, and can also help stabilize soil structure (Morcillo & Manzanera, 2021). Numerous studies have confirmed the effect of EPS in alleviating salt stress through Na^+ chelation, making it inaccessible to the root. In soybeans, EPS produced by *B. japonicum* and *B. subtilis* alleviated the effects of salinity stress by sodium uptake restriction (Han & Lee, 2005). Similarly, EPS produced by *Pseudomonas* sp. AK-1 was able to bind free Na^+ from the soil, making Na^+ unavailable to soybeans and maintaining normal plant growth (Kasotia et al., 2016). It has been suggested that EPS may trigger different plant antioxidant mechanisms to alleviate drought stress. For instance, the activity of the antioxidant enzyme superoxide dismutase (SOD) was higher in maize plants inoculated with *P. aeruginosa* Pa2 in plants under drought-stress conditions (Naseem & Bano, 2014). Sandhya et al. (2009) reported that plants treated with bacteria-producing EPS exhibited more excellent drought resistance (Sandhya et al., 2009). Applying a six-microbe mixture comprising *Paenibacillus polymyxa*, *Bacillus megaterium*, *P. putida*, *P. aeruginosa*, *P. fluorescens*, and *T. harzianum* enhanced the ability of tomato crops to withstand drought. Additionally, it led to improved crop production, increased soil nutrient content, and heightened activity of antioxidant enzymes in the tomato plants (Krishna et al., 2022).

Regarding hydric stress, it is well known that water deficit represents a significant agricultural threat that limits crop growth and productivity. Several mechanisms have been proposed for PGPR-mediated drought stress tolerance in plants, which include root morphology modification (Egamberdieva & Kucharova, 2009), plant hormonal regulation (Liu et al., 2013), ACC deaminase activity (Glick, 2004), VOCs (Zhang et al., 2008), EPS production (Morcillo & Manzanera, 2021), and accumulation of osmolytes and antioxidants (Sandhya et al., 2011). Some PGPR strains can generate hormones, such as gibberellin, cytokinin, and ABA, of which production has been related to plant drought protection (Liu et al., 2013; Seleiman et al., 2021). Also, beneficial bacteria produce IAA, which increases root growth and formation of lateral roots, helping plants enhance water and nutrient uptake (Buragohain et al.,

2024). During water deficit, plants synthesize ethylene to regulate plant homeostasis, reducing root and shoot growth (Binder, 2020). ET, a stress-response hormone, is produced excessively during stress conditions, and ACC is a direct precursor. The action of bacterial ACC deaminase degrades plant ET, reducing the deleterious effect of this hormone, ameliorating plant stress, and promoting plant growth (Gupta et al., 2021). Mayak et al. (2004) demonstrated that the *Achromobacter piechaudii* ARV8 strain produces ACC deaminase, which induces tolerance against drought and salinity in pepper and tomato plants (Mayak et al., 2004). Other microbial strains also produce ACC deaminase, which disrupts ethylene production; these microbes mitigate the stress effects by degrading ACC and promoting healthy plant growth (Shahid et al., 2023). Microbial consortia modulate plant hormonal pathways, promoting the production of phytohormones such as ABA and auxins, which are essential for stress adaptation and growth regulation (Bhat et al., 2022). Moreover, these microorganisms mitigate oxidative stress by reducing reactive oxygen species (ROS) accumulation, which causes cellular damage under extreme stress conditions (Dong et al., 2024).

Microorganisms can also enhance the production of antioxidants that reduce oxidative stress in plants, protecting cells from damage. Stress conditions lead to oxidative damage, so antioxidants, such as SOD, catalase, and glutathione, help in neutralizing ROS. Microbial strains produce cytokinins and antioxidants, stimulate the accumulation of abscisic acid (ABA), and promote the degradation of reactive oxygen species (ROS), thereby enhancing plant tolerance to environment stresses (Choudhary et al., 2016). *P. aeruginosa, T. harzianum*, and *B. subtilis* were activated the phenylpropanoid pathway and antioxidant defence systems, regulated reactive oxygen species (ROS) dynamics, and modulated the oxidative burst and increasing H_2O_2 production in pea plants (Jain et al., 2015a). When plants experience stress, their photosynthetic system becomes disrupted, causing a significant increase in ROS and a detoxification process triggered by photorespiration. One potential consequence of water deficit is the elevated generation of various ROS, including hydrogen peroxide, hydroxyl radical, superoxide radical, and singlet oxygen. These ROS can disrupt standard plant metabolic functions and potentially cause cell death by inducing oxidative stress on macromolecules such as proteins and associated lipids (Fadiji et al., 2022).

3.5.2 Protection against Biotic Stress

Under biotic stress conditions, such as pathogens infection, herbivory, or competition with other plants, microbial consortia play critical role in enhancing plant defense mechanisms. These beneficial microbes act through multiple pathways to protect plants and mitigate damage caused by biotic agents (Nunes et al., 2024) (Figure 3.2). The mechanism of stimulation of the plant immune system by microbial consortia involves a series of complex interactions between beneficial microorganisms and plants. These interactions modulate plant physiology and activate defense responses, including the production of antimicrobial compounds, such as phenolic compounds and flavonoids, which can inhibit the growth of phytopathogens (Vishwakarma et al., 2020). Bacteria can detect chemical signals from plants, which are released into the rhizosphere to attract them. In turn, bacteria can release volatile compounds and siderophores, which can directly affect the immune system.

PGPRs and/or their elicitors usually prime the host to stimulate defense mechanisms but only after pathogen challenge. These responses activate the plant immune system, producing defense-related compounds, such as PR proteins, phenolics, and phytoalexins (El-Saadony et al., 2022). The expression of PR genes is often part of the broader defense response. The beneficial bacteria can induce it by activating signaling pathways such as JA and ET (Zhu et al., 2022). The production of PR proteins includes enzymes like chitinases, glucanases, peroxidases, and defensins that can break down the cell walls of pathogens or inhibit their ability to grow. SA is mainly associated with resistance against biotrophic pathogens (Roychowdhury et al., 2024). PGPR activation of the SA pathway is one of the most common mechanisms for the induction of PR proteins. Also, JA and ET are associated with resistance to necrotrophic pathogens (Yu et al., 2022).

PR proteins have various functions, depending on their type, but they are generally involved in defense against microbial infections (Dos Santos & Franco, 2023). PR proteins are grouped into different classes, each with specific functions in plant defense: PR-1, proteins with antifungal and antibacterial activity. They act by inhibiting the proliferation of pathogens, such as PR-2, β-glucanases that degrade the cell walls of fungi and, therefore, hinder their growth. This type of protein is essential in the defense against necrotrophic fungi; PR-3, chitinases that have activity against fungi, as they break down chitin, a component of the fungal cell wall; PR-4, proteins considered as defensins, which have antibacterial and antifungal properties; PR-5, proteins include chitinases and endoglucanases, which are effective in defending against fungal pathogens. Additionally, some of these proteins have antioxidant functions; PR-10 is a protein that is produced in response to oxidative stress. They can also protect plants from the damaging effects of free radicals generated during infection or environmental stress (Kamle et al., 2020; Mashabela et al., 2022).

When beneficial bacteria stimulate the immune system of plants, phenolic compounds can have direct antimicrobial effects, inhibiting the growth of phytopathogens (Wallis & Galarneau, 2020). Flavonoids and other phenolic compounds have a few key roles in plant defense. Flavonoids are secondary metabolites produced in plants in response to various stressors, including pathogen attacks (Shomali et al., 2022). These compounds can interfere with the cell membrane of pathogens, alter their metabolism, or even inhibit enzymes essential for their survival (Seth et al., 2023). For example, some flavonoids can inhibit the activity of chitinases or key proteins of bacterial pathogens. Flavonoids can also affect the metabolic pathways of pathogens, such as inhibiting toxin production or inactivating extracellular enzymes secreted by phytopathogens, such as proteases or lipases, which are necessary for the invasion of plant cells (Falcone Ferreyra et al., 2012). By reducing the ability of pathogens to produce toxic substances or inhibiting enzymes, flavonoids reduce the pathogenicity of these organisms (Kumar et al., 2023a).

Stimulation of plants to produce phytoalexins through interaction with microbial consortia represents an effective mechanism for protection against a wide variety of phytopathogens (Yu et al., 2022). Phytoalexins are low-molecular weight compounds, generally phenolic, that plants produce rapidly and locally in response to pathogen infections or other stress factors. These molecules are produced by activating specific biosynthetic pathways, often involving SA or JA signaling, depending on the nature of the microbial interaction. For example, in soybean (*Glycine max*), phytoalexins such as glyceollins are activated in response to *Rhizobium* or PGPR bacteria (Jahan et al., 2020).

PGPR elicitors also enhance cell wall reinforcement, a structural barrier to pathogen invasion via callose deposition or accumulation of phenolic compounds and lignin (Sun et al., 2024). Using these molecular strategies, *Aspergillus spinulosporus* and *B. amyloliquefaciens* enhanced the defense system in rice against *Xanthomonas oryzae* pv. *oryzae*, the causative agent of bacterial blight and bacterial leaf streak, which are among the most severe rice diseases (Jain et al., 2020). *Solanum tuberosum*, *Trichoderma viride*, *B. subtilis*, and *P. fluorescens* consortia activate phenylpropanoid derivatives and the redox homeostasis pathway, functioning as a biocontrol agent against *A. solani* while enhancing the overall plant growth (Kumar et al., 2023b). Microbial consortia are critical in improving plant disease resistance by inducing the ISR and increasing the production of PR proteins, phenolic compounds, and phytoalexins. These consortia strengthen their natural defense mechanisms, which not only prevent infection by pathogens but also help reduce dependence on pesticides and chemicals in agriculture.

3.6 CONTRIBUTION OF MOLECULAR TOOLS FOR STUDYING MICROBIAL CONSORTIA AND PLANT INTERACTIONS

Microbe consortia plant interactions occur in the rhizosphere, a soil region near plant roots. This rhizosphere is a dynamic environment driven by microbial community interaction, root exudates, nutrient availability, and soil physical-chemical properties; therefore, studying the rhizosphere is a cumbersome task (Lakshmanan et al., 2014). First attempts to achieve insights into these complex interactions relied on microbiological, biochemical, and microscope techniques and identification of soil-cultured isolates (Franco-Duarte et al., 2019). Determining microbial community interactions may be achieved by classic microbiological methods, such as co-cultivating microorganisms and evaluating their behavior or treating cultures with supernatant or cell extracts (Baichman-Kass et al., 2023). Still, results are limited and do not reflect reality within complex communities. These methods are time-consuming and labor-intensive and sometimes unable to identify organisms at the species level and unculturable organisms (Bochner, 2009).

Alternative biochemical methods include enzyme-linked immunosorbent assay (ELISA); mass spectrometry–based methods such as gas chromatography and matrix-assisted laser desorption ionization time-of-flight mode (MALDI-TOF) (Arvizu et al., 2016). Microbial consortia may protect phytopathogens, enhancing plant defense responses. The consortia comprising *B. subtilis* BHHU100, *P. aeruginosa* PJHU15, and *T. harzianum* TNHU27 were used in the pea rhizosphere to observe proteome-level changes upon *S. sclerotiorum* challenge (Jain et al., 2013). MALDI-TOF MS, MS/MS, and two-dimensional gel electrophoresis (2-DE) gels methods detected the protein differences between increase or decrease in proteins involved in phenylpropanoid metabolism, carbohydrate and nitrogen metabolism, photosynthesis, respiration, stress regulation, and disease/defense-related processes. According to this, the presence of microbial consortia highlights the protection that can be provided by these microorganisms in challenged plants (Jain et al., 2015b).

On the other hand, nucleic acid–based methodologies yield faster results and allow the identification of microorganisms, mostly at species or strain level, regardless of whether organisms are culturable. As every organism has a specific DNA signature, DNA fingerprinting techniques are valuable tools to identify organisms composing communities of interest (Fakruddin et al., 2013). Such tools include single-strand conformation polymorphism, denaturing gradient gel electrophoresis, amplicon length heterogeneity polymerase chain reaction, ribosomal intergenic spacer analysis (RISA) and automated RISA, and terminal restriction fragment length polymorphism (Johnston-Monje & Lopez Mejia, 2020).

The Omics Era, which includes metagenomics, transcriptomics, proteomics, and metabolomics, is producing an unprecedented amount of information, not only on community composition but also on determining complete genomes from environmental samples, genetic diversity, and species evolution (Handelsman, 2005; Moreno Andrade et al., 2018). Furthermore, it enables the identification of gene expression patterns, novel genes and proteins, and protein–protein interactions, thereby providing comprehensive insights into the composition, structure and metabolic interactions within microbial communities (Hettich et al., 2013; Kleiner, 2019). Metagenomics has revolutionized our understanding of the complex associations between microorganisms and their environments, including their plant interactions (Handelsman, 2005). Investigating these microbial communities' genetic composition and functional potential makes it possible to identify key microbial players and pathways that promote plant health and productivity (Trivedi et al., 2010). For example, in wheat, sorghum, and maize, the most detected bacterial genera are *Pantoea*, *Pseudomonas*, *Rhizobium*, *Bacillus*, *Sphingomonas*, and *Stenotrophomonas* (Kavamura et al., 2021; Mehta et al., 2020). Furthermore, transcriptome sequencing technology was used to study the changes in gene expression profiles of wheat roots and leaves under salt stress after inoculation with *B. subtilis* HG-15 and *B. velezensis* JC-K3 to further understand the mechanisms of bacterial consortia from the perspective of plants (Ji et al., 2023). Molecular tools have revolutionized research on the interaction between microbial consortia and plants. Through advanced techniques such as genomics, transcriptomics, and metagenomics, researchers can identify and characterize the functions of microorganisms present in the rhizosphere, analyze gene expression patterns in plants, and understand the ecological interactions that occur at the molecular level.

3.7 MICROBIAL CONSORTIA IN AGRICULTURE

The impact of using microbial consortia is significant in terms of benefits to plants, activating various biological mechanisms that contribute to improved plant quality and growth, which results in higher agricultural yields while potentially reducing the environmental effect through decreased use of agrochemicals (Negi et al., 2024; Tabacchioni et al., 2021). Combining microorganisms with different nutritional requirements, growth characteristics, and modes of action is required, mainly if synergistic interactions occur (Mamun et al., 2024). However, successfully implementing microbial consortia faces numerous challenges, including competition from native organisms, incompatibility between microorganisms, and adaptation to the habitat (Arif et al., 2020). For this reason, investigating specific microbial inter-relationships is crucial for establishing a functionally successful consortium (Minchev et al., 2021).

Microbial consortia are a powerful and effective strategy for improving soil health, promoting plant growth, and controlling pathogens in various crops (Moradtalab et al., 2020). Application examples in tomato, rice, corn, wheat, and soybean crops, among others, highlight several key benefits: improving the crop yield, reducing dependence on chemical fertilizers through nitrogen fixation, and biological control of phytopathogens, which reduces the need for synthetic pesticides (Table 3.1). Additionally, they promote the tolerance of plants against abiotic factors such as salinity and water stress. In recent years, numerous studies have examined microbial consortia in agriculture. Among the benefits of microbial consortia selection for agriculture include plant disease control (Ram et al., 2022; Upamanya et al., 2020) or plant growth promotion (Kaur et al., 2022; Yadav et al., 2022). In addition, microbial consortia can also be used to improve soils and mitigate abiotic stresses (Anshu et al., 2022; Petrillo et al., 2022).

Common genera in such consortia include *Bacillus* (Salazar et al., 2023), *Pseudomonas* (Kaur et al., 2022), *Metarhizium* (Upamanya et al., 2020), *Azospirillum* (Barua et al., 2023), *Trichoderma* (Oljira et al., 2020), and *Beauveria* (Upamanya et al., 2020). A variety of consortia are summarized in Table 3.1, along with their synergistic or additive effects.

Additional studies demonstrate the potential of microbial consortia to mitigate plant abiotic stress (Basu & South, 2022). Examples include enhanced tolerance to drought (Ashry et al., 2022), salt (Ellouzi et al., 2023; Kumar et al., 2020), and heavy metal stresses (Shilev et al., 2020). The use of microbial consortia in agriculture is a reality. The potential benefits can result from carefully selecting microbial mixtures, allowing for improvements in consortia applications.

3.8 CONCLUSION

Microbial consortia are critical and complex in protecting and promoting plant growth and establishing molecular interactions that benefit microorganisms and plants. Microorganisms in consortia facilitate the uptake of nutrients, such as nitrogen, phosphorus, and potassium, through biological nitrogen fixation or phosphorus solubilization processes. It allows better efficiency in using soil nutrients and reduces the need for chemical fertilizers. Additionally, contact with microbial consortia activates the defense mechanisms against several phytopathogens, resulting in disease resistance. This phenomenon is associated with the induction of immune responses at the molecular level, such as the accumulation of phenolic compounds, the production of antimicrobial proteins, and the modulation of defense-related gene expression. Microbial consortia can help plants survive under adverse conditions, such as prolonged droughts, saline soils, or extreme temperatures. Some microorganisms can produce compounds that protect plant cells from oxidative damage or help maintain water balance, which minimizes the adverse effects of these stressors. Thus, microbial consortia act integrated through mechanisms promoting plant nutrition, protecting against pathogens, inducing immune defenses, and improving tolerance to abiotic stresses. These interactions enhance agricultural performance under challenging conditions and contribute to crop sustainability. Although these consortia are fully effective, challenges related to the stability and persistence of microorganisms in the soil, adaptation to different types of crops, climatic conditions, and management of their diversity in the agricultural ecosystem must be overcome. Additionally, large-scale application is limited by competition with native soil flora, availability of suitable strains, lack of trained personnel, and a lack of awareness among farmers.

There are still several unresolved aspects regarding the development and application of microbial consortia. However, it is undoubtedly that microbial consortia in agriculture offer great potential to improve sustainability by reducing dependence on chemical fertilizers and helping plants cope with abiotic stressors. With advances in omics technologies, it will be possible to identify the most effective microorganisms in different environmental and agricultural conditions with greater precision. These advances will allow improved soil quality and optimized agricultural practices through biotechnology. Therefore, the future of sustainable agriculture depends on understanding the molecular interactions between plants and microorganisms, which will allow for more efficient use of natural resources and open the door to new forms of agroecological management, optimizing crop health and reducing the environmental impact of agricultural activity.

REFERENCES

Abbasi, S., AlipourKafi, S., Karimi, E. & Sadeghi, A. (2022) *Streptomyces* consortium improved quality attributes of bell pepper fruits, induced plant defense priming, and changed microbial communities of rhizosphere under commercial greenhouse conditions. *Rhizosphere*, 23, 100570.

Afzal, A., Khan, M. Y., Zahir, Z. A., Asghar, H. N., Muhmood, A., Rashid, M., Aslam, Z., Javed, S. A. & Nadeem, S. M. (2023) Plant growth-promoting bacterial consortia improved the physiology and growth of maize by regulating osmolytes and antioxidants balance under salt-affected field conditions. *Heliyon*, 9(7), e17816.

Ahemad, M. & Kibret, M. (2014) Mechanisms and applications of plant growth promoting rhizobacteria: Current perspective. *J King Saud Univ Sci*, 26, 1–20.

Ali, S. A. M., Sayyed, R. Z., Mir, M. I., Khan, M. Y., Hameeda, B., Alkhanani, M. F., Haque, S., Mohammad Al Tawaha, A. R. & Poczai, P. (2022) Induction of systemic resistance in maize and antibiofilm activity of surfactin from *Bacillus velezensis* MS20. *Front Microbiol*, 13, 879739.

Anshu, A., Agarwal, P., Mishra, K., Yadav, U., Verma, I., Chauhan, S., Srivastava, P. K. & Singh, P. C. (2022) Synergistic action of *Trichoderma koningiopsis* and *T. asperellum* mitigates salt stress in paddy. *Physiol Mol Biol Plants*, 28(5), 987–1004.

Arif, I., Batool, M. & Schenk, P. M. (2020) Plant microbiome engineering: Expected benefits for improved crop growth and resilience. *Trends Biotechnol*, 38(12), 1385–1396.

Aravindharajan, S. T. M., Shukla, L., Vijaysri, D., Singh, S. K., Kumar, R., Kumar, R., Rudra, S. G., Chawla, G. & Kumar, A. (2025) Comparative paddy straw degradation potential of Psychrotrophic Fungal Consortia and Pusa Decomposer at low temperatures. *Bioresour Technol Rep*, 32, 102314.

Arvizu, H., I., Hernandez-Flores, J. L., Romero, G., S., Cruz-Hernandez, A., Saldaña, G., C., Pastrana, M., X., Jones, G. H. & Campos-Guillén, J. (2016) Identification by MALDI-TOF Mass Spectrometry of bacteria in air samples in a biosafety level 2 laboratory. *Mod Environ Sci Eng*, 2, 238–245.

Ashry, N. M., Alaidaroos, B. A., Mohamed, S. A., Badr, O. A. M., El-Saadony, M. T. & Esmael, A. (2022) Utilization of drought-tolerant bacterial strains isolated from harsh soils as a plant growth-promoting rhizobacteria (PGPR). *Saudi J Biol Sci*, 29(3), 1760–1769.

Auchtung, J. M., Lee, C. A. & Grossman, A. D. (2006) Modulation of the ComA-dependent quorum response in *Bacillus subtilis* by multiple Rap proteins and Phr peptides. *J Bacteriol*, 188(14), 5273–5285.

Audenaert, K., Pattery, T., Cornelis, P. & Hofte, M. (2002) Induction of systemic resistance to *Botrytis cinerea* in tomato by *Pseudomonas aeruginosa* 7NSK2: Role of salicylic acid, pyochelin, and pyocyanin. *Mol Plant Microbe Interact*, 15(11), 1147–1156.

Azeem, M., Haider, M. Z., Javed, S., Saleem, M. H. & Alatawi, A. (2022) Drought stress amelioration in maize (*Zea mays* L.) by inoculation of *Bacillus* spp. strains under sterile soil conditions. *Agriculture*, 12(1), 50.

Baichman-Kass, A., Song, T. & Friedman, J. (2023) Competitive interactions between culturable bacteria are highly non-additive. *Elife*, 12, e83398.

Barua, N., Clouse, K. M., Ruiz Diaz, D. A., Wagner, M. R., Platt, T. G. & Hansen, R. R. (2023) Screening the maize rhizobiome for consortia that improve *Azospirillum brasilense* root colonization and plant growth outcomes. *Front Sustain Food Syst*, 7, 1–15.

Basu, D. & South, P. F. (2022) Design and analysis of native photorespiration gene motifs of promoter untranslated region combinations under short term abiotic stress conditions. *Front Plant Sci*, 13, 828729.

Begum, N., Wang, L., Ahmad, H., Akhtar, K., Roy, R., Khan, M. I. & Zhao, T. (2022) pCo-inoculation of arbuscular mycorrhizal fungi and the plant growth-promoting rhizobacteria improve growth and photosynthesis in Tobacco under drought stress by up-regulating antioxidant and mineral nutrition metabolism. *Microb Ecol*, 83(4), 971–988.

Bhat, B. A., Tariq, L., Nissar, S., Islam, S. T., Islam, S. U., Mangral, Z., Ilyas, N., Sayyed, R. Z., Muthusamy, G., Kim, W. & Dar, T. U. H. (2022) The role of plant-associated rhizobacteria in plant growth, biocontrol and abiotic stress management. *J Appl Microbiol*, 133(5), 2717–2741.

Biessy, A. & Filion, M. (2018) Phenazines in plant-beneficial *Pseudomonas* spp.: Biosynthesis, regulation, function and genomics. *Environ Microbiol*, 20(11), 3905–3917.

Binder, B. M. (2020) Ethylene signaling in plants. *J Biol Chem*, 295(22), 7710–7725.

Blanco Nouche, C., Paris, C., Dhalleine, T., Oger, P., Turpault, M.-P., Uroz, S. & Spear, J. R. (2023) The non-ribosomal peptide synthetase-independent siderophore (NIS) rhizobactin produced by *Caballeronia mineralivorans* PML1(12) confers the ability to weather minerals. *Appl Environ Microbiol*, 89, e0045323.

Bo, T., Kong, C., Zou, S., Mo, M. & Liu, Y. (2022) *Bacillus nematocida* B16 enhanced the rhizosphere colonization of *Pochonia chlamydosporia* ZK7 and controlled the efficacy of the root-knot nematode *Meloidogyne incognita*. *Microorganisms*, 10(2), 218.

Bochner, B. R. (2009) Global phenotypic characterization of bacteria. *FEMS Microbiol Rev*, 33(1), 191–205.

Bouchard-Rochette, M., Machrafi, Y., Cossus, L., Nguyen, T. A., Antoun, T., Droit, H. & Tweddell, A. (2022) *Bacillus pumilus* PTB180 and *Bacillus subtilis* PTB185: Production of lipopeptides, antifungal activity, and biocontrol ability against *Botrytis cinerea*. *Biol Control*, 170, Article 104925.

Buragohain, K., Tamuly, D., Sonowal, S. & Nath, R. (2024) Impact of drought stress on plant growth and its management using plant growth promoting rhizobacteria. *Indian J Microbiol*, 64(2), 287–303.

Butcher, J., Sarvan, S., Brunzelle, J. S., Couture, J. F. & Stintzi, A. (2012) Structure and regulon of *Campylobacter jejuni* ferric uptake regulator Fur define apo-Fur regulation. *Proc Natl Acad Sci USA*, 109(25), 10047–10052.

Calderon, R. B. & Dangi, S. R. (2024) Arbuscular mycorrhizal fungi and *Rhizobium* improve nutrient uptake and microbial diversity relative to dryland site-specific soil conditions. *Microorganisms*, 12(4), 667.

Carroll, C. S. & Moore, M. M. (2018) Ironing out siderophore biosynthesis: A review of non-ribosomal peptide synthetase (NRPS)-independent siderophore synthetases. *Crit Rev Biochem Mol Biol*, 53(4), 356–381.

Cawoy, H., Mariutto, M., Henry, G., Fisher, C., Vasilyeva, N., Thonart, P., Dommes, J. & Ongena, M. (2014) Plant defense stimulation by natural isolates of *Bacillus* depends on efficient surfactin production. *Mol Plant Microbe Interact*, 27(2), 87–100.

Chandler, S., Van Hese, N., Coutte, F., Jacques, P., Höfte, M. & De Vleesschauwer, D. (2015) Role of cyclic lipopeptides produced by *Bacillus subtilis* in mounting induced immunity in rice (*Oryza sativa* L.). *Physiol Mol Plant P*, 91, 20–30.

Che, S., Xu, Y., Qin, X., Tian, S., Wang, J., Zhou, X., Cao, Z., Wang, D., Wu, M., Wu, Z., Yang, M., Wu, L. & Yang, X. (2024) Building microbial consortia to enhance straw degradation, phosphorus solubilization, and soil fertility for rice growth. *Microb Cell Fact*, 23(1), 232.

Chi, S. I., Akuma, M., Xu, R., Plante, V., Hadinezhad, M. & Tambong, J. T. (2024) Phenazines are involved in the antagonism of a novel subspecies of *Pseudomonas chlororaphis* strain S1Bt23 against *Pythium ultimum*. *Sci Rep*, 14(1), 20517.

Chompa, S. S., Zuan, A. T. K., Amin, A. M., Hun, T. G., Ghazali, A. H. A., Sadeq, B. M., Akter, A., Rahman, M. E. & Rashid, H. O. (2024) Growth and protein response of rice plant with plant growth-promoting rhizobacteria inoculations under salt stress conditions. *Int Microbiol*, 27(4), 1151–1168.

Choudhary, D. K., Kasotia, A., Jain, S., Vaishnav, A., Kumari, S., Sharma, K. P. & Varma, A. (2016) Bacterial-mediated tolerance and resistance to plants under abiotic and biotic stresses. *J Plant Growth Regul*, 35, 276–300.

Cirillo, V., Romano, I., Woo, S. L., Di Stasio, E., Lombardi, N., Comite, E., Pepe, O., Ventorino, V. & Maggio, A. (2023) Inoculation with a microbial consortium increases soil microbial diversity and improves agronomic traits of tomato under water and nitrogen deficiency. *Front Plant Sci*, 14, 1304627.

Curatti, L., Brown, C. S., Ludden, P. W. & Rubio, L. M. (2005) Genes required for rapid expression of nitrogenase activity in *Azotobacter vinelandii*. *Proc Natl Acad Sci USA*, 102(18), 6291–6296.

Dai, Z., Liu, G., Chen, H., Chen, C., Wang, J., Ai, S., Wei, D., Li, D., Ma, B., Tang, C., Brookes, P. C. & Xu, J. (2020) Long-term nutrient inputs shift soil microbial functional profiles of phosphorus cycling in diverse agroecosystems. *ISME J*, 14(3), 757–770.

De Vleesschauwer, D., Djavaheri, M., Bakker, P. A. & Hofte, M. (2008) *Pseudomonas fluorescens* WCS374r-induced systemic resistance in rice against *Magnaporthe oryzae* is based on pseudobactin-mediated priming for a salicylic acid-repressible multifaceted defense response. *Plant Physiol*, 148(4), 1996–2012.

Devi, N. O., Tombisana Devi, R. K., Debbarma, M., Hajong, M. & Thokchom, S. (2022) Effect of endophytic *Bacillus* and arbuscular mycorrhiza fungi (AMF) against Fusarium wilt of tomato caused by *Fusarium oxysporum* f. sp. lycopersici. *Egypt J Biol Pest Control*, 32, 1–14.

Dobrzyński, J., Kulkova, I., Jakubowska, Z. & Wróbel, B. (2025) Non-native PGPB consortium consisting of *Pseudomonas* sp. G31 and *Azotobacter* sp. PBC2 promoted winter wheat growth and slightly altered the native bacterial community. *Sci Rep*, 15, 3248.

Dong, H., Wang, Y., Di, Y., Qiu, Y., Ji, Z., Zhou, T., Shen, S., Du, N., Zhang, T., Dong, X., Guo, Z., Piao, F. & Li, Y. (2024) Plant growth-promoting rhizobacteria *Pseudomonas aeruginosa* HG28-5 improves salt tolerance by regulating Na(+)/K(+) homeostasis and ABA signaling pathway in tomato. *Microbiol Res*, 283, 127707.

Dos Santos, C. & Franco, O. L. (2023) Pathogenesis-Related Proteins (PRs) with enzyme activity activating plant defense responses. *Plants*, 12(11), 2226.

Dos Santos, P. C., Fang, Z., Mason, S. W., Setubal, J. C. & Dixon, R. (2012) Distribution of nitrogen fixation and nitrogenase-like sequences amongst microbial genomes. *BMC Genom*, 13, 162.

Dutta, P., Mahanta, M., Singh, S. B., Thakuria, D., Deb, L., Kumari, A., Upamanya, G. K., Boruah, S., Dey, U., Mishra, A. K., Vanlaltani, L., VijayReddy, D., Heisnam, P. & Pandey, A. K. (2023) Molecular interaction between plants and *Trichoderma* species against soil-borne plant pathogens. *Front Plant Sci*, 14, 1145715.

Egamberdieva, D. & Kucharova, Z. (2009) Selection for root colonizing bacteria stimulating wheat growth in saline soils. *Biol Fert Soil*, 45, 561–573.

El-Saadony, M. T., Saad, A. M., Soliman, S. M., Salem, H. M., Ahmed, A. I., Mahmood, M., El-Tahan, A. M., Ebrahim, A. A. M., Abd El-Mageed, T. A., Negm, S. H., Selim, S., Babalghith, A. O., Elrys, A. S., El-Tarabily, K. A. & AbuQamar, S. F. (2022) Plant growth-promoting microorganisms as biocontrol agents of plant diseases: Mechanisms, challenges and future perspectives. *Front Plant Sci*, 13, 923880.

Elhaissoufi, W., Ghoulam, C., Barakat, A., Zeroual, Y. & Bargaz, A. (2022) Phosphate bacterial solubilization: A key rhizosphere driving force enabling higher P use efficiency and crop productivity. *J Adv Res*, 38, 13–28.

Ellouzi, H., Zorrig, W., Amraoui, S., Oueslati, S., Abdelly, C., Rabhi, M., Siddique, K. H. M. & Hessini, K. (2023) Seed priming with salicylic acid alleviates salt stress toxicity in barley by suppressing ROS accumulation and improving antioxidant defense systems, compared to halo- and gibberellin priming. *Antioxidants*, 12(9), 1779.

Etesami, H. & Alikhani, H. A. (2016) Co-inoculation with endophytic and rhizosphere bacteria allows reduced application rates of N-fertilizer for rice plant. *Rhizosphere*, 2, 5–12.

Fadiji, A. E., Santoyo, G., Yadav, A. N. & Babalola, O. O. (2022) Efforts towards overcoming drought stress in crops: Revisiting the mechanisms employed by plant growth-promoting bacteria. *Front Microbiol*, 13, 962427.

Fakruddin, M., Mannan, K. S. B., Mazumdar, R. M., Chowdhury, A. & Hossain, N. (2013) Identification and characterization of microorganisms: DNA-fingerprinting methods. *Songklanakarin J Sci Technol*, 35(4), 397–404.

Falcone Ferreyra, M. L., Rius, S. P. & Casati, P. (2012) Flavonoids: Biosynthesis, biological functions, and biotechnological applications. *Front Plant Sci*, 3, 222.

Farzand, A., Moosa, A., Zubair, M., Khan, A. R., Massawe, V. C., Tahir, H. A. S., Sheikh, T. M. M., Ayaz, M. & Gao, X. (2019) Suppression of *Sclerotinia sclerotiorum* by the induction of systemic resistance and regulation of antioxidant pathways in tomato using fengycin produced by *Bacillus amyloliquefaciens* FZB42. *Biomolecules*, 9(10), 613.

Fazle-Rabbee, M. & Baek, K. H. (2020) Antimicrobial activities of lipopeptides and polyketides of *Bacillus velezensis* for agricultural applications. *Molecules*, 25(21), 4973.

Fiodor, A., Ajijah, N., Dziewit, L. & Pranaw, K. (2023) Biopriming of seed with plant growth-promoting bacteria for improved germination and seedling growth. *Front Microbiol*, 14, 1142966.

Fortunato, S., Nigro, D., Lasorella, C., Marcotuli, I., Gadaleta, A. & de Pinto, M. C. (2023) The role of glutamine synthetase (GS) and glutamate synthase (GOGAT) in the improvement of Nitrogen use efficiency in cereals. *Biomolecules*, 13(12), 1771.

Fouda, A., Hassan, S., Eid, A. & Ewais, E. (2019) The interaction between plants and bacterial endophytes under salinity stress. In: Jha, S. (ed.) *Endophytes and Secondary Metabolites*. Reference Series in Phytochemistry. Springer, Cham, Switzerland, pp. 1–17.

Franco-Duarte, R., Cernakova, L., Kadam, S., Kaushik, K. S., Salehi, B., Bevilacqua, A., Corbo, M. R., Antolak, H., Dybka-Stepien, K., Leszczewicz, M., Relison Tintino, S., Alexandrino de Souza, V. C., Sharifi-Rad, J., Coutinho, H. D. M., Martins, N. & Rodrigues, C. F. (2019) Advances in chemical and biological methods to identify microorganisms-from past to present. *Microorganisms*, 7(5), 130.

Garcia-Sanchez, M., Bertrand, I., Barakat, A., Zeroual, Y., Oukarroum, A. & Plassard, C. (2023) Improved rock phosphate dissolution from organic acids is driven by nitrate assimilation of bacteria isolated from nitrate and $CaCO_3$-rich soil. *PLoS One*, 18(3), e0283437.

Garcia, D., Gonzalez-Almario, A. & Cotes, A. M. (2023) Controlling Fusarium wilt of cape gooseberry by microbial consortia. *Lett Appl Microbiol*, 76(7), ovad072.

Geudens, N. & Martins, J. C. (2018) Cyclic lipodepsipeptides from *Pseudomonas* spp.-Biological swiss-army knives. *Front Microbiol*, 9, 1867.

Glick, B. R. (2004) Bacterial ACC deaminase and the alleviation of plant stress. *Adv Appl Microbiol*, 56, 291–312.

Gonzalez, A., Angarica, V. E., Sancho, J. & Fillat, M. F. (2014) The FurA regulon in *Anabaena* sp. PCC 7120: In silico prediction and experimental validation of novel target genes. *Nucleic Acids Res*, 42(8), 4833–46.

Goswami, D., Thakker, J. N. & Dhandhukia, P. C. (2016) Portraying mechanics of plant growth promoting rhizobacteria (PGPR): A review. *Cogent Food Agric*, 2, 1127500.

Gotze, S., Arp, J., Lackner, G., Zhang, S., Kries, H., Klapper, M., Garcia-Altares, M., Willing, K., Gunther, M. & Stallforth, P. (2019) Structure elucidation of the syringafactin lipopeptides provides insight in the evolution of nonribosomal peptide synthetases. *Chem Sci*, 10(48), 10979–10990.

Gotze, S. & Stallforth, P. (2020) Structure, properties, and biological functions of nonribosomal lipopeptides from pseudomonads. *Nat Prod Rep*, 37(1), 29–54.

Grahovac, J., Pajcin, I. & Vlajkov, V. (2023) *Bacillus* VOCs in the context of biological control. *Antibiotics*, 12(3), 581.

Gupta, A., Bano, A., Rai, S., Kumar, M., Ali, J., Sharma, S. & Pathak, N. (2021) ACC deaminase producing plant growth promoting rhizobacteria enhance salinity stress tolerance in *Pisum sativum. 3 Biotech*, 11(12), 514.

Han, H. S. & Lee, K. D. (2005) Physiological responses of soybean-inoculation of *Bradyrhizobium japonicum* with PGPR in saline soil conditions. *Res J Agric Biol Sci*, 1, 216–221.

Handelsman, J. (2005) Metagenomics: Application of genomics to uncultured microorganisms. *Microbiol Mol Biol Rev*, 68(4), 669–685.

Hartmann, A., Klink, S. & Rothballer, M. (2021) Importance of N-Acyl-homoserine lactone-based Quorum Sensing and Quorum Quenching in pathogen control and plant growth promotion. *Pathogens*, 10(12), 1561.

Hasan, A., Tabassum, B., Hashim, M. & Khan, N. (2024) Role of plant growth promoting rhizobacteria (PGPR) as a plant growth enhancer for sustainable agriculture: A review. *Bacteria*, 3(2), 59–75.

Hasnain, A. & Kanwal, S. (2020) Biological uses and importance of iron regulation. *MARKHOR (J. Zool.)*, 1, 11–13.

Hedden, P. (2020) The current status of research on gibberellin biosynthesis. *Plant Cell Physiol*, 61(11), 1832–1849.

Hernandez-Reyes, C., Schenk, S. T., Neumann, C., Kogel, K. H. & Schikora, A. (2014) N-acyl-homoserine lactones-producing bacteria protect plants against plant and human pathogens. *Microb Biotechnol*, 7(6), 580–588.

Hettich, R. L., Pan, C., Chourey, K. & Giannone, R. J. (2013) Metaproteomics: Harnessing the power of high performance mass spectrometry to identify the suite of proteins that control metabolic activities in microbial communities. *Anal Chem*, 85(9), 4203–4214.

Houida, S., Yakkou, L., Kaya, L. O., Bilen, S., Fadil, M., Raouane, M., El Harti, A. & Amghar, S. (2022) Biopriming of maize seeds with plant growth-promoting bacteria isolated from the earthworm *Aporrectodea molleri*: Effect on seed germination and seedling growth. *Lett Appl Microbiol*, 75(1), 61–69.

Jahan, M. A., Harris, B., Lowery, M., Infante, A. M., Percifield, R. J. & Kovinich, N. (2020) Glyceollin transcription factor GmMYB29A2 regulates soybean resistance to *Phytophthora sojae*. *Plant Physiol*, 183(2), 530–546.

Jain, A., Chatterjee, A. & Das, S. (2020) Synergistic consortium of beneficial microorganisms in rice rhizosphere promotes host defense to blight-causing *Xanthomonas oryzae* pv. oryzae. *Planta*, 252(6), 106.

Jain, A., Singh, A., Singh, S., Sarma, B. K. & Singh, H. B. (2015a) Biocontrol agents-mediated suppression of oxalic acid induced cell death during *Sclerotinia sclerotiorum*-pea interaction. *J Basic Microbiol*, 55(5), 601–606.

Jain, A., Singh, A., Singh, S. & Singh, H. B. (2013) Microbial consortium-induced changes in oxidative stress markers in pea plants challenged with *Sclerotinia sclerotiorum*. *J Plant Growth Regul*, 32, 388–398.

Jain, A., Singh, A., Singh, S., Singh, V. & Singh, H. B. (2015b) Comparative proteomic analysis in pea treated with microbial consortia of beneficial microbes reveals changes in the protein network to enhance resistance against *Sclerotinia sclerotiorum*. *J Plant Physiol*, 182, 79–94.

Ji, C., Liang, Z., Cao, H., Chen, Z., Kong, X., Xin, Z., He, M., Wang, J., Wei, Z., Xing, J., Li, C., Zhang, Y., Zhang, H., Sun, F., Li, J. & Li, K. (2023) Transcriptome-based analysis of the effects of compound microbial agents on gene expression in wheat roots and leaves under salt stress. *Front Plant Sci*, 14, 1109077.

Johnston-Monje, D. & Lopez Mejia, J. (2020) Botanical microbiomes on the cheap: Inexpensive molecular fingerprinting methods to study plant-associated communities of bacteria and fungi. *Appl Plant Sci*, 8(4), e11334.

Kadi, N., Oves-Costales, D., Barona-Gomez, F. & Challis, G. L. (2007) A new family of ATP-dependent oligomerization-macrocyclization biocatalysts. *Nat Chem Biol*, 3(10), 652–656.

Kamle, M., Borah, R., Bora, H., Jaiswal, A. K., Singh, R. K. & Kumar, P. (2020) Systemic acquired resistance (SAR) and Induced Systemic Resistance (ISR): Role and mechanism of action against phytopathogens. In: Hesham, A. L., Upadhyay, R., Sharma, G., Manoharachary, C., Gupta, V. (eds.) *Fungal Biotechnology and Bioengineering*. Fungal Biology. Springer, Cham, 457–470.

Kapadia, C., Sayyed, R. Z., El Enshasy, H. A., Vaidya, H., Sharma, D., Patel, N., Malek, R. A., Syed, A., Elgorban, A. M., Ahmad, K. & Zuan, A. T. K. (2021) Halotolerant microbial consortia for sustainable mitigation of salinity stress, growth promotion, and mineral uptake in tomato plants and soil nutrient enrichment. *Sustainability*, 13(15), 8369.

Kasotia, A., Varma, A., Tuteja, N. & Choudhary, D. K. (2016) Amelioration of soybean plant from saline-induced condition by exopolysaccharide producing *Pseudomonas*-mediated expression of high affinity K+-transporter (HKT1) gene. *Curr Sci*, 12, 1961–1967.

Kaspar, F., Neubauer, P. & Gimpel, M. (2019) Bioactive secondary metabolites from *Bacillus subtilis*: A comprehensive review. *J Nat Prod*, 82(7), 2038–2053.

Kaur, T., Devi, R., Kumar, S., Sheikh, I., Kour, D. & Yadav, A. N. (2022) Microbial consortium with nitrogen fixing and mineral solubilizing attributes for growth of barley (*Hordeum vulgare* L.). *Heliyon*, 8(4), e09326.

Kavamura, V. N., Mendes, R., Bargaz, A. & Mauchline, T. H. (2021) Defining the wheat microbiome: Towards microbiome-facilitated crop production. *Comput Struct Biotechnol J*, 19, 1200–1213.

Keswani, C., Singh, S. P., Garcia-Estrada, C., Mezaache-Aichour, S., Glare, T. R., Borriss, R., Rajput, V. D., Minkina, T. M., Ortiz, A. & Sansinenea, E. (2022) Biosynthesis and beneficial effects of microbial gibberellins on crops for sustainable agriculture. *J Appl Microbiol*, 132(3), 1597–1615.

Khan, M., Khan, A. U., Hasan, M. A., Yadav, K. K., Pinto, M. M. C., Malik, N., Yadav, V. K., Khan, A. H., Islam, S. & Sharma, G. K. (2021) Agro-nanotechnology as an emerging field: A novel sustainable approach for improving plant growth by reducing biotic stress. *Appl Sci*, 11(5), 2282–2292.

Khan, M. Y., Nadeem, S. M., Sohaib, M., Waqas, M. R., Alotaibi, F., Ali, L., Zahir, Z. A. & Al-Barakah, F. N. I. (2022) Potential of plant growth promoting bacterial consortium for improving the growth and yield of wheat under saline conditions. *Front Microbiol*, 13, 958522.

Khourchi, S., Elhaissoufi, W., Loum, M., Ibnyasser, A., Haddine, M., Ghani, R., Barakat, A., Zeroual, Y., Rchiad, Z., Delaplace, P. & Bargaz, A. (2022) Phosphate solubilizing bacteria can significantly contribute to enhance P availability from polyphosphates and their use efficiency in wheat. *Microbiol Res*, 262, 127094.

Kleiner, M. (2019) Metaproteomics: Much more than measuring gene expression in microbial communities. *mSystems*, 4(3), 10–128.

Kour, D., Rana, K., Kaur, T., Yadav, N., Yadav, A., Kumar, M., Kumar, V., Dhaliwal, H. S. & Saxena, A. K. (2021) Biodiversity, current developments and potential biotechnological applications of phosphorus-solubilizing and-mobilizing microbes: A review. *Pedosphere*, 31, 43–75.

Kramer, J., Ozkaya, O. & Kummerli, R. (2020) Bacterial siderophores in community and host interactions. *Nat Rev Microbiol*, 18(3), 152–163.

Krishna, R., Jaiswal, D. K., Ansari, W. A., Singh, S., Soumia, P. S., Singh, A. K. & Verma, J. P. (2022) Potential microbial consortium mitigates drought stress in tomato (*Solanum lycopersicum* L.) plant by up-regulating stress-responsive genes and improving fruit yield and soil properties. *J Soil Sci Plant Nut*, 22(4), 4598–4615.

Kumar, A., Singh, S., Gaurav, A. K., Srivastava, S. & Verma, J. P. (2020) Plant growth-promoting bacteria: Biological tools for the mitigation of salinity stress in plants. *Front Microbiol*, 11, 1216.

Kumar, A. & Verma, J. P. (2018) Does plant-Microbe interaction confer stress tolerance in plants: A review? *Microbiol Res*, 207, 41–52.

Kumar, G. A., Kumar, S., Bhardwaj, R., Swapnil, P., Meena, M., Seth, C. S. & Yadav, A. (2023a) Recent advancements in multifaceted roles of flavonoids in plant-rhizomicrobiome interactions. *Front Plant Sci*, 14, 1297706.

Kumar, M., Giri, V. P., Pandey, S., Gupta, A., Patel, M. K., Bajpai, A. B., Jenkins, S. & Siddique, K. H. M. (2021) Plant-growth-promoting rhizobacteria emerging as an effective bioinoculant to improve the growth, production, and stress tolerance of vegetable crops. *Int J Mol Sci*, 22(22), 12245.

Kumar, S., Chandra, R., Behera, L., Sudhir, I., Meena, M., Singh, S. & Keswani, C. (2023b) Microbial consortium mediated acceleration of the defense response in potato against *Alternaria solani* through prodigious inflation in phenylpropanoid derivatives and redox homeostasis. *Heliyon*, 9, 11–21.

Lakshmanan, V., Selvaraj, G. & Bais, H. P. (2014) Functional soil microbiome: Belowground solutions to an aboveground problem. *Plant Physiol*, 166(2), 689–700.

Li, Y., Heloir, M. C., Zhang, X., Geissler, M., Trouvelot, S., Jacquens, L., Henkel, M., Su, X., Fang, X., Wang, Q. & Adrian, M. (2019) Surfactin and fengycin contribute to the protection of a *Bacillus subtilis* strain against grape downy mildew by both direct effect and defence stimulation. *Mol Plant Pathol*, 20(8), 1037–1050.

Liang, J. L., Liu, J., Jia, P., Yang, T. T., Zeng, Q. W., Zhang, S. C., Liao, B., Shu, W. S. & Li, J. T. (2020) Novel phosphate-solubilizing bacteria enhance soil phosphorus cycling following ecological restoration of land degraded by mining. *ISME J*, 14(6), 1600–1613.

Liu, F., Xing, S., Ma, H., Du, Z. & Ma, B. (2013) Cytokinin-producing, plant growth-promoting rhizobacteria that confer resistance to drought stress in *Platycladus orientalis* container seedlings. *Appl Microbiol Biotechnol*, 97(20), 9155–9164.

Liu, L., Wang, W., Wu, S. & Gao, H. (2022) Recent advances in the siderophore biology of *Shewanella*. *Front Microbiol*, 13, 823758.

Liu, N., Sun, H., Tang, Z., Zheng, Y., Qi, G. & Zhao, X. (2023) Transcription factor Spo0A regulates the biosynthesis of difficidin in *Bacillus amyloliquefaciens*. *Microbiol Spectr*, 11(4), e0104423.

Lozano-Gonzalez, J. M., Valverde, S., Montoya, M., Martin, M., Rivilla, R., Lucena, J. J. & Lopez-Rayo, S. (2023) Evaluation of siderophores generated by *Pseudomonas* bacteria and their possible application as Fe biofertilizers. *Plants*, 12(23), 4054.

Ma, Z., Hua, G. K. H., Ongena, M. & Hofte, M. (2016) Role of phenazines and cyclic lipopeptides produced by *Pseudomonas* sp. CMR12a in induced systemic resistance on rice and bean. *Environ Microbiol Rep*, 8(5), 896–904.

Maheshwari, D. K., Dubey, R. C., Agarwal, M., Dheeman, S., Aeron, A. & Bajpaik, V. K. (2015) Carrier based formulations of biocoenotic consortia of disease suppressive *Pseudomonas aeruginosa* KRP1 and *Bacillus licheniformis* KRB1. *Ecol Eng*, 81, 272–277.

Mamun, A. A., Neumann, G., Moradtalab, N., Ahmed, A., Dupuis, B., Darbon, G., Nawaz, F., Declerck, S., Mai, K., Vogt, W., Ludewig, U. & Weinmann, M. (2024) Microbial consortia versus single-strain Inoculants as drought stress protectants in potato affected by the form of N supply. *Horticulturae*, 10, 102–115.

Mashabela, M. D., Piater, L. A., Dubery, I. A., Tugizimana, F. & Mhlongo, M. I. (2022) Rhizosphere tripartite interactions and PGPR-mediated metabolic reprogramming towards ISR and plant priming: A metabolomics review. *Biology*, 11(3), 346.

Mayak, S., Tirosh, T. & Glick, B. R. (2004) Plant growth-promoting bacteria confer resistance in tomato plants to salt stress. *Plant Physiol Biochem*, 42(6), 565–572.

Mehta, S., Singh, B., Patra, A., Tripathi, A., Easwaran, M., Choudhary, J. R., Choudhary, M. & Aggarwal, S. K. (2020) *Microbiomes and Plant Health: Panoply and Their Applications*. Elsevier, Amsterdam, The Netherlands. Maize Microbiome: Current Insights for the Sustainable Agriculture, pp. 267–297.

Metwally, R. A., Azab, H. S., Al-Shannaf, H. M. & Rabie, G. H. (2022) Prospective of mycorrhiza and *Beauvaria bassiana* silica nanoparticles on *Gossypium hirsutum* L. plants as biocontrol agent against cotton leafworm, *Spodoptera littoralis*. *BMC Plant Biol*, 22(1), 409.

Meziane, H., Sluis, I., Loon, L. C., Hofte, M. & Bakker, P. A. (2005) Determinants of *Pseudomonas putida* WCS358 involved in inducing systemic resistance in plants. *Mol Plant Pathol*, 6(2), 177–185.

Mhlongo, M. I., Piater, L. A., Steenkamp, P. A., Labuschagne, N. & Dubery, I. A. (2021) Metabolomic evaluation of tissue-specific defense responses in tomato plants modulated by PGPR-priming against *Phytophthora capsici* infection. *Plants*, 10(8), 1530.

Minchev, Z., Kostenko, O., Soler, R. & Pozo, M. J. (2021) Microbial consortia for effective biocontrol of root and foliar diseases in tomato. *Front Plant Sci*, 12, 756368.

Moradtalab, N., Ahmed, A., Geistlinger, J., Walker, F., Hoglinger, B., Ludewig, U. & Neumann, G. (2020) Synergisms of microbial consortia, N forms, and micronutrients alleviate oxidative damage and stimulate hormonal cold stress adaptations in maize. *Front Plant Sci*, 11, 396.

Morcillo, R. J. L. & Manzanera, M. (2021) The effects of plant-associated bacterial exopolysaccharides on plant abiotic stress tolerance. *Metabolites*, 11(6), 337.

Moreno Andrade, V. D., Saldana Gutierrez, C., Calvillo Medina, R. P., Cruz Hernandez, A., Vazquez Cruz, M. A., Torres Ruiz, A., Romero Gomez, S., Ramos Lopez, M. A., Alvarez-Hidalgo, E., Lopez-Gaytan, S. B., Ramirez, N. S., Jones, G. H., Hernandez-Flores, J. L. & Campos-Guillen, J. (2018) Microbial diversity in commercial bee pollen from Europe, Chile, and Mexico, based on 16S rRNA gene amplicon metagenome sequencing. *Genome Announc*, 6(20), 10–1128.

Muthuraja, R. & Muthukumar, T. (2022) Co-inoculation of halotolerant potassium solubilizing *Bacillus licheniformis* and *Aspergillus violaceofuscus* improves tomato growth and potassium uptake in different soil types under salinity. *Chemosphere*, 294, 133718.

Naseem, H. & Bano, A. (2014) Role of plant growth-promoting rhizobacteria and their exopolysaccharide in drought tolerance of maize. *J Plant Interact*, 9, 689–701.

Negi, R., Sharma, B., Jan, T., Kaur, T., Chowdhury, S., Kapoor, M., Singh, S., Kumar, A., Rai, A. K., Rustagi, S., Shreaz, S., Kour, D., Ahmed, N., Kumar, K. & Yadav, A. N. (2024) Microbial consortia: Promising tool as plant bioinoculants for agricultural sustainability. *Curr Microbiol*, 81(8), 222.

Niewiadomska, A., Płaza, A., Wolna-Maruwka, A., Budka, A., Głuchowska, K., Rudziński, R. & Kaczmarek, T. (2023) Consortia of plant growth-promoting rhizobacteria and selected catch crops for increasing microbial activity in soil under spring barley grown as an organic farming system. *Appl Sci*, 13(8), 5120.

Novello, G., Bona, E., Nasuelli, M., Massa, N., Sudiro, C., Campana, D. C., Gorrasi, S., Hochart, M. L., Altissimo, A., Vuolo, F. & Gamalero, E. (2024) The impact of Nitrogen-fixing bacteria-based biostimulant alone or in combination with commercial inoculum on tomato native rhizosphere microbiota and production: An open-field trial. *Biology*, 13(6), 400.

Nunes, P. S. O., Lacerda-Junior, G. V., Mascarin, G. M., Guimaraes, R. A., H., M. F., Arthurs, S. & Bettiol, W. (2024) Microbial consortia of biological products: Do they have a future? *Biol Control*, 188, 105439.

Oljira, A. M., Hussain, T., Waghmode, T. R., Zhao, H., Sun, H., Liu, X., Wang, X. & Liu, B. (2020) *Trichoderma* enhances net photosynthesis, water use efficiency, and growth of wheat (*Triticum aestivum* L.) under salt stress. *Microorganisms*, 8(10), 1565.

Ongena, M., Jourdan, E., Adam, A., Paquot, M., Brans, A., Joris, B., Arpigny, J. L. & Thonart, P. (2007) Surfactin and fengycin lipopeptides of *Bacillus subtilis* as elicitors of induced systemic resistance in plants. *Environ Microbiol*, 9(4), 1084–1090.

Oshunsanya, S. O., Nwosu, N. J. & Li, Y. (2019) Abiotic stress in agricultural crops under climatic conditions. In: Jhariya, M., Banerjee, A., Meena, R., Yadav, D. (eds.) *Sustainable Agriculture, Forest and Environmental Management*. Springer, Singapore, vol. 1, pp. 1–16.

Panda, S. K. & Das, S. (2024) Potential of plant growth-promoting microbes for improving plant and soil health for biotic and abiotic stress management in mangrove vegetation. *Rev Environ Sci Biotechnol*, 23(3), 801–837.

Park, J. Y., Oh, S. A., Anderson, A. J., Neiswender, J., Kim, J. C. & Kim, Y. C. (2011) Production of the antifungal compounds phenazine and pyrrolnitrin from *Pseudomonas chlororaphis* O6 is differentially regulated by glucose. *Lett Appl Microbiol*, 52(5), 532–537.

Perez-Moncada, U. A., Santander, C., Ruiz, A., Vidal, C., Santos, C. & Cornejo, P. (2024) Design of microbial consortia based on arbuscular mycorrhizal fungi, yeasts, and bacteria to improve the biochemical, nutritional, and physiological status of strawberry plants growing under water deficits. *Plants*, 13(11), 1556.

Petrillo, C., Vitale, E., Ambrosino, P., Arena, C. & Isticato, R. (2022) Plant growth-promoting bacterial consortia as a strategy to alleviate drought stress in *Spinacia oleracea*. *Microorganisms*, 10(9), 1798.

Prajapati, K., Sharma, M. & Modi, H. J. I. J. A. S. (2013) Growth promoting effect of potassium solubilizing microorganisms on *Abelmoscus esculantus*. *Int J Agric Sci*, 3(1), 181–188.

Prsic, J. & Ongena, M. (2020) Elicitors of plant immunity triggered by beneficial bacteria. *Front Plant Sci*, 11, 594530.

Puranik, S., Shukla, L., Kundu, A., Kamil, D., Paul, S., Venkadasamy, G., Salim, R., Singh, S.K., Kumar, D. and Kumar, A., 2023. Exploring potent fungal isolates from sanitary landfill soil for in vitro degradation of dibutyl phthalate. *J Fungi*, 9(1), 125.

Qin, H., Liu, Y., Cao, X., Jiang, J., Lian, W., Qiao, D., Xu, H. & Cao, Y. (2020) RpoS is a pleiotropic regulator of motility, biofilm formation, exoenzymes, siderophore and prodigiosin production, and trade-off during prolonged stationary phase in *Serratia marcescens*. *PLoS One*, 15(6), e0232549.

Ram, R. M., Debnath, A., Negi, S. & Singh, H. B. (2022) Use of microbial consortia for broad spectrum protection of plant pathogens: Regulatory hurdles, present status and future prospects. *Biopesticides*, 2, 319–335.

Rawat, P., Das, S., Shankhdhar, D. & Shankhdhar, S. (2021) Phosphate-solubilizing microorganisms: Mechanism and their role in phosphate solubilization and uptake. *J Soil Sci Plant Nutr*, 21, 49–68.

Raza, W., Ling, N., Liu, D., Wei, Z., Huang, Q. & Shen, Q. (2016) Volatile organic compounds produced by *Pseudomonas fluorescens* WR-1 restrict the growth and virulence traits of *Ralstonia solanacearum*. *Microbiol Res*, 192, 103–113.

Robson, R. L. & Postgate, J. R. (1980) Oxygen and hydrogen in biological nitrogen fixation. *Annu Rev Microbiol*, 34, 183–207.

Romero, D., de Vicente, A., Rakotoaly, R. H., Dufour, S. E., Veening, J. W., Arrebola, E., Cazorla, F. M., Kuipers, O. P., Paquot, M. & Perez-Garcia, A. (2007) The iturin and fengycin families of lipopeptides are key factors in antagonism of *Bacillus subtilis* toward Podosphaera fusca. *Mol Plant Microbe Interact*, 20(4), 430–440.

Roychowdhury, R., Mishra, S., Anand, G., Dalal, D., Gupta, R., Kumar, A. & Gupta, R. (2024) Decoding the molecular mechanism underlying salicylic acid (SA)-mediated plant immunity: An integrated overview from its biosynthesis to the mode of action. *Physiol Plant*, 176(3), e14399.

Rungsirivanich, P., Parlindungan, E., O'Connor, P. M., Field, D., Mahony, J., Thongwai, N. & van Sinderen, D. (2021) Simultaneous production of multiple antimicrobial compounds by *Bacillus velezensis* ML122-2 isolated From Assam Tea Leaf [*Camellia sinensis* var. assamica (J.W.Mast.) Kitam.]. *Front Microbiol*, 12, 789362.

Salazar, B., Ortiz, A., Keswani, C., Minkina, T., Mandzhieva, S., Pratap Singh, S., Rekadwad, B., Borriss, R., Jain, A., Singh, H. B. & Sansinenea, E. (2023) *Bacillus* spp. as bio-factories for antifungal secondary metabolites: Innovation beyond whole organism formulations. *Microb Ecol*, 86(1), 1–24.

Saleem, M., Nawaz, F., Hussain, M. B. & Ikram, R. M. (2021) Comparative effects of individual and consortia plant growth promoting bacteria on physiological and enzymatic mechanisms to confer drought tolerance in maize (*Zea mays* L.). *J Soil Sci Plant Nutr*, 21, 3461–3476.

Samain, E., Ernenwein, C., Aussenac, T. & Selim, S. (2022) Effective and durable systemic wheat-induced resistance by a plant-growth-promoting rhizobacteria consortium of *Paenibacillus* sp. strain B2 and *Arthrobacter* spp. strain AA against *Zymoseptoria tritici* and drought stress. *Physiol Mol Plant Pathol*, 119, 101830.

Sandhya, V., Ali, S. Z., Grover, M., Reddy, G. & Venkateswarlu, B. (2011) Drought-tolerant plant growth promoting *Bacillus* spp.: Effect on growth, osmolytes, and antioxidant status of maize under drought stress. *J Plant Inter*, 6, 1–14.

Sandhya, V. Z. A. S., SK. Z, A., Grover, M., Reddy, G. & Venkateswarlu, B. S. S. S. (2009) Alleviation of drought stress effects in sunflower seedlings by the exopolysaccharides producing *Pseudomonas putida* strain GAP-P45. *Biol Fertil Soils*, 46, 17–26.

Santi, C., Bogusz, D. & Franche, C. (2013) Biological nitrogen fixation in non-legume plants. *Ann Bot*, 111(5), 743–67.

Santos, R., Batista, B. B. & da Silva Neto, J. F. (2020) Ferric uptake regulator Fur coordinates siderophore production and defense against iron toxicity and oxidative stress and contributes to virulence in *Chromobacterium violaceum*. *Appl Environ Microbiol*, 86(21), e01620-20.

Santoyo, G., Guzmán-Guzmán, P., Parra-Cota, F. I., Santos-Villalobos, S. D. L., Orozco-Mosqueda, M. D. C. & Glick, B. R. (2021) Plant growth stimulation by microbial consortia. *Agronomy*, 11(2), 219.

Santoyo, G., Orozco-Mosqueda, M. D. C., Afridi, M. S., Mitra, D., Valencia-Cantero, E. & Macias-Rodriguez, L. (2024) *Trichoderma* and *Bacillus* multifunctional allies for plant growth and health in saline soils: Recent advances and future challenges. *Front Microbiol*, 15, 1423980.

Sathya, A., Rehman, V., Srinivas, V., Kudapa, H. & Gopalakrishnan, S. (2024) Efficacy of two different microbial consortia on salinity tolerance in chickpea: An in-planta evaluation on biochemical, histochemical, and genomic aspects. *3 Biotech*, 14(11), 285.

Schikora, A., Schenk, S. T. & Hartmann, A. (2016) Beneficial effects of bacteria-plant communication based on quorum sensing molecules of the N-acyl homoserine lactone group. *Plant Mol Biol*, 90(6), 605–612.

Schikora, A., Schenk, S. T., Stein, E., Molitor, A., Zuccaro, A. & Kogel, K. H. (2011) N-acyl-homoserine lactone confers resistance toward biotrophic and hemibiotrophic pathogens via altered activation of AtMPK6. *Plant Physiol*, 157(3), 1407–1418.

Schmitz, S. & Rosenbaum, M. A. (2020) Controlling the production of *Pseudomonas* phenazines by modulating the genetic repertoire. *ACS Chem Biol*, 15(12), 3244–3252.

Schulz-Bohm, K., Martin-Sanchez, L. & Garbeva, P. (2017) Microbial volatiles: Small molecules with an important role in intra- and inter-kingdom interactions. *Front Microbiol*, 8, 2484.

Seleiman, M. F., Al-Suhaibani, N., Ali, N., Akmal, M., Alotaibi, M., Refay, Y., Dindaroglu, T., Abdul-Wajid, H. H. & Battaglia, M. L. (2021) Drought stress impacts on plants and different approaches to alleviate its adverse effects. *Plants*, 10(2), 259.

Seth, K., Vyas, P., Deora, S., Gupta, A. K., Meena, M., Swapnil, P. & Harish (2023) Understanding plant-plant growth-promoting rhizobacteria (PGPR) interactions for inducing plant defense. In: Swapnil, P., Meena, M., Harish, Marwal, A., Vijayalakshmi, S., Zehra, A. (eds.) *Plant-Microbe Interaction—Recent Advances in Molecular and Biochemical Approaches*. Elsevier: Academic Press, vol. 2, pp. 201–226.

Shahid, M., Singh, U. B., Khan, M. S., Singh, P., Kumar, R., Singh, R. N., Kumar, A. & Singh, H. V. (2023) Bacterial ACC deaminase: Insights into enzymology, biochemistry, genetics, and potential role in amelioration of environmental stress in crop plants. *Fron. Microbiol*, 14, 1132770.

Shi, W., Xing, Y., Zhu, Y., Gao, N. & Ying, Y. (2022) Diverse responses of pqqC- and phoD-harbouring bacterial communities to variation in soil properties of Moso bamboo forests. *Microb Biotechnol*, 15(7), 2097–2111.

Shilev, S., Babrikova, I. & Babrikov, T. (2020) Consortium of plant growth-promoting bacteria improves spinach (*Spinacea oleracea* L.) growth under heavy metal stress conditions. *J Chem Technol Biotechnol*, 95, 932–939.

Shomali, A., Das, S., Arif, N., Sarraf, M., Zahra, N., Yadav, V., Aliniaeifard, S., Chauhan, D. K. & Hasanuzzaman, M. (2022) Diverse physiological roles of flavonoids in plant environmental stress responses and tolerance. *Plants*, 11(22), 3158.

Siddika, A., Rashid, A. A., Khan, S. N., Khatun, A., Karim, M. M., Prasad, P. V. V. & Hasanuzzaman, M. (2024) Harnessing plant growth-promoting rhizobacteria, *Bacillus subtilis* and *B. aryabhattai* to combat salt stress in rice: A study on the regulation of antioxidant defense, ion homeostasis, and photosynthetic parameters. *Front Plant Sci*, 15, 1419764.

Singh, D., Geat, N., Jadon, K. S., Verma, A., Sharma, R., Rajput, L. S., Mahla, H. R. & Kakani, R. K. (2024) Isolation and characterization of biocontrol microbes for development of effective microbial consortia for managing *Rhizoctonia bataticola* Root Rot of cluster bean under hot arid climatic conditions. *Microorganisms*, 12(11), 2331.

Singh, D., Shukla, L., Kamil, D., Singh, S. K. & Kumar, A. (2025) Evaluating a fungal consortium for efficient rice stubble degradation. *Journal of Pure & Applied Microbiology*, 19(1). https://doi.org/10.22207/JPAM.19.1.30

Singh, S., Mitra, N., G., K., S. R., Yaduwanshi, B., Singh, S. & Soni, K. (2021) Effect of microbial consortia on available nutrients and microbial population in soil of soybean in vertisol. *BFIJ*, 13, 451–458.

Soltero-Rios, M. A., Medina-Orozco, A. Y., Valenzuela-Yerena, M. A., Andrade-Robles, N. M., Mendoza-Flores, A. A., Morelia-Jiménez, J. A. Y., Ochoa-Castañeda, M. Y., Robles-Hernández, A. & González-Estrada, R. R. (2024) Microbial consortia: An approach in plant growth promotion and plant diseases management. In: Kumar, A., Solanki, M. K. (eds.) *Microbial Biocontrol Techniques*. Microorganisms for Sustainability, vol. 54. Springer, Singapore, 1–17.

Soumare, A., Diedhiou, A. G., Thuita, M., Hafidi, M., Ouhdouch, Y., Gopalakrishnan, S. & Kouisni, L. (2020) Exploiting biological nitrogen fixation: A route towards a sustainable agriculture. *Plants*, 9(8), 1011.

Srivastava, S., Tyagi, R. & Sharma, S. (2024) Seed biopriming as a promising approach for stress tolerance and enhancement of crop productivity: A review. *J Sci Food Agric*, 104(3), 1244–1257.

Sruthy, K. S., Shukla, L., Kundu, A., Singh, S. K., Abdulrahman Alodaini, H., Hatamleh, A. A., Santoyo, G. & Kumar, A. (2023). Effect of microbial consortium constructed with lignolytic ascomycetes fungi on degradation of rice stubble. *J Fungi*, 9(5), 567.

Sruthy, K. S., Puranik, S., Kumar, V., Kaushik, A., Vikram, K. V., Manoj, M., Shukla, L., Singh, S. K. & Kumar, A. (2025) Recent developments in microbial degradation of crop residues: A comprehensive review. *Int J Environ Sci Technol*, 22(7), 6247–6264.

Steil, L., Hoffmann, T., Budde, I., Volker, U. & Bremer, E. (2003) Genome-wide transcriptional profiling analysis of adaptation of *Bacillus subtilis* to high salinity. *J Bacteriol*, 185(21), 6358–6370.

Stummer, B. E., Zhang, X., Yang, H. & Harvey, P. R. (2022) Co-inoculation of *Trichoderma gamsii* A5MH and *Trichoderma harzianum* Tr906 in wheat suppresses in planta abundance of the crown rot pathogen *Fusarium pseudograminearum* and impacts the rhizosphere soil fungal microbiome. *Biol Control*, 165, Article 104809.

Suleimanova, A. D., Beinhauer, A., Valeeva, L. R., Chastukhina, I. B., Balaban, N. P., Shakirov, E. V., Greiner, R. & Sharipova, M. R. (2015) Novel glucose-1-phosphatase with High phytase activity and unusual metal Ion activation from soil bacterium *Pantoea* sp. Strain 3.5.1. *Appl Environ Microbiol*, 81(19), 6790–6799.

Suleman, M., Yasmin, S., Rasul, M., Yahya, M., Atta, B. M. & Mirza, M. S. (2018) Phosphate solubilizing bacteria with glucose dehydrogenase gene for phosphorus uptake and beneficial effects on wheat. *PLoS One*, 13(9), e0204408.

Sun, J., Liu, Y., Lin, F., Lu, Z. & Lu, Y. (2021) CodY, ComA, DegU and Spo0A controlling lipopeptides biosynthesis in *Bacillus amyloliquefaciens* fmbJ. *J Appl Microbiol*, 131(3), 1289–1304.

Sun, W., Shahrajabian, M. H. & Soleymani, A. (2024) The roles of Plant-Growth-Promoting Rhizobacteria (PGPR)-based biostimulants for agricultural production yystems. *Plants*, 13(5), 613.

Tabacchioni, S., Passato, S., Ambrosino, P., Huang, L., Caldara, M., Cantale, C., Hett, J., Del Fiore, A., Fiore, A., Schluter, A., Sczyrba, A., Maestri, E., Marmiroli, N., Neuhoff, D., Nesme, J., Sorensen, S. J., Aprea, G., Nobili, C., Presenti, O., Giovannetti, G., Giovannetti, C., Pihlanto, A., Brunori, A. & Bevivino, A. (2021) Identification of beneficial microbial consortia and bioactive compounds with potential as plant biostimulants for a sustainable agriculture. *Microorganisms*, 9(2), 426.

Tang, J., Li, Y., Zhang, L., Mu, J., Jiang, Y., Fu, H., Zhang, Y., Cui, H., Yu, X. & Ye, Z. (2023) Biosynthetic pathways and functions of Indole-3-Acetic Acid in microorganisms. *Microorganisms*, 11(8), 2077.

Theatre, A., Cano-Prieto, C., Bartolini, M., Laurin, Y., Deleu, M., Niehren, J., Fida, T., Gerbinet, S., Alanjary, M., Medema, M. H., Leonard, A., Lins, L., Arabolaza, A., Gramajo, H., Gross, H. & Jacques, P. (2021) The surfactin-like lipopeptides from *Bacillus* spp.: Natural biodiversity and synthetic biology for a broader application range. *Front Bioeng Biotechnol*, 9, 623701.

Thomas, G., Withall, D. & Birkett, M. (2020) Harnessing microbial volatiles to replace pesticides and fertilizers. *Microb Biotechnol*, 13(5), 1366–1376.

Timofeeva, A. M., Galyamova, M. R. & Sedykh, S. E. (2023) Plant growth-promoting bacteria of soil: Designing of consortia beneficial for crop production. *Microorganisms*, 11(12), 2864.

Timofeeva, A. M., Galyamova, M. R. & Sedykh, S. E. (2024) How do plant growth-promoting bacteria use plant hormones to regulate stress reactions? *Plants*, 13(17), 2371.

Toribio, A. J., Suarez-Estrella, F., Jurado, M. M., Lopez-Gonzalez, J. A., Martinez-Gallardo, M. R. & Lopez, M. J. (2022) Design and validation of cyanobacteria-rhizobacteria consortia for tomato seedlings growth promotion. *Sci Rep*, 12(1), 13150.

Travençoli-Rossetim, M. F., Vargas-Motta, A. C., Rocha-Kondo, Y., Santos-Ruthes, B. E., Hungria, M., Falcão-Salles, J. & Kaschuk, G. (2025) Enhancing soybean yield through inoculation of multifunctional microbial consortia. *Int J Agron*, 1, 1–13.

Trivedi, P., Duan, Y. & Wang, N. (2010) Huanglongbing, a systemic disease, restructures the bacterial community associated with citrus roots. *Appl Environ Microbiol*, 76(11), 3427–3436.

Tsavkelova, E. A., Klimova, S., Cherdyntseva, T. A. & Netrusov, A. I. (2006) Microbial producers of plant growth stimulators and their practical use: A review. *Prikl Biokhim Mikrobiol*, 42(2), 133–143.

Tsukanova, K. A., Chebotar, V. K., Meyer, J. J. M. & Bibikova, T. N. (2017) Effect of plant growth-promoting Rhizobacteria on plant hormone homeostasis. *S Afr J Bot*, 113, 91–102.

Tyagi, J., Mishra, A., Kumari, S., Singh, S., Agarwal, H., Pudake, R. N. & Joshi, N. C. (2023) Deploying a microbial consortium of *Serendipita indica*, *Rhizophagus intraradices*, and *Azotobacter chroococcum* to boost drought tolerance in maize. *EEB*, 206, 105142.

Upamanya, G. K., Bhattacharyya, A. & Dutta, P. (2020) Consortia of entomo-pathogenic fungi and bio-control agents improve the agro-ecological conditions for brinjal cultivation of Assam. *3 Biotech*, 10(10), 450.

Vishwakarma, K., Kumar, N., Shandilya, C., Mohapatra, S., Bhayana, S. & Varma, A. (2020) Revisiting plant-microbe interactions and microbial consortia application for anhancing austainable agriculture: A review. *Front Microbiol*, 11, 560406.

Wallis, C. M. & Galarneau, E. R. (2020) Phenolic compound induction in plant-microbe and plant-insect interactions: A Meta-analysis. *Front Plant Sci*, 11, 580753.

Wan, W., Qin, Y., Wu, H., Zuo, W., He, H., Tan, J., Wang, Y. & He, D. (2020) Isolation and characterization of phosphorus solubilizing bacteria with multiple Phosphorus sources utilizing capability and their potential for lead immobilization in soil. *Front Microbiol*, 11, 752.

Wang, Q., Liu, J. & Zhu, H. (2018) Genetic and molecular mechanisms underlying symbiotic specificity in legume-Rhizobium interactions. *Front Plant Sci*, 9, 313.

Wang, Y., Dall'Agnol, R. F., Bertani, I., Bez, C. & Venturi, V. (2024a) Identification of synthetic consortia from a set of plant-beneficial bacteria. *Microb Biotechnol*, 17(2), e14330.

Wang, Y., Pei, Y., Wang, X., Dai, X. & Zhu, M. (2024b) Antimicrobial metabolites produced by the plant growth-promoting rhizobacteria (PGPR): *Bacillus* and *Pseudomonas*. *Adv Agrochem*, 3, 206–221.

Wang, Z. J., Li, X., Wang, J. H., Qi, S. S., Dai, Z. C. & Du, D. L. (2022) Effect of nitrogen-fixing bacteria on resource investment of the root system in an invasive clonal plant under low nutritional environment. *Flora*, 297, 152166.

Wehner, G., Kopahnke, D., Richter, K., Kecke, S., Schikora, A. & Ordon, F. (2019) Priming is a suitable strategy to enhance resistance towards leaf rust in barley. *Phytob J*, 3, 46–51.

Wei, X., Huang, X., Tang, L., Wu, D. & Xu, Y. (2013) Global control of GacA in secondary metabolism, primary metabolism, secretion systems, and motility in the rhizobacterium *Pseudomonas aeruginosa* M18. *J Bacteriol*, 195(15), 3387–400.

Wilder, C. N., Diggle, S. P. & Schuster, M. (2011) Cooperation and cheating in *Pseudomonas aeruginosa*: The roles of the las, rhl and pqs quorum-sensing systems. *ISME J*, 5(8), 1332–1343.

Wilmowicz, E., Kucko, A., Bogati, K., Wolska, M., Swidzinski, M., Burkowska-But, A. & Walczak, M. (2022) *Glomus* sp. and *Bacillus* sp. strains mitigate the adverse effects of drought on maize *(Zea $_{may}$s* L.). *Front Plant Sci*, 13, 958004.

Wu, X., Pan, L., Liu, Y., Jiang, P., Lee, I., Drevensek-Olenik, I., Zhang, X. & Xu, J. (2013) Cell-cell communication induces random spikes of spontaneous calcium oscillations in multi-BV-2 microglial cells. *Biochem Biophys Res Commun*, 431(4), 664–669.

Xiao, J., Guo, X., Qiao, X., Zhang, X., Chen, X. & Zhang, D. (2021) Activity of fengycin and iturin A isolated from *Bacillus subtilis* Z-14 on *Gaeumannomyces graminis* Var. tritici and soil microbial diversity. *Front Microbiol*, 12, 682437.

Xie, B., Wei, X., Wan, C., Zhao, W., Song, R., Xin, S. & Song, K. (2024a) Exploring the biological pathways of siderophores and their multidisciplinary applications: A comprehensive review. *Molecules*, 29(10), 2318.

Xie, P., Xu, Y., Tang, J., Wu, S. & Gao, H. (2024b) Multifaceted regulation of siderophore synthesis by multiple regulatory systems in *Shewanella oneidensis*. *Commun Biol*, 7, 498.

Xu, P. & Wang, E. (2023) Diversity and regulation of symbiotic nitrogen fixation in plants. *Curr Biol*, 33(11), R543–R559.

Xu, Z., Pehlivan, N., Ghorbani, A. & Wu, C. (2022) Effects of *Azorhizobium caulinodans* and *Piriformospora indica* co-inoculation on growth and fruit quality of tomato (*Solanum lycopersicum* L.) under salt stress. *Horticulturae*, 8, 302–312.

Yadav, R., Ror, P., Beniwal, R., Kumar, S. & Ramakrishna, W. (2022) *Bacillus* sp. and arbuscular mycorrhizal fungi consortia enhance wheat nutrient and yield in the second-year field trial: Superior performance in comparison with chemical fertilizers. *J Appl Microbiol*, 132(3), 2203–2219.

Yaduwanshi, B., Sahu, R. K., Mitra, N. G. & Dwivedi, B. S. (2021) Impact of microbial consortia on microbial population and available nutrients in soil under soybean crop. *Indian Soc Soil Sci*, 69(2), 187–194.

Yamanaka, K., Oikawa, H., Ogawa, H. O., Hosono, K., Shinmachi, F., Takano, H., Sakuda, S., Beppu, T. & Ueda, K. (2005) Desferrioxamine E produced by *Streptomyces griseus* stimulates growth and development of *Streptomyces tanashiensis*. *Microbiology*, 151 (Pt 9), 2899–2905.

Yang, B., Zheng, M., Dong, W., Xu, P., Zheng, Y., Yang, W., Luo, Y., Guo, J., Niu, D., Yu, Y. & Jiang, C. (2023) Plant disease resistance-related pathways recruit beneficial bacteria by remodeling root exudates upon *Bacillus cereus* AR156 Treatment. *Microbiol Spectr*, 11(2), e0361122.

Yim, B., Heider, M. A., Bloem, E., Vetterlein, D., Behr, J. H., Babin, D. & Smalla, K. (2025) Exploring the potential of seed inoculation with microbial consortia to mitigate drought stress in maize plants under greenhouse conditions. *Plant Soil*, 1, 1–15.

Yu, Y., Gui, Y., Li, Z., Jiang, C., Guo, J. & Niu, D. (2022) Induced systemic resistance for improving plant immunity by beneficial microbes. *Plants*, 11(3), 386.

Yuan, L., Zhang, S., Wang, Y., Li, Y., Wang, X. & Yang, Q. (2018) Surfactin inhibits membrane fusion during invasion of epithelial cells by enveloped viruses. *J Virol*, 92(21), 10–1128.

Zamioudis, C. & Pieterse, C. M. (2012) Modulation of host immunity by beneficial microbes. *Mol Plant Microbe Interact*, 25(2), 139–150.

Zayed, O., Hewedy, O. A., Abdelmoteleb, A., Ali, M., Youssef, M. S., Roumia, A. F., Seymour, D. & Yuan, Z. C. (2023) Nitrogen journey in plants: From uptake to metabolism, stress response, and microbe interaction. *Biomolecules*, 13(10), 1443.

Zhang, H., Kim, M. S., Sun, Y., Dowd, S. E., Shi, H. & Pare, P. W. (2008) Soil bacteria confer plant salt tolerance by tissue-specific regulation of the sodium transporter HKT1. *Mol Plant Microbe Interact*, 21(6), 737–744.

Zhang, W., Chen, Y., Huang, K., Wang, F. & Mei, Z. (2023) Molecular mechanism and agricultural application of the NifA-NifL system for nitrogen fixation. *Int J Mol Sci*, 24(2), 907.

Zhang, Y., Teper, D., Xu, J. & Wang, N. (2019) Stringent response regulators (p)ppGpp and DksA positively regulate virulence and host adaptation of *Xanthomonas citri*. *Mol Plant Pathol*, 20(11), 1550–1565.

Zhen, C., Ge, X. F., Lu, Y. T. & Liu, W. Z. (2023) Chemical structure, properties and potential applications of surfactin, as well as advanced strategies for improving its microbial production. *AIMS Microbiol*, 9(2), 195–217.

Zhu, L., Huang, J., Lu, X. & Zhou, C. (2022) Development of plant systemic resistance by beneficial rhizobacteria: Recognition, initiation, elicitation and regulation. *Front Plant Sci*, 13, 952397.

Zhu, Z., Zhang, H., Leng, J., Niu, H., Chen, X. & Liu, D. (2020) Isolation and characterization of plant growth-promoting rhizobacteria and their effects on the growth of *Medicago sativa* L. under salinity conditions. *Int J General Molec Microbiol* (*Antonie van Leeuwenhoek*), 113, 1263–1278.

4 Elucidating Possibilities of Grasses and Their Allied Beneficial Microorganisms for Sustainable Agricultural Production

Shriniketan Puranik,
Meghalamane Krishnamurthy Murali,
Murtheppa Manoj, Priyanka Mulage,
Revanna Mythra, Sandeep Kumar Singh,
and Livleen Shukla

4.1 INTRODUCTION

Grasses are among the most ecologically essential and versatile plant taxa, serving as a critical foundation for ecosystem functionality, restoration, and sustainable agriculture due to their unique physiological, morphological, and biochemical adaptations that facilitate the regulation of soil nutrient dynamics; improvement of biodiversity; mitigation of air, soil, and water pollution; and enhancement of climate resilience across diverse landscapes. Their extensive and deep-rooted systems not only stabilize the soil, preventing erosion and improving aggregation, but also act as crucial channels for the uptake, storage, recycling, and gradual return of major macronutrients like nitrogen, phosphorus, and potassium via the processes of organic matter decomposition, root exudation, and active interactions with soil microorganisms and legumes, ensuring continual enrichment and robust fertility even in marginal or degraded soils, as demonstrated in semiarid regions around the world. Grasses are integral for accumulation and cycling of soil organic matter (SOM), that increases soil structure, boosts water retention, and provides a nurturing environment for diverse microbial and fungal communities including arbuscular mycorrhizal fungi (AMF), which further boosts phosphorus and micronutrient uptake and contribute to resilient agroecosystems by supporting plant stress tolerance and maintaining soil function against drought, nutrient depletion, and other stressors (dos Santos Lucas et al., 2022).

 DOI: 10.1201/9781003664451-4

Innovative grass-based practices, such as using vetiver and switchgrass for mulch decomposition or prairie strips and grassed waterways for controlling runoff and preventing sediment loss, underscore the multifaceted value of grasses as both natural engineers of their environment and vital assets in sustainable nutrient management, showing quantifiable improvements in nutrient cycling ability, C-accumulation, and microbial activity compared to bare or degraded land. In addition to their roles belowground, grasses mitigate air pollution by using dense foliage and high leaf surface area to capture airborne particulates and filter volatile organic compounds (VOCs), with supporting evidence showing that species such as *Sansevieria trifasciata*, *Tradescantia pallida*, and *Chlorophytum comosum* can reduce levels of PM_{10}, $PM_{2.5}$, and atmospheric VOCs by huge margins, highlighting their contribution to improved air quality in agricultural and urban environments (Vitaliano et al., 2024; Desiderio et al., 2025). Grasses and associated vegetation form effective biofilters that intercept a significant proportion of pesticide residues and airborne chemicals prior to soil penetration or atmospheric dispersion, with grass cover and windbreaks reducing erosion-induced dust and minimizing drift of agrochemicals, thus protecting both the health of the ecosystem and the broader community. In soil pollutant mitigation, grasses exhibit remarkable phytoremediation potential through strategies like phytoextraction and phytostabilization of heavy metals and hazardous substances. Species including ryegrass, vetiver, finger millet, sorghum, and pearl millet have shown the ability to accumulate, stabilize, and prevent the leaching of toxic elements like cadmium, lead, arsenic, and chromium, especially when grown on contaminated or landfill soils and when paired with organic amendments like compost and biochar, sometimes reducing available soil cadmium by more than 60% in a single growing season while also supporting fodder or bioenergy production. These same grass species have been successfully applied to restore polluted sites and marginal lands, simultaneously providing agronomic benefits and environmental services by binding soil, improving aggregation, enhancing microbial communities, and reducing erosion and runoff by up to 90% relative to non-grassed controls, all of which are key for land reclamation and long-term ecosystem recovery.

In water quality improvement, grasses employed as buffer strips, constructed wetlands, and vegetative barriers function as living filters that slow surface runoff, trap sediments, sequester nutrients, and absorb agricultural pollutants, enabling dramatic reductions in nutrient export (especially nitrogen and phosphorus), biological oxygen demand (BOD), and chemical oxygen demand (COD) with removal efficiencies for heavy metals like cadmium, lead, and zinc and for excess nutrients often exceeding 70% in engineered wetland systems (Marzec et al., 2018; Sladkovska et al., 2022). The resilience of grasses in fluctuating hydrological regimes, including repeated wet–dry cycles, allows them to maintain efficient filtration of pollutants and consistent bioremediation, whether in natural riparian habitats or artificial stormwater management systems, thereby stabilizing water quality even under climatic stress or high-intensity land use. Such interventions not only provide immediate reductions in pollution loads but also create habitat and foster biodiversity, reinforcing ecosystem structure and resilience through the creation of heterogeneous plant and microbial communities. Grasses thereby represent a sustainable, low-input solution for environmental stewardship, integrating seamlessly with conservation agriculture, restoration ecology, and landscape design to enhance both ecological integrity and productivity; their systemic contribution to soil carbon sequestration and climate regulation through photosynthesis, deep storage of organic carbon, and reduction of greenhouse gas emissions further align them with strategies for climate change mitigation, presenting cost-effective and scientifically substantiated pathways for realizing resilient, multifunctional, and productive agricultural and ecological systems worldwide. Thus, the multifunctional roles of grasses spanning biogeochemical cycling, pollution remediation, habitat creation, and climate stabilization are indispensable for addressing contemporary environmental pressures, and their wise integration into land management strategies holds the promise of revitalizing landscapes for both current and future generations, ensuring both environmental health and sustainable productivity in an era of mounting global challenges.

4.2 ROLE OF GRASSES IN ECOSYSTEMS

Grasses, among the most ecologically prevalent and versatile plant taxa, deliver crucial ecosystem services regulating soil nutrient dynamics and remediating pollution across air, soil, and water compartments. Their rapid growth, extensive root architecture, and adaptability to stress underpin these roles (Cheng et al., 2021). Four key functions of grasses are regulation of soil nutrients, and mitigation of air, soil, and water pollution, which are discussed in the following subsections.

4.2.1 Regulation of Soil Nutrients

Grasses are essential for maintaining soil nutrient balance in farming systems, as they help improve soil fertility, structure, and long-term ecosystem health. They are integral to the regulation of soil nutrients in agricultural landscapes. Through their root systems, contribution to organic matter, specialized nutrient acquisition mechanisms, and interactions with legumes and soil microorganisms, grasses enhance soil fertility and structure (Wei et al., 2022). The inclusion of grasses in agricultural practices is crucial for maintaining sustainable soil nutrient cycles and securing the long-term productivity of cropping systems.

Grasses influence soil nutrient dynamics largely through their well-developed root systems. These roots stabilize soil structure, reduce erosion, and play an important role in absorbing and recycling essential macronutrients, including nitrogen (N), phosphorus (P), and potassium (K) (Nascimento et al., 2024). These nutrients re-enter the soil mainly through the decomposition of plant litter, the release of root exudates, and the death of plant tissues (Guan et al., 2023). Nitrogen is returned mainly in organic forms such as amino acids and proteins, which are later mineralized to ammonium (NH_4^+) and nitrate (NO_3^-) (Li et al., 2024a). Phosphorus is released predominantly in organic forms like phytates and orthophosphates during decomposition (Su et al., 2021). Potassium, being more soluble, is typically returned in ionic form (K^+), readily available for plant uptake. Thus, grasses play a crucial role in nutrient cycling by contributing both organic and inorganic forms of essential elements back to the soil system. During growth and subsequent decomposition, grasses return accumulated nutrients to the soil, thereby improving soil quality and ensuring their availability to succeeding plants. This repeated process of nutrient uptake and recycling forms the foundation of long-term soil fertility (Halli et al., 2022). Moreover, grasses are key contributors to increasing SOM. The decomposition of their residues adds organic carbon into the soil matrix, strengthening soil structure, improving its capacity to retain water, and enhancing nutrient retention. Higher SOM levels also sustain a diverse microbial population responsible for the mineralization and release of nutrients essential for plant growth (Gerke, 2022).

Grasses play an essential role in regulating soil nutrient dynamics through the fixation, uptake, and recycling of essential macronutrients such as NPK. Quantitative studies have shown that different grass species vary in their capacity to fix and return these nutrients to the soil. For instance, in a study on vetiver grass, mulch decomposition over 125 days released approximately 12.3 kg N, 6.0 kg P, and 34.0 kg K per hectare, derived from initial contents of 32.9 kg N, 15.9 kg P, and 90.6 kg K per hectare (da Rocha et al., 2020). In rocky karst regions, grasslands contribute to annual returns ranging from 17.9 to 55.9 kg N, 1.8 to 4.7 kg P, and 6.6 to 37.0 kg K per hectare through litter and root turnover (Wang et al., 2024). Similarly, in semiarid regions, the dominant grass species *Leymus chinensis* has shown an annual nitrogen release of approximately 3.4 kg/ha, assuming 1 Mg/ha of litter with a decomposition rate constant (k) of 0.34 yr^{-1} (Zhu et al., 2021).

Typha species play a significant role in nutrient cycling within wetland and aquatic ecosystems by accumulating, storing, and releasing macronutrients such as NPK through their biomass and decomposition processes. Quantitative studies have shown that in *Typha glauca*, approximately 45% of the nitrogen and phosphorus accumulated in aboveground shoots during the growing season are translocated to rhizomes before senescence, while potassium is rapidly lost from tissues during early decomposition stages (Shapiro, 2022). In decomposition studies of submerged *Typha latifolia* litter, Mahmud et al. (2021) observed that most of the K and Na were lost within the first 20 days, over 50% of calcium and phosphorus were lost early, and nitrogen loss accelerated after 95 days of decomposition. These findings underscore the dynamic and species-specific patterns of nutrient storage and recycling in *Typha* grasses, highlighting their importance in wetland and their potential application in phytoremediation and ecological restoration.

In addition to these benefits, certain grass species have developed specialized mechanisms to improve nutrient acquisition. For instance, under conditions of micronutrient deficiency, some grasses release phytosiderophores, organic compounds that bind and solubilize nutrients such as iron (Fe), increasing their availability for plant uptake, which is especially advantageous in calcareous soils where the availability of micronutrients is naturally low (Prasad et al., 2022). As an indirect strategy, the co-cultivation of grasses with legumes in mixed cropping systems further enhances soil nutrient regulation. Legumes convert atmospheric nitrogen into a usable form through their symbiotic association with rhizobia, thereby increasing the soil's nitrogen content. When grown alongside grasses, this fixed nitrogen becomes accessible to the grasses, promoting their growth and contributing to a balanced nutrient profile in the soil. Such synergistic interactions between grasses and legumes are instrumental in ensuring sustainable nutrient management strategies (Bousselin et al., 2021). Additionally, grasses promote the proliferation of beneficial soil microbes such as AMF.

These fungi establish symbiotic relationships with plant roots, increasing the functional root area and thereby enhancing nutrient absorption, particularly of phosphorus. The presence of AMF in grass-dominated systems facilitates efficient nutrient cycling and contributes to the resilience of the agroecosystem (Li et al., 2024b). Common AMF species that perform a vital role in grass-dominated systems include *Glomus intraradices* (reclassified as *Rhizophagus intraradices*), *Glomus mosseae* (*Funneliformis mosseae*), *Claroideoglomus etunicatum*, and *Acaulospora laevis* (Smilauer et al., 2021). These fungi enhance the solubilization and mobilization of phosphorus and other micronutrients such as zinc and copper, especially in nutrient-poor soils (Bhantana et al., 2021). The presence of AMF in grass-dominated systems facilitates efficient nutrient cycling and contributes to the resilience and productivity of the agroecosystem by improving plant stress tolerance and maintaining soil structure.

Grasses contribute significantly to soil fertility, especially in degraded or nutrient-poor environments by cycling nutrients and promoting organic matter buildup. A recent field study in semiarid India compared grass-covered plots to unplanted controls. It has been reported that grasses enhance nutrient cycling by 1.1–1.6 times compared to fallow conditions, and their presence can double the soil's carbon accumulation capacity (Patidar et al., 2023). Furthermore, mechanistically, roots adhere to soil particles, reduce erosion, and enhance aggregation and microbial habitat. The vertical penetration of grass root systems can sustain nutrient availability in deeper horizons, supporting sustained biomass productivity. Such findings underscore the idea that grasses are crucial in revitalizing marginal and nutrient-depleted lands, providing durable soil function across moisture regimes (Halli et al., 2022).

4.2.2 Mitigation of Air Pollution

Grasses mitigate air pollution within agricultural systems through multiple mechanisms. Their dense foliage effectively captures airborne particulate matter (PM), reducing the concentration of pollutants such as PM_{10} and $PM_{2.5}$ in the atmosphere. This is mainly beneficial in arid and semiarid regions where wind erosion is prevalent (Ogwu et al., 2024). Besides these grasses, other plant types such as dendrophytes (woody perennials like trees and shrubs) and certain herbaceous species also play a complementary role in filtering air pollutants (Sladkovska et al., 2022). For instance, snake plants (*S. trifasciata*), known for their vertical leaf orientation and waxy surface, are highly efficient at trapping dust and absorbing VOCs (Permana et al., 2022). Similarly, plants like *Ficus benjamina* and *Azadirachta indica* (neem) are often planted in and around agricultural fields for their proven ability to absorb gaseous pollutants like sulphur dioxide (SO_2), nitrogen oxides (NO_x), and ozone (O_3), while also serving as windbreaks to reduce particulate resuspension. Together, grasses and associated vegetative barriers create a biologically active filtration system that improves air quality and contributes to healthier agricultural environments.

Grasses and associated vegetation significantly contribute to the reduction of airborne pollutants and pesticide residues within agricultural systems through interception and phytoremediation mechanisms (Jia et al., 2023). Quantitative research findings suggest that certain plant species can reduce VOCs by up to 90%–100% in controlled environments (Bao et al., 2023). For instance, indoor experiments using plants like *T. pallida* and *C. comosum* have demonstrated a 76%–92% reduction in formaldehyde within the first 15 minutes, reaching nearly 100% removal within 24 hours (Talaie and Ponraj, 2024). Although direct field-scale studies on VOC reduction by grasses are limited, these results indicate that vegetation with high leaf surface area can play a substantial role in atmospheric VOC mitigation. Regarding pesticide reduction, dense grass cover and vegetative barriers have been found to intercept up to 60% of applied pesticides before they reach the soil or are lost through drift (Braun et al., 2024). These findings underline the practical value of integrating grasses and vegetative buffers into agricultural landscapes not only for improving air quality through VOC absorption but also for minimizing pesticide dispersion and enhancing overall agroecosystem sustainability.

Moreover, grasses contribute significantly to carbon sequestration. Photosynthesis by aerial parts and extensive root systems store carbon in the soil, enhancing SOM and reducing atmospheric CO_2 levels. This mechanism contributes to climate change mitigation while simultaneously improving soil health and overall fertility (da Silva et al., 2022). Additionally, certain grass species can absorb VOCs emitted from fertilizers and pesticides, thereby improving air quality (Zheng et al., 2023). Beyond fertilizers and pesticides, grasses also play a role in mitigating other agrochemical wastes such as herbicide residues (e.g., glyphosate, atrazine), fungicides, and residual antibiotics from animal manure. Some grass species, like *Vetiveria zizanioides* (vetiver) and *Pennisetum purpureum* (Napier grass), have shown potential in phytoremediation by absorbing or degrading these agrochemical pollutants from soil and water (Dorafshan et al., 2023). These grasses can take up organic pollutants through their roots and facilitate their breakdown via rhizodegradation or microbial action in the rhizosphere. Thus, grasses not only serve as a tool for nutrient cycling and carbon capture but also contribute to detoxifying agricultural landscapes contaminated with a broad spectrum of agrochemical residues. Grasses also harbour some microbes in their aerial parts that can aid in airborne contaminants degradation. Sorkhoh et al. (2011) studied hydrocarbon degradation in American grass (*Cynodon* sp.) under microcosm conditions. They found an 81% reduction in hydrocarbons (4.1 μg/cm^3) as compared to surface-sterilized plants as control (21.8 μg/cm^3). This reduction was attributed to phyllospheric microbes like *Rhodococcus erythreus* and *Pseudomonas stutzeri*. Also, a study showed higher colonization of formaldehyde-degrading *Bacillus cereus* in shoots of *Zamioculcas zamiifolia* when introduced to the contaminant aerially (Khaksar et al., 2017). Thus, grass–microbe interactions can be underscored as a sustainable strategy for air pollution mitigation.

4.2.3 Mitigation of Soil Pollution

Grasses significantly help reduce soil pollution in agriculture by using their natural ability to clean up contaminants through phytoremediation. Species including ryegrass (*Lolium perenne*) have been utilized to stabilize heavy metals like cobalt, mercury, molybdenum, and nickel in contaminated soils, especially when mixed with soil amendments like compost and iron oxides, effectively reducing their bioavailability and environmental hazards (Masotla et al., 2023). Similarly, turfgrass species, such as bermudagrass (*Cynodon dactylon*), tall fescue (*Festuca arundinacea*), and perennial ryegrass, have demonstrated effectiveness in accumulating and stabilizing heavy metals including chromium, copper, zinc, and lead, making them suitable for phytoremediation efforts in contaminated agricultural lands (Satognon, 2024). In one study, *L. perenne* reduced the bioavailable fraction of cadmium in amended soils by up to 64% over a 90-day period (Yang et al., 2022). Likewise, turf-grass species including Bermuda grass (*C. dactylon*), tall fescue (*Festuca arundinacea*), and perennial ryegrass have been reported to sequester and immobilize heavy metals such as chromium, copper, zinc, and lead, highlighting their potential use in phytoremediation of polluted environments (Liu et al., 2019).

In addition to these, several cereal grasses and millets including finger millet (*Eleusine coracana*), sorghum (*Sorghum bicolor*), and pearl millet (*Pennisetum glaucum*) have also shown phytoremediation potential. These crops can accumulate trace elements like cadmium, arsenic, and lead in their root and shoot tissues and are often used in phytostabilization strategies. Their robust root systems, high biomass production, and tolerance to abiotic stress make them ideal candidates for reclaiming polluted agricultural lands while simultaneously supporting food and fodder production (Gorelova et al., 2022). Finger millet has shown tolerance and accumulation capacity for heavy metals like cadmium and lead, especially in soils amended with biochar and organic compost (Rakkammal et al., 2024). Sorghum, with its deep root system and high biomass, is effective at extracting and stabilizing heavy metals such as arsenic, zinc, and chromium, making it widely employed in the phytoremediation of industrially contaminated soils (Gorelova et al., 2023). Pearl millet and foxtail millet (*Setaria italica*) have also demonstrated the ability to tolerate and immobilize toxic metals like nickel and lead through root accumulation and restricted translocation to shoots (Awan et al., 2023).

Another promising species, proso millet (*Panicum miliaceum*) has shown the ability to grow in selenium-contaminated soils, aiding in selenium stabilization and preventing its leaching (Mushtaq et al., 2022). Moreover, Napier grass (*P. purpureum*) and vetiver grass (*Chrysopogon zizanioides*) are extensively used for phytostabilization of landfill sites and polluted irrigation areas due to their exceptional tolerance to heavy metals and agrochemical residues (Masinire et al., 2021). These grasses and millet species, apart from contributing to soil remediation, also provide fodder, grain, and biomass, offering an integrated approach to sustainable agriculture and environmental health. The dense root systems of these grasses not only aid in pollutant uptake but also improve soil structure and stimulate microbial activity, thereby supporting overall soil health. Moreover, the use of grasses in practices like prairie strips has been shown to lower soil erosion by up to 95% and decrease nutrient runoff, thereby preventing further soil degradation (Athar et al., 2022). These findings underscore the importance of integrating grasses into agricultural practices as a sustainable strategy for soil pollution mitigation. Among cereal grasses and millets, finger millet has shown a 47% reduction in lead content in contaminated soils when grown with biochar and compost amendments (Kang et al., 2022). *S. bicolor*, known for its deep root system and high biomass, is effective in stabilizing arsenic, zinc, and chromium. In a controlled field trial, sorghum reduced soil arsenic levels by 38% and increased microbial biomass carbon by 29%, indicating improvement in soil quality alongside metal stabilization (Kumar et al., 2024). Additionally, proso millet (*P. miliaceum*) has demonstrated the capacity to reduce selenium leaching by over 50% in selenium-contaminated soils, acting as a stabilizer while maintaining acceptable biomass production (Rasool et al., 2023). Likewise, Napier grass (*P. purpureum*) and vetiver grass (*C. zizanioides*) are widely used for landfill reclamation and wastewater treatment zones due to their ability to tolerate and stabilize high concentrations of lead and cadmium. For instance, vetiver has been reported to remove up to 75% of cadmium and 67% of lead from polluted sites over a single growing season (Masinire et al., 2021). Microbes associated to grasses also contribute to soil remediation. Arora et al. (2016) reported that AM fungi (arbuscular mycorrhizal fungi) and *Azospirillum* harboured by *Panicum virgatum* (switchgrass) had Pb and Cd degradation capacity in soil. Iqbal et al. (2024) assessed petroleum degradation in soil containing Bermuda grass (*C. dactylon*) and consortium of *Arthrobacter oxydans* and *Pseudomonas* sp., isolated from Italian ryegrass (*Lolium multiflorum*). They found 75% degradation of the contaminant and attributed it to increase in alkane hydroxylase gene (*alkB*) expression. These findings highlight the pivotal role of grass–microbe interaction for mitigating soil pollution, offering both ecological restoration and agronomic advantages.

4.2.4 Mitigation of Water Pollution

Grasses function as natural filters, stabilizers, and nutrient sinks, thereby playing a vital role in reducing water pollution within agricultural ecosystems. Their integration into farming systems through practices such as buffer strips, prairie strips, grassed waterways, and vegetative barriers has proven effective in reducing runoff, trapping sediments, and absorbing excess nutrients, thereby enhancing water quality (Plunge et al., 2022). Grassed waterways are another vital practice, involving the establishment of grass-covered channels to manage surface water flow. These structures effectively reduce sediment and nutrient losses by stabilizing soil in areas of concentrated flow and promoting water infiltration. Properly designed and maintained grassed waterways can significantly mitigate the transport of pollutants to adjacent streams and rivers, thereby protecting aquatic ecosystems. The vetiver system, utilizing *C. zizanioides*, is renowned for its deep root system and ability to form dense hedge rows. When planted along contours, vetiver hedge rows act as physical barriers that slow down runoff, trap sediments, and enhance water infiltration. Switchgrass (*P. virgatum*) barriers have also demonstrated efficacy in conserving soil and water. These dense, stiff-stemmed grasses, when established perpendicular to field slopes, obstruct runoff, reduce its velocity, and facilitate sediment deposition. Additionally, switchgrass barriers enhance soil quality by boosting organic carbon levels and porosity, thereby improving water infiltration and nutrient-holding capacity while reducing the likelihood of water contamination (Hu et al., 2022a, 2022b). Integrating grasses into farming systems contributes to the mitigation of water pollution while simultaneously enhancing soil quality, promoting biodiversity, and strengthening resilience to climatic fluctuations. With the growing emphasis on sustainable agriculture, the targeted use of grasses represents an efficient and eco-friendly method for maintaining water quality and improving ecosystem functioning.

Grasses also function effectively as phytobarriers and filter plants in riparian zones, stormwater systems, and constructed wetlands (Kurade et al., 2021). While much research relies on aquatic macrophytes, engineered systems increasingly use grasses such as *Phragmites* and *Typha* species in combination with soil and microbial processes. These grasses remove nutrients (nitrates, phosphates) and trace metals from runoff and wastewater via uptake, adsorption, sedimentation, and microbial transformation. For instance, constructed wetlands planted with grasses show >70% removal efficiencies for heavy metals such as Cd, Pb, and Zn, and significant reductions in BOD and COD (Karam et al., 2022). Grass coverage enhances soil stabilization along water margins, reducing erosion and sediment-associated pollutants. Their survival under fluctuating moisture conditions wet–dry cycles, while maintaining filtration capacity, makes them reliable bioremediation agents in both rural and urban environments (Singh et al., 2022).

Though much research focuses on aquatic macrophytes, engineered wetlands increasingly use grasses such as *Phragmites australis* and *T. latifolia* in combination with soil and microbial processes. Constructed wetlands planted with these grasses have demonstrated removal efficiencies exceeding 70% for heavy metals (Cd, Pb, Zn), 65% for nitrates, and 50%–75% for total phosphorus (Thakur et al., 2023). Moreover, reductions in BOD and COD are substantial. Alquwaizany et al. (2022) reported an average 72% reduction in BOD and 68% reduction in COD over a treatment period of 15 days in grass-planted constructed wetlands. These grass-covered systems enhance soil stabilization along water margins, reducing erosion and associated pollutant loads. Their resilience to fluctuating moisture conditions allows consistent performance under variable hydrological regimes. A study by Yadav et al. (2025) found that during alternating wet–dry cycles, *Typha* and *Phragmites*-based wetlands maintained over 60% efficiency in BOD and COD reduction, highlighting the grasses' reliability in both rural and urban bioremediation efforts.

Some microbes associated with grasses also aid in remediation in water matrix. Yasin et al. (2021) observed 97.5% degradation of sodium dodecyl sulphate in water, in the presence of para grass (*Brachiaria mutica*) and bacterial consortium developed from grasses (*Acinetobacter* sp. strain BRSI56, *Acinetobacter junii* strain TYRH47, and *Acinetobacter* sp. strain CYRH21). Colour, ammonia nitrogen, and tannins removal from oil mill effluent was reported in the presence of *P. purpureum.* They observed that microbial groups belonging to *Anaerolineaceae*, *Cyanobacteria*, *Acidobacteria*, and *Nitrosomonadaceae* were responsible for remediation of effluent (Ujang et al., 2021). Thus, it is evident that grass–microbe interactions are important for decontamination of water sources.

4.3 ROLE OF MICROORGANISMS IN REGULATION OF VARIOUS NUTRIENT CYCLES

Microorganisms are tiny, often overlooked, yet incredibly powerful and are the unsung heroes that drive the planet's most fundamental processes. Without these microbial engineers, the Earth's ecosystems would collapse, as nutrients would become locked away, unavailable to plants and animals. So, microorganisms are the heart of biogeochemical cycles, which are the pathways by which elements move through the living (biotic) and non-living (abiotic) components of the ecosystem. The major cycles include carbon, nitrogen, phosphorus, and sulphur cycles, each crucial for ecosystem productivity and stability (Rousk and Bengtson, 2014; Bayu, 2024).

4.3.1 C Cycle and Soil Organic Matter

The carbon cycle is a fundamental Earth process, linking the atmosphere, biosphere, and geosphere through the movement of carbon in various forms. SOM has a central role, acting as both a reservoir and a dynamic participant in carbon fluxes, and understanding the relationship between SOM and the carbon cycle is crucial for managing soil health, agricultural productivity, and climate change alleviation. So, the C cycle involves the exchange of carbon between the atmosphere, plants, animals, soil, and oceans. Plants sequester CO_2 via photosynthesis, synthesizing organic compounds that form plant tissues. When plants and animals die, their residues return to the soil, where they are decomposed by microorganisms, releasing CO_2 back into the atmosphere or stabilizing it as SOM (Horwath, 2007).

Soil microbes, including bacteria, fungi, and actinomycetes, decompose plant residues (such as grass litter and root exudates) and transform organic matter through diverse metabolic pathways (Wu et al., 2024). In grassland ecosystem, a substantial proportion of organic carbon inputs from both aboveground litter and belowground rhizodeposition is rapidly processed by soil microbes. This leads to the release of carbon dioxide via microbial respiration, and synthesis of more stable organic compounds that contribute to long-term carbon storage. Mycorrhizal fungi, which are prevalent in grasslands, form symbiotic associations with grass roots. These fungi receive plant-derived carbon and in turn, enhance nutrient uptake and influence the flow and stabilization of carbon in the soil (Gougoulias et al., 2014). Microbial necromass (dead microbial biomass) is a significant and persistent component of SOM, contributing to the long-term sequestration of carbon in soils (Sokol et al., 2022).

Microbial decomposition transforms plant-derived carbon into humus and other stable forms of SOM, which are critical for soil fertility, structure, and water retention. The efficiency with which soil microbes convert plant carbon into microbial biomass (microbial carbon use efficiency) directly affects the amount of carbon retained in soils versus the amount released as CO_2 (Tao et al., 2023). Grassland management practices that promote diverse and active microbial communities, such as reduced tillage, organic amendments, and maintaining plant cover, can enhance soil carbon sequestration by supporting microbial processes that favour SOM stabilization. The interplay between grasses and soil microorganisms regulates both carbon influx (via plant growth and rhizodeposition) and carbon efflux (via microbial respiration and decomposition), influencing the role of grassland soils as carbon sinks or sources (Azevedo et al., 2024).

The study across ten alpine ecosystems on the Tibetan Plateau found that the Eastern and Northeastern alpine meadows were strong CO_2 sinks (~200 g C m^{-2} yr^{-1}), while western grasslands were weak sinks or sources. Soil temperature mainly controlled carbon fluxes in cold, humid zones, whereas soil moisture dominated in dry western areas, where higher soil water content increased carbon sink strength by 20%–30%. Overall, carbon flux variability followed longitudinal and altitudinal gradients (Wang et al., 2021). Henneron et al. (2020) investigated how plants regulate soil organic carbon (SOC) dynamics via rhizodeposition using continuous ^{13}C labelling across 12 grassland species. All species stimulated native SOC mineralization, with the rhizosphere priming effect exhibiting an 11-fold variation, ranging from +26% to +295% relative to unplanted soils.

Carbon is the base moiety of ecosystem energy flow and biogeochemical cycling, several grassy ecosystem services are dependent on C cycling between plants, soil, microorganisms, and environmental factors. Factors like soil moisture, temperature, pH, and nutrient availability modulate microbial activity and thus, the rates of carbon cycling and SOM formation (Wu et al., 2024). Microbial activity and community composition are essential to cycle C regulation, as mushrooms and bacteria are the main degradation factors of the soil. The capacity of microbes to mineralize carbon inputs to plants or decompose organic matter has major influence on C storage capacity of soil (Dignac et al., 2017).

Soil microbes are very crucial in maintaining soil C storage and turnover. Soil microorganisms are essential for the decomposition of grass-derived organic matter, the cycling and storage of carbon, and the maintenance of soil health in grassland ecosystems. Their activity determines whether soils act as carbon sinks or sources, with profound implications for climate regulation and ecosystem sustainability (Gougoulias et al., 2014). Effective grassland management that promotes microbial diversity and activity can therefore enhance SOC stabilization, improve soil fertility, and contribute to climate change mitigation.

4.3.2 N Cycle

Nitrogen is an essential element for all living organisms as it forms a key component of proteins, nucleic acids, and other vital biomolecules. Despite its abundance in the atmosphere, most organisms cannot directly utilize atmospheric nitrogen (N_2). N is necessary in soil particles–water–microbe–plants network. Among its various forms, inorganic nitrogen plays a vital role for plant growth. However, the plant-available forms, ammonium (NH_4^+) and nitrate (NO_3^-), are usually constrained in soils. Since a huge portion of nitrogen exists in organic form, it must first be mineralized into inorganic N before absorption by plants. Organic N compounds are degraded by enzymes secreted by microbes to release ammonia (NH_3), thereby increasing the pool of plant-available N in soils (Mahala et al., 2021). Microorganisms are central to the nitrogen cycle, transforming nitrogen into bioavailable forms like NO_3^- and NH_4^+ and maintaining ecosystem productivity and balance. The nitrogen cycle is a complex biogeochemical process driven primarily by microbes, which transform nitrogen into various chemical forms essential for life. Microbial communities play central roles in six major nitrogen transformation processes like nitrogen fixation, ammonification, nitrification, denitrification, anaerobic ammonium oxidation (anammox), and assimilation. Each of these steps is catalysed by specific groups of bacteria and archaea, whose activity is tightly regulated by environmental factors such as temperature, oxygen, organic matter, and nutrient availability (Zhao et al., 2024).

Biological nitrogen fixation is the reduction of atmospheric N_2 to NH_3 by diazotrophs, using the nitrogenase enzyme complex (MoFe protein and Fe protein), requiring 8 electrons and 16 ATP (adenosine triphosphate) molecules. It occurs through different modes: symbiotic fixers (e.g., *Rhizobium* with legumes, *Frankia* with trees, *Anabaena* with *Azolla*), free-living fixers (e.g., *Azotobacter, Clostridium*), and associative fixers (e.g., *Azospirillum, Gluconacetobacter*). converting atmospheric nitrogen (N_2) into ammonia (NH_3), making nitrogen accessible to plants. Recent discoveries have expanded the known diversity of these nitrogen-fixing microbes, including both free-living and symbiotic bacteria in terrestrial and aquatic ecosystems (Mattoo, 2023). Rice seedlings inoculated with *Pseudomonas fluorescens* (Mp) or co-inoculated with mixture of *Azospirillum brasilense* and *P. fluorescens* (Mbp) showed significantly higher rhizosphere ammonification than the control (M0), especially at higher N levels. Compared with N0, ammonification increased by 60.6%–63.1% at N90, 148.1%–161.4% at N180, and 267.0%–309.0% at N270 (Zhang et al., 2021).

Microorganisms are important in the degradation of complex polymers like cellulose and lignin, which higher organisms cannot directly utilize. During this process, part of the organic nitrogen is assimilated into microbial biomass, while the remaining fraction is released (mineralized) as inorganic nitrogen. Liu et al. (2025) reported that the intercropping enhanced net N mineralization by 1.6 times, with soil properties and enzyme activities driving fungal variation as compared to monocropping of maize, and under N180 treatment, mineralized N was positively linked to *Glomeromycota* fungal population. Ammonium (NH_4^+) is the most common form of nitrogen present in the biosphere and readily taken up by plants. However, under aerobic conditions, ammonia is unstable and undergoes transformation by soil nitrifying bacteria into nitrite (NO_2^-) and subsequently nitrate (NO_3^-), thereby occupying a central role in the nitrogen cycle (Chamoli et al., 2024). This conversion is called nitrification, which involves the oxidation of ammonia to nitrite and then nitrate, mediated by ammonia- and nitrite-oxidizing bacteria, with key functional genes such as *amoA* and *nxr* serving as indicators of this process (Wang et al., 2025).

Long-term grazing significantly influences nitrifying microbial communities in grassland soils. Among ammonia-oxidizing bacteria (AOB), *Nitrosomonas* was dominant; there will be decrease in the *Nitrosomonas* community by 2% as compared to ungrazed soils. These findings highlight that grazing intensity strongly regulates nitrifying communities through changes in soil physicochemical characteristics (Chen et al., 2022). Jin et al. (2023) observed that alpine grassland degradation on the Tibetan Plateau reduced soil nitrification. Gross nitrification in <5 cm soil under *Kobresia pygmaea* was 1.67 times higher than in bare subsoil patches, while net nitrification rates were negative across all stages, highlighting strong N limitation, especially in degraded soils.

Biological nitrate elimination can be attained via denitrification. This process destroys good ozone in stratosphere, global warming, and volatilization to N_2. Denitrification is the microbial process that reduces nitrate (NO_3^-) and nitrite (NO_2^-) to dinitrogen gas (N_2), returning bioavailable nitrogen to the atmosphere by using these oxides as terminal electron acceptors under oxygen-limited conditions. Key microorganisms involved include *Thiobacillus denitrificans, Micrococcus denitrificans*, and *Pseudomonas denitrificans*, possessing genes such as *nirS*, *nirK*, and *nosZ* (Zhao et al., 2024). Nitrous oxide (N_2O), a potent greenhouse gas and ozone-depleting compound, is primarily generated through denitrification in agroecosystems. A study in Australian cereal systems revealed that N_2O emissions decreased with increasing temperature and moisture, while denitrification was predominant at 70% soil moisture (Pan et al., 2025). Anammox is another microbial process that occurs naturally in aquatic environments, where specialized anammox bacteria convert ammonia (NH_3^+) into nitrogen gas (N_2) under oxygen-limited conditions via unique membrane-bound organelle called the *anammoxosome*, which provides the anaerobic environment necessary for this conversion. Representative genera include *Anammoxoglobus* and *Brocadia* (Wang et al., 2025). Sediments along the Peru margin (100–3240 m) contained dense *Thioploca* mats that exhibited anammox rates ranging from 1.8 to 44.2 nmol N $cm^{-3}h^{-1}$, with an average relative contribution of 34.2%. Further, *Candidatus scalindua* was identified, and the high abundance of *hzo* genes indicated a substantial role of anammox in nitrogen removal across these sediments (Rich et al., 2020). The impact of increased nitrogen availability on grasses is substantial. Recent studies show that nitrogen fertilization can increase aboveground biomass in grasses by up to 79% compared to unfertilized controls, and significantly raise the nitrogen content of grass tissues. This enrichment is a direct consequence of enhanced microbial-mediated nitrogen transformations in the soil, which increase the pool of bioavailable nitrogen for plant uptake. However, excessive nitrogen inputs can disrupt microbial community balance and ecosystem health, emphasizing the need for sustainable management of nitrogen resources (Lin et al., 2023). So, the microorganisms are the engines of the nitrogen cycle, enabling the transformation and flow of nitrogen through ecosystems. Their activity not only regulates soil fertility and plant productivity in grasses, where nitrogen enrichment can boost growth of grasses by nearly 80%, but also influences global nitrogen budgets and environmental health (Wang et al., 2025).

Recent research in grassland ecosystems has shown that the composition and activity of these microbial communities are influenced by grass species, nitrogen fertilization, and environmental factors. For example, moderate nitrogen fertilization increases the abundance and activity of AOB and enhances nitrification potential, thereby raising the pool of plant-available nitrogen. However, excessive nitrogen can suppress beneficial microbial populations and alter the balance of nitrogen transformations (Zhao et al., 2024).

The net effect is that nitrogen-cycling microorganisms directly influence how much nitrogen is available for grasses, impacting their growth, tissue nitrogen content, and overall productivity. In practical terms, studies have documented that nitrogen fertilization—mediated through microbial transformations—can increase the aboveground nitrogen content in grasses by up to 79% compared to unfertilized controls, provided microbial communities remain active and balanced. Further, fertilization rate and type of grass species influence soil N transformation by microbes in N-limited C_4 grasslands. Switchgrass increased population of AOB and nitrite reducers in soil. Thus, the diversity and function of nitrogen-cycling microorganisms are essential for sustaining grassland fertility, optimizing grass nitrogen levels, and supporting ecosystem productivity (Mattoo, 2023).

4.3.3 P Cycle

The phosphorus cycle is a vital biogeochemical process in both terrestrial and aquatic ecosystems, and microorganisms are at its core, orchestrating the transformation, mobilization, and recycling of phosphorus (P). Phosphorus is essential for all living organisms, as it is a key component of DNA, RNA, ATP, and cellular membranes. Despite its abundance in soils, up to 95–99% of soil phosphorus is immobilized in forms unavailable to plants, making phosphorus a limiting nutrient in many ecosystems (Qin et al., 2024). Microorganisms, particularly bacteria and fungi, act as "phosphorus gatekeepers," controlling how phosphorus is accessed, cycled, and ultimately made available to plants, including grasses.

Microbial processes in the phosphorus cycle include the solubilization of inorganic phosphate, mineralization of organic phosphorus, uptake and transport of phosphate, and adaptive responses to phosphorus scarcity. Mineralization of organic phosphorus is the conversion of organic P into insoluble inorganic P by enzymes such as phosphatase, phytase, and P lyase, mediated by bacteria (*Bacillus*, *Pseudomonas*) and fungi (*Aspergillus*, *Penicillium*). Phosphate-solubilizing microorganisms (PSMs), such as species of *Pseudomonas*, *Bacillus*, *Aspergillus*, and *Penicillium*, produce organic acids and siderophores that convert insoluble phosphates into plant-available forms (Tian et al., 2021). Genes like *phoD* (mineralization), *gcd* (solubilization), and *pst* (transport) are key molecular markers of these microbial functions (Qin et al., 2024). Uptake of soluble inorganic phosphate can be enhanced by mycorrhizal fungi, which form mutualistic associations with plant roots and act as efficient P mobilizers (Etesami et al., 2021). After uptake, the assimilated phosphorus is incorporated into microbial nucleic acids, phospholipids, and other protoplasmic substances. In phosphorus-limited soils, microbes may also adapt by replacing membrane phospholipids with non-phosphorus lipids and optimizing phosphorus use efficiency. In aquatic systems, recent research has uncovered even more dynamic microbial phosphorus cycling, including the use and production of reduced phosphorus compounds and novel metabolic pathways that influence the availability of phosphorus throughout the water column (Duhamel, 2025; Sosa, 2018).

In grasses, phosphorus availability is directly influenced by the activity of these microorganisms. When phosphate-solubilizing microbes are active, they increase the pool of bioavailable phosphorus, which in turn enhances grass growth, tissue phosphorus content, and overall productivity. Recent studies show that microbial interventions or phosphorus fertilization can increase phosphorus content in grasses by approximately 15%–40% compared to unfertilized controls, depending on the grass species, soil type, and the effectiveness of the microbial community (Tian et al., 2021). In phosphorus-deficient grassland soils of KwaZulu-Natal, phosphate solubilizers treatment showed higher enzyme activities than control treatment, with β-glucosaminidase, acid phosphatase, and alkaline phosphatase increasing by 21.4%, 47.3%, and 40.4%, respectively. *Pseudomonas nitroreducens* and *Variovorax paradoxus* exhibited maximum P-solubilization under acidic conditions (pH 4.2–4.9), with acid (45.3 U/mL) and alkaline phosphatase (94.0 U/mL) activities after 10 days, highlighting their potential in improving nutrient bioavailability (Magadlela et al., 2023). According to Torres-Cuesta et al. (2023), PSB inoculation combined with rock phosphate (RP) fertilization significantly enhanced acid phosphomonoesterase activity in Kikuyu grass pastures, showing a 77.7% increase after six months compared to the uninoculated control. This improvement in enzymatic activity was associated with a 54.2% increase in forage yield. Mythra et al. (2022) demonstrated that co-inoculation of *Bacillus megaterium* or *Aspergillus* sp. with mycorrhizae and RP markedly enhanced phosphorus uptake in maize. P uptake increased to 0.9 mg/plant and 0.8 mg/plant, respectively, compared to 0.09 mg/plant in the control, representing more than a tenfold improvement. This enhanced nutrient acquisition was also associated with improved maize growth.

So, microorganisms are indispensable to the phosphorus cycle, mediating the release, transformation, and recycling of phosphorus in soils and water. Their activity is crucial for making phosphorus available to grasses, with microbial-driven increases in phosphorus content supporting higher productivity and ecosystem health (Qin et al., 2024).

4.3.4 S Cycle

Sulphur is an essential element cycling through Earth's spheres in diverse chemical forms; its versatility underpins numerous biological and abiotic processes, making it central to ecosystem functions, climate modulation, and industrial applications. Microorganisms including sulphur-oxidizers and -reducers drive the redox transformations that sustain the global sulphur cycle, shaping both ancient environments and modern technologies. Sulphur cycle is the microbial conversion between reduced and oxidized forms. In aerobic or microoxic conditions, sulphur oxidizers (*Beggiatoa*, *Thiobacillus*) employ pathways such as Sox to transform sulphide, thiosulphate, and elemental sulphur into sulphate (Twible et al., 2024). Conversely, in anaerobic zones sediments, wetlands, or engineered bioreactors, sulphate-reducing microbes (SRMs) reduce sulphate to hydrogen sulphide, often producing isotopically light sulphide. Disproportionation reactions, for example, of elemental sulphur or thiosulphate, further diversify sulphur transformations in low-redox habitats (Joo et al., 2022). These microbial activities have profound environmental and biotechnological implications. SRMs contribute to acid mine drainage and malodorous urban streams, whereas sulphur-oxidizing bacteria are integral to biodesulphurization, metal bioleaching, *etc.* (Gupta et al., 2024). In constructed wetlands, seasonal redox oscillations determine the efficiency of sulphur cycling and contaminant removal, and such insights inform design improvements (Mateos et al., 2023).

In terrestrial ecosystems, sulphate generated by weathering (pyrite oxidation) may not always reach oceans; retention through microbial reduction and burial forms a cryptic sulphur sink on land. This sink played a key role during Earth's history, modulating atmospheric oxygen and ocean chemistry in response to geological and biological changes. At the molecular scale, recent studies are uncovering electron transfer mechanisms underpinning microbial sulphur redox reactions. Zhuang et al. (2024) provide a detailed look at biogeochemical electron pathways that drive sulphur transformations. Metagenomic work from Himalayan hot springs expands our understanding of microbial diversity and the evolutionary distribution of *dsr* genes across oxidative and reductive pathways (Nagar et al., 2024). Metagenomic analysis of a Northern Himalayan sulphur hot spring revealed diverse S-cycling microbes, mainly Proteobacteria, with conserved *dsrAB* genes ($n = 178$) across all habitats. Sulphur-oxidizing and sulphate-reducing bacteria showed key adaptations, while unassigned DsrAB proteins were linked to *Desulfovibrio* and *Archaeoglobus fulgidus*, indicating evolutionary significance and lateral gene transfer in S cycling (Nagar et al., 2022).

The evolutionary origins of key sulphur enzymes reflect Earth's changing oxygen levels. Phylogenomic analyses indicate that sulphide oxidation evolved in the Archean, with thiosulphate-utilizing enzymes and organic sulphur cycling becoming widespread only after the Great Oxidation Event and in the Mid-Proterozoic, respectively (Mateos et al., 2023). These developments parallel shifts in global redox balance, deepening our understanding of Earth's atmospheric and biospheric evolution. The sulphur cycle interconnects geochemistry, microbiology, climate science, and environmental engineering across scales from enzymatic mechanics to planetary processes. Recent advances, spanning isotope geochemistry to genomics and biotechnological applications, reveal a system more complex and dynamic than ever. Moving forward, integrating these insights is critical for managing pollution, mitigating climate impact, and understanding life's role in Earth's redox evolution.

4.4 IMPACT OF INTERACTION BETWEEN GRASSES AND MICROORGANISMS ON CROP YIELD AND PRODUCTIVITY

The complex interactions between grasses and soil microbes are fundamental in influencing both agricultural productivity and long-term sustainability. Grasses, encompassing both wild species and cultivated cereals, form symbiotic associations with diverse microbial communities, including bacteria, fungi, and archaea. Through their extensive root systems, exude a variety of organic compounds that serve as substrates for microbial communities, fostering a dynamic environment conducive to influence nutrient cycling, plant health, and stress resilience, thereby directly impacting crop yield and productivity. Microorganisms, in turn, boost the soil fertility through key processes like nitrogen fixation, phosphorus solubilization, and the synthesis of growth-promoting compounds (Laishram et al., 2024). Recent studies have accentuated the significance of these interactions, emphasizing their role in sustainable agriculture and the potential to lower reliance on chemical fertilizers. Understanding the mechanisms behind grass–microbe interactions is thus essential for developing strategies aimed at improving crop yield and ensuring food security.

The symbiotic relationship between grasses and soil microorganisms is a keystone of agroecosystem functionality. One of the primary mechanisms through which grasses influence soil microorganisms is via root exudation. Grasses secrete diverse organic compounds into the rhizosphere, including sugars, amino acids, and phenolic molecules, which function as substrates supporting microbial proliferation. These root exudates can preferentially promote beneficial microorganisms, such as plant growth-promoting rhizobacteria (PGPRs), which subsequently enhance nutrient uptake by plants. PGPRs, especially nitrogen-fixing microbes, convert atmospheric nitrogen (N_2) into ammonia, making it bioavailable to plants. Unlike legumes, cereals and millets do not form classical nodules but can engage in associative symbioses with nitrogen-fixing bacteria such as *Azospirillum*, *Herbaspirillum*, and *Gluconacetobacter*. These bacteria colonize the rhizosphere and can significantly improve nitrogen uptake in grasses. In a study, inoculating maize with *A. brasilense* improved root mass by 12.1%, leaf N concentration by 4.3%, and grain yield by 5.4% (Barbosa et al., 2022). Similarly, an increase in wheat grain yield by 37.95% was seen when inoculated with 75% inorganic N fertilizer along with *A. brasilense* (UASDAZO-29) compared with 100% inorganic N fertilizer alone (Murali and Navi, 2024). PGPRs can solubilize phosphorus (P) where bioavailability of insoluble P is enhanced by PSMs. Phosphate-solubilizing actinomycetes, such as *Streptomyces cellostaticus* S18, enhance phosphorus availability by secreting organic acids and phosphatases. In greenhouse trials with finger millet, inoculation with strain S18 increased plant height from 56.21 ± 2.86 cm to 72.89 ± 1.43 cm, and shoot P uptake was found significantly higher with strain S18+75% inorganic P fertilizer (30.15 ± 6.31 mg/pot) when compared with 100% inorganic P alone (12.86 ± 1.90 mg/pot) (Babu et al., 2025) (Table 4.1). PGPRs also enhance the availability of potassium (K), a vital nutrient for plant growth, by converting non-available mineral-bound forms into soluble K^+ ions accessible to plant roots through secretion of organic acids, ion exchange, and enzyme-mediated mineral transformation that lower rhizosphere pH and dissolve K-bearing silicate minerals like feldspar, mica, and illite (Khan et al., 2024). A study on K-solubilizing rhizobacterial strains like *B. cereus* applied to Rhodes grass found an increase of 52.4% in shoot length and 72.7% in root length when compared to control, showcasing potential benefits for forage grass productivity (Javaid et al., 2023). Similarly, the application of PGPR consortia has been shown to notably improve maize yield by 27.81% under rainfed conditions by improving nutrient uptake and stress tolerance (Manjunath et al., 2023). Fodder grasses and PGPR interaction plays a pivotal role in enhancing crop yield and productivity, primarily through the modulation of phytohormone levels. PGPRs are known to synthesize a range of phytohormones, including indole-3-acetic acid (IAA), gibberellins, cytokinins, and abscisic acid, which collectively influence plant growth and stress responses. The IAA produced by PGPR stimulates root expansion and branching, ultimately supporting better nutrient and water acquisition needed for healthy plant growth. PGPRs including *Providencia rettgeri* and other strains producing high levels of IAA were tested on barley (*Hordeum vulgare*) under drought stress. These IAA-producing bacteria significantly enhanced root growth parameters, viz., length, biomass, and root/shoot ratios. The improved root architecture resulted in better water and nutrient uptake, supporting higher shoot biomass and grain yield under dry conditions (Ferioun et al., 2025). Also, the 1-aminocyclopropane-1-carboxylate (ACC) deaminase enzyme produced by PGPR converts ACC into ammonia (NH_3) and α-ketobutyrate, thereby enhancing plant growth and increasing tolerance to stress under both normal and adverse conditions. ACC deaminase-positive *Bacillus amyloliquefaciens* MMR04, isolated from pearl millet, improved seed germination (91.7%) and vigour (1,214), while control seedlings showed 85% and 858.35 of seed germination and seedling vigour, respectively, under drought stress. The bacterium also elevated antioxidant enzyme activity (superoxide dismutase: SOD, ascorbate peroxidase: APX), reduced stress ethylene (via ACC deaminase), and repressed stress-responsive genes, resulting in drought tolerance (Murali et al., 2021). Moreover, phytohormone

TABLE 4.1
Recent Studies on Interaction of Grasses and Microbes on Crop Growth and Productivity

Grass	Microbe(s)	Type of Treatment	Experiment Details	% Change in Plant Biometrics/Yield Parameters	Mechanism	Reference
Barley	*P. fluorescens*	Seed biopriming	Green house study Stress: salinity (50, 100, and 150 mM NaCl) Inoculum population: 10^8 cfu/mL Duration of study: 120 days	Increase in grains/spike: 12%–14% at all salinity levels	Increase in ACC deaminase and decrease in ethylene level	Azadikhah et al. (2019)
Wheat	*Azospirillum brasilense* + *Rhizobium pisi*	Seed biopriming	Wire house study Inoculum population: 10×10^8 cfu/mL Duration of study: 120 days	Upsurge in yield (36%), no. of grains per plant (11%), and seed index (17%)	Biological N_2 fixation and phytohormone synthesis	Zaheer et al. (2019)
Pearl millet	*B. amyloliquefaciens*	Seed biopriming	Seedling cup study Stress: drought (well-watered and watering on alternate days) Inoculum population: 10^8 cfu/mL Duration of study: 40 days	Enhancement in seed germination (0.14-fold), shoot fresh weight (0.53-fold), ascorbate peroxidase (APX) (0.62-fold), superoxide dismutase (SOD) (0.73-foldA), and decrease in melondialdehyde (2.3-fold)	Synthesis of ACC deaminase enzyme, antioxidant enzyme activity like APX and SOD	Murali et al. (2021)
Rice	*B. subtilis* and *B. megaterium*	Seedling dip method + soil application	Field study Stress: drought (100% and 80% deficit irrigation) Inoculum load: 250 mL (seedling dip) 20 mL/seedling (field application) Duration of study: 2 years	Increased plant height (9.4%), no. of tillers (15.3%), no. of panicles (18.6%), and shoot dry weight (29.6%) under deficit irrigation	Production of antioxidant enzymes, IAA, salicylic acid, and exopolysaccharide	El-Mageed et al. (2022)

(Continued)

TABLE 4.1 (*Continued*)
Recent Studies on Interaction of Grasses and Microbes on Crop Growth and Productivity

Grass	Microbe(s)	Type of Treatment	Experiment Details	% Change in Plant Biometrics/Yield Parameters	Mechanism	Reference
Maize	*P. stutzeri* 1568/pVA3 (engineered ammonium-excreting strain, nitrogen-fixing)	Seedling dip method + soil application	Both pot and field studies Inoculum load: 25 mL/kg seed Inoculum population: 8.8×10^8 cfu/mL Duration of study: 210 days	Improvement in plant biomass (3.3%) and N_2 fixation (39.4%)	Biological nitrogen fixation, ammonium excretion, and enhanced nitrogenase activity	Jiang et al. (2022)
Little millet	*Burkholderia* sp. (FKK5) and *Azotobacter chroococcum*	Seed biopriming	Greenhouse study Inoculum population: 10^9 cfu/mL Duration of study: 90 days	Increase in grain yield (28.14% and 20.43%), and total biomass (23.08% and 19.09%)	N_2 fixation, solubilization of complex phosphates, IAA and siderophore production	Mankar et al. (2022)
Pearl millet	*Enterobacter* sp. MN17 and *Stenotrophomonas maltophilia* FA-9	Seed biopriming	Field study Different N sources; organic farm yard manure and inorganic (urea) Inoculum population: 10^8 cfu/mL Duration of study: 2 years	Increased soil organic carbon (34.88%), no. of grains per ear (23.63%), seed index (7.85%), and grain yield (25.0%)	Improved nitrogen use efficiency, P-solubilization, and phytohormone synthesis	Dawood et al. (2023)
Rice	*Delftia tsuruhatensis* strain D9 + *Delftia* sp. strain MS2As2 + *Bacillus* sp.	Seed biopriming + soil application	Pot trial experiment Inoculum load: 10 L/ha Inoculum population: 10^8 cfu/mL Duration of study: 150 days	Increase in no. of tillers (11.84%), no. of panicles (27.78%), and relative agronomic effectiveness (63.02%)	Nitrogenase activity. Synthesis of IAA and organic acids	Harahap et al. (2023)
Rhodes grass	*B. cereus* SM2	Seed biopriming	Pot trial experiment Inoculum population: $OD_{600} = 0.70$ Duration of study: 4 weeks	Enhanced shoot length (52.4%) and root length (72.7%)	Mineral (P, K, and Mn) solubilization and IAA production	Javaid et al. (2023)

(*Continued*)

TABLE 4.1 (*Continued*)
Recent Studies on Interaction of Grasses and Microbes on Crop Growth and Productivity

Grass	Microbe(s)	Type of Treatment	Experiment Details	% Change in Plant Biometrics/Yield Parameters	Mechanism	Reference
Maize	Microbial consortia 1 (MC1) (*P. putida* P7 + *Paenibacillus favisporus* B30) and Microbial consortia 2 (MC2) (*P. putida* P45 + *B. amyloliquefaciens* B17)	Seed + soil application	Field study Inoculum load: Seed treatment—30 g/kg seed Soil application— 2.5 kg/ha Inoculum population: 10^8 cfu/mL Duration of study: 3 years	Increase in grain yield: MC1—27.78% and MC2—23.21%	Enhanced soil enzymatic activity and rain water use efficiency	Manjunath et al. (2023)
Bermuda grass	*Paenibacillus riograndensis* (DH44) + *Bacillus pumilus* strains (AP7 and AP18) + *Bacillus sphaericus* (AP282)	Soil application	Field study Different N dosages; full rate (112 kg/ha) and half rate (56 kg/ha) Inoculum population: 10^6 cfu/mL each Duration of study: 2 years	No difference in mean forage height (~21.9 cm) and biomass (89.24 g) when consortia + 50% N applied compared to 100% N alone	Enhanced nitrogenase and soil enzymatic activities	Sullins et al. (2023)
Sorghum	*B. subtilis* (PG152) and *Enterobacter* sp. (PG197)	Seed coating with talc-based biofertilizer	Greenhouse study Inoculum population: 10^9 cfu/mL Duration of study: 90 days	Increase in plant height (47.32% and 38.83%), shoot dry weight (88.23% and 82.70%), and grain yield (35.47% and 23.19%)	Indole acetic acid, gibberellic acid, and HCN production	Chiranjeevi et al. (2024)

(*Continued*)

TABLE 4.1 (*Continued*)
Recent Studies on Interaction of Grasses and Microbes on Crop Growth and Productivity

Grass	Microbe(s)	Type of Treatment	Experiment Details	% Change in Plant Biometrics/Yield Parameters	Mechanism	Reference
Little millet	*Bacillus albus* LRS2	Seed biotization	*In vitro* study Stress: drought (−0.2 MPa to −3.6 MPa) Inoculum population: 10^8 cfu/mL Duration of study: 7 days	Elevated germination percentage (33%), shoot (13.5%), and root length (25.2%) under induced drought stress (−0.45 MPa)	High ACC deaminase activity, exopolysaccharide production, solubilization of P, K, and Zn, and synthesis of IAA	Rajkumar et al. (2024)
Maize (silage)	*Methylobacterium symbioticum*	Foliar spray	Both pot and field study Different N dosages (0%, 40%, 80%, and 160%) Inoculum load: 333 g/ha diluted to 100 L water Duration of study: 2 years	Enhancement in dry matter yield (~20.2%) and plant nitrogen concentration (~8.16%) at all N dosages	Biological N fixation by utilizing phyllospheric exudates like C_1 sources (methanol), organic acids, soluble carbohydrates, etc.	Rodrigues et al. (2024)
Finger millet	*Streptomyces cellostaticus* S18	Root dip method	Pot trial experiment Varied P dosage (50%, 75%, and 100%) Inoculum population: $OD_{600} = 1.42$ Duration of study: 120 days	Increase in plant height (28.57%) and shoot P uptake (150%) in S18+75% P than 100% P alone	P-solubilization and IAA production	Babu et al. (2025)

biosynthesis mediated by PGPR contributes to the attenuation of abiotic stresses, such as salinity and drought, by modulating plant stress-responsive signalling pathways and enhancing physiological adaptation mechanisms (Kumar et al., 2023).

The benefits of grass–microbe interactions extend beyond nutrient management. These interactions can also increase plant tolerance to abiotic stresses, such as drought and salinity, by stimulating systemic resistance, improve water uptake, and modulate stress-related hormones, thereby increasing plant resilience. Plant probiotics, including species like *Bacillus subtilis*, *A. brasilense*, and *P. stutzeri*, have been reported to sustain stress resilience in crops like rice, wheat, and maize (Zaheer et al., 2019; El-Mageed et al., 2022; Jiang et al., 2022). Also, inoculation with *Bacillus* sp. has been reported to enhance root growth and development in wheat under drought conditions by maintaining soil moisture of 75% water holding capacity, leading to improved water and nutrient uptake (Malik et al., 2024). In forage pearl millet, the application of liquid microbial inoculants containing strains like *Burkholderia seminalis* and *Stenotrophomonas maltophilia* led to a significant increase in green fodder and dry matter yield (Arora et al., 2024). These PGPR strains exhibited tolerance to various abiotic stresses and maintained their plant growth-promoting traits across diverse environmental conditions.

The integration of grasses into cropping systems, such as through intercropping or cover cropping, has been shown to enrich microbial diversity and activity. In a study conducted in maize fields, intercropping maize with legumes like soybean and peanut significantly increased soil microbial biomass carbon and nitrogen, along with the activities of soil enzymes such as urease, phosphatase, and catalase (Xiao et al., 2023). These changes were associated with enhanced nutrient cycling and soil fertility. Moreover, intercropping systems exhibited higher bacterial and fungal diversity compared to monocropping systems, indicating a more robust and resilient soil microbial community. Intercropping grasses with legumes further amplify these benefits. Leguminous plants enrich the soil with nitrogen through symbiotic associations with nitrogen-fixing bacteria, which, in turn, benefits the companion grass crops. Studies have demonstrated that intercropping systems, such as maize–soybean combinations, not only enhance overall biomass but also improve soil microbial diversity and function (Guo et al., 2025). This diversification of the microbial community enhances nutrient cycling and suppresses soil-borne pathogens, leading to healthier crops and higher yields. Furthermore, the role of grasses in shaping the soil microbiome extends over nutrient dynamics. Grasses influence the composition and function of microbial communities, thereby affecting soil structure, organic matter content, and overall ecosystem services. In walnut orchards, intercropping with grasses like ryegrass and hairy vetch altered the soil microbial community, enhancing functions related to nitrogen cycling and carbohydrate metabolism (Wang et al., 2023). Such modifications contribute to improved soil health and consequently, better crop performance. Advancements in nanotechnology have introduced novel approaches to augment grass–microbe interactions. The development of nanobiofertilizers, which integrate nanomaterials with plant probiotics, offers an assuring strategy for agricultural sustainability. These formulations facilitate the controlled release of nutrients and support microbial sustainability, leading to improved plant growth and resilience against environmental stress (Upadhayay et al., 2023).

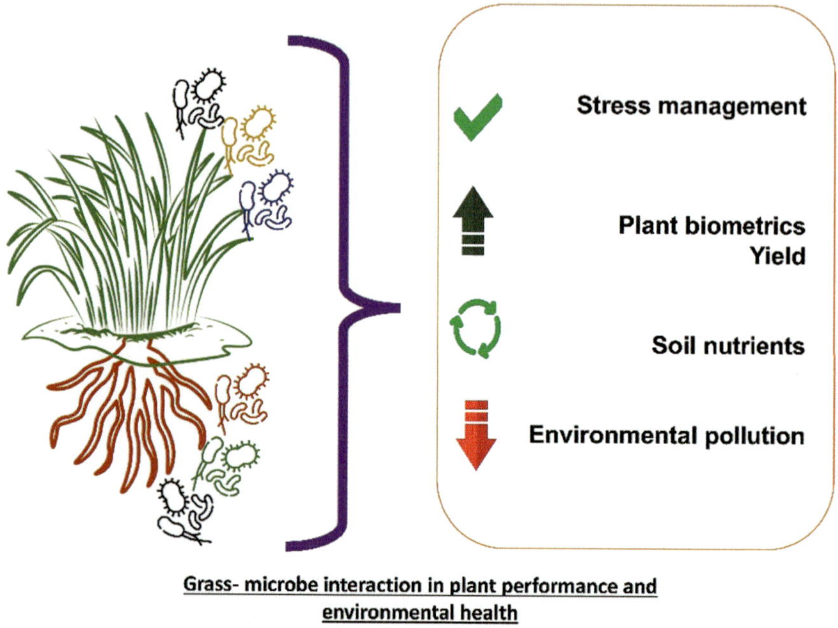

FIGURE 4.1 Role of grass–microbe interaction in plant performance and environmental health.

The integration of grasses and beneficial microorganisms into cropping systems holds significant potential for enhancing agricultural productivity (Figure 4.1). Through mutualistic associations, grasses and soil microbes ease nutrient cycling, enhance soil structure, and promote plant health. Experimental evidence from recent studies highlights the benefit of integrating PGPR, intercropping strategies, organic amendments, and nanotechnology-based solutions to optimize these interactions. Such approaches contribute to sustainable agricultural practices by increasing resource use efficiency, reducing environmental impacts, and ensuring food security. As the global need for agricultural products continues to rise, harnessing the synergistic potential of grass–microbe interactions will be instrumental in developing resilient and productive cropping systems.

4.5 CONCLUSION

Grasses are major contributors to agricultural production. However, with increasing productivity, quantum of agriculture input usage is also spiking. The chemical inputs often remain in environment or convert to toxic forms, thereby polluting environment. Thus, microorganisms that can convert unavailable nutrients to available form, and neutralize toxic contaminants are important in grass production. Further, microorganisms degrade SOM into a form that is useful for plants. This enhances soil fertility by enhancing nutrient availability in the soil. These microbes can act as grass endophytes or are present on rhizosphere, rhizoplane, phyllosphere, or phylloplane, contributing in various ways, thereby maximizing yield. This synergistic association also decontaminate xenobiotic compounds in matrices like soil, water, and air. Thus, integrating efficient microorganisms in grasses ecosystem not only improves the yield, but also aids in cleaning the environment. This strategy will ultimately lead us towards achieving SDG 13, that emphasizes on mitigating climate change and its consequences.

REFERENCES

Alquwaizany, A. S., Hussain, G., & Al-Zarah, A. I. (2022). Changes in physico-chemical composition of wastewater by growing *Phragmites australis* and *Typha latifolia* in an arid environment in Saudi Arabia. *Environmental Science and Pollution Research*, *29*(26), 39838–39846.

Arora, K., Sharma, S., & Monti, A. (2016). Bio-remediation of Pb and Cd polluted soils by switchgrass: A case study in India. *International Journal of Phytoremediation*, *18*(7), 704–709.

Arora, V., Pandove, G., Brar, S. K., Oberoi, H. K., & Kalia, A. (2024). Evaluating plant growth-promoting rhizobacteria to improve the productivity of forage pearl millet. *Current Microbiology*, *81*(7), 172.

Athar, T., Fatima, H., Khan, M. K., Pandey, A., Shahid, S., Hamurcu, M., & Gezgin, S. (2022). Potential role of turfgrasses and understorey vegetation for mitigating the changing climate. In M. A. Dervash, & A. A. Wani (Eds.), *Climate Change Alleviation for Sustainable Progression* (pp. 347–367). CRC Press.

Awan, S. A., Khan, I., Rizwan, M., Irshad, M. A., Xiaosan, W., Zhang, X., & Huang, L. (2023). Reduction in the cadmium (Cd) accumulation and toxicity in pearl millet (*Pennisetum glaucum* L.) by regulating physio-biochemical and antioxidant defense system via soil and foliar application of melatonin. *Environmental Pollution*, *328*, 121658.

Azadikhah, M., Jamali, F., Nooryazdan, H. R., & Bayat, F. (2019). Growth promotion and yield enhancement of barley cultivars using ACC deaminase producing *Pseudomonas fluorescens* strains under salt stress. *Spanish Journal of Agricultural Research*, *17*(1), e0801–e0801.

Azevedo, L. C. B., Bertini, S. C. B., Ferreira, A. S., Rodovalho, N. S., Ferreira, L. F. R., & Kumar, A. (2024). Microbial contribution to the carbon flux in the soil: A literature review. *Revista Brasileira de Ciência do Solo*, *48*, e0230065.

Babu, P. M., Panda, N., Nayak, R. K., Sethi, D., Biswal, S., Mishra, M. K., … & Pattanayak, S. K. (2025). Isolation, characterization and screening of phosphate (P) solubilizing actinomycetes and exploring its potency in finger millet (*Eleusine coracana* L.). *BMC Plant Biology*, *25*(1), 362.

Bao, X., Zhou, W., Xu, L., & Zheng, Z. (2023). A meta-analysis on plant volatile organic compound emissions of different plant species and responses to environmental stress. *Environmental Pollution*, *318*, 120886.

Barbosa, J. Z., de Almeida Roberto, L., Hungria, M., Corrêa, R. S., Magri, E., & Correia, T. D. (2022). Meta-analysis of maize responses to *Azospirillum brasilense* inoculation in Brazil: Benefits and lessons to improve inoculation efficiency. *Applied Soil Ecology*, *170*, 104276.

Bayu, T. (2024). Systematic review on the role of microbial activities on nutrient cycling and transformation implication for soil fertility and crop productivity. *bioRxiv*, 209–217.

Bhantana, P., Rana, M. S., Sun, X. C., Moussa, M. G., Saleem, M. H., Syaifudin, M., … & Hu, C. X. (2021). Arbuscular mycorrhizal fungi and its major role in plant growth, zinc nutrition, phosphorous regulation and phytoremediation. *Symbiosis*, *84*, 19–37.

Bousselin, X., Cassagne, N., Baux, A., Valantin-Morison, M., Herrera, J. M., Lorin, M., … & Fustec, J. (2021). Interactions between plants and plant-soil in functionally complex mixtures including grass pea, faba bean and niger, intercropped with oilseed rape. *Agronomy*, *11*(8), 1493.

Braun, R. C., Mandal, P., Nwachukwu, E., & Stanton, A. (2024). The role of turfgrasses in environmental protection and their benefits to humans: Thirty years later. *Crop Science*, *64*(6), 2909–2944.

Chamoli, A., Bhambri, A., Karn, S. K., & Raj, V. (2024). Ammonia, nitrite transformations and their fixation by different biological and chemical agents. *Chemistry and Ecology*, *40*(2), 166–199.

Chen, Q., Shang, Y., Zhu, R., Bao, Q., & Lin, S. (2022). Long-term enclosure at heavy grazing grassland affects soil nitrification via ammonia-oxidizing bacteria in Inner Mongolia. *Scientific Reports*, *12*(1), 21464.

Cheng, K., Xu, X., Cui, L., Li, Y., Zheng, J., Wu, W., … & Pan, G. (2021). The role of soils in regulation of freshwater and coastal water quality. *Philosophical Transactions of the Royal Society B*, *376*(1834), 20200176.

Chiranjeevi, M., Goudar, G. D., Pu, K., & Yalavarthi, N. (2024). Study on effect of plant growth promoting rhizobacteria on sorghum (*Sorghum bicolor* L.) under gnotobiotic conditions. *Frontiers in Microbiology*, *15*, 1374802.

da Rocha, J. G., Coelho, F. C., Lelis, R. T., de Souza Santos, G. C., & da Costa Jaeggi, M. E. P. (2020). Mulch decomposition and nitrogen, phosphorus and potassium release in vetiver grass at different plant densities. *Pesquisa Agropecuária Tropical*, *50*, e64603.

da Silva, H. M., Júnior, J. C. D., Silveira, M. L., Junior, M. A. L., Cardoso, A. S., & Vendramini, J. M. (2022). Greenhouse gas mitigation and carbon sequestration potential in humid grassland ecosystems in Brazil: A review. *Journal of Environmental Management*, *323*, 116269.

Dawood, A., Majeed, A., Ul-Allah, S., Naveed, M., Farooq, S., Sarwar, N., & Hussain, M. (2023). Combined application of organic and inorganic nitrogen and seed inoculation with rhizobacteria (*Stenotrophomonas maltophilia* FA-9) improved productivity, nitrogen use efficiency, and economic returns of pearl millet. *Sustainability*, *15*(10), 8248.

Desiderio, G. A., da Silva Cavalcanti, P. V., de Andrade, A. M. L., Manuel, M. I. Z., Desiderio, A. C. L., & Aiub, C. A. F. (2025). Air quality biomonitoring in the Guaiúba Chemical Park using *Tradescantia pallida* var. *purpurea*: Genotoxic effects assessed by micronucleus and stamen hair mutation bioassays associated with climate data. *Caderno Pedagógico*, *22*(9), e18528–e18528.

Dignac, M. F., Derrien, D., Barré, P., Barot, S., Cécillon, L., Chenu, C., ... & Basile-Doelsch, I. (2017). Increasing soil carbon storage: Mechanisms, effects of agricultural practices and proxies. A review: Soil C storage: Mechanisms, practices and proxies. *Agronomy for Sustainable Development*, *37*(2), 14.

Dorafshan, M. M., Abedi-Koupai, J., Eslamian, S., & Amiri, M. J. (2023). Vetiver grass (*Chrysopogon zizanioides* L.): A hyper-accumulator crop for bioremediation of unconventional water. *Sustainability*, *15*(4), 3529.

dos Santos Lucas, L., Neto, A. R., de Moura, J. B., de Souza, R. F., Santos, M. E. F., de Moura, L. F., … & Dutra e Silva, S. (2022). Mycorrhizal fungi arbuscular in forage grasses cultivated in Cerrado soil. *Scientific Reports*, *12*(1), 3103.

Duhamel, S. (2025). The microbial phosphorus cycle in aquatic ecosystems. *Nature Reviews Microbiology*, *23*(4), 239–255.

El-Mageed, A., Taia, A., El-Mageed, A., Shimaa, A., El-Saadony, M. T., & Abdelaziz, S. (2022). Plant growth-promoting rhizobacteria improve growth, morph-physiological responses, water productivity, and yield of rice plants under full and deficit drip irrigation. *Rice*, *15*(1), 16.

Etesami, H., Jeong, B. R., & Glick, B. R. (2021). Contribution of arbuscular mycorrhizal fungi, phosphate–solubilizing bacteria, and silicon to P uptake by plant. *Frontiers in Plant Science*, *12*, 699618.

Ferioun, M., Zouitane, I., Bouhraoua, S., Elouattassi, Y., Belahcen, D., Errabbani, A., … & El Ghachtouli, N. (2025). Applying microbial biostimulants and drought-tolerant genotypes to enhance barley growth and yield under drought stress. *Frontiers in Plant Science*, *15*, 1494987.

Gerke, J. (2022). The central role of soil organic matter in soil fertility and carbon storage. *Soil Systems*, *6*(2), 33.

Gorelova, S. V., Kolbas, A. P., Muratova, A. Y., Frontasyeva, M. V., Zinicovscaia, I., & Okina, O. I. (2023). Prospects for the use of *sorghum bicolor* for phytoremediation of soils contaminated with heavy metals in temperate climates. In L. Newman, A. A. Ansari, S. S. Gill, M. Naeem, & Gill, R. (Eds.), *Phytoremediation: Management of Environmental Contaminants, Volume 7* (pp. 263–301). Cham: Springer International Publishing.

Gorelova, S. V., Muratova, A. Y., Zinicovscaia, I., Okina, O. I., & Kolbas, A. (2022). Prospects for the use of *Echinochloa frumentacea* for Phytoremediation of soils with multielement anomalies. *Soil Systems*, *6*(1), 27.

Gougoulias, C., Clark, J. M., & Shaw, L. J. (2014). The role of soil microbes in the global carbon cycle: Tracking the below-ground microbial processing of plant-derived carbon for manipulating carbon dynamics in agricultural systems. *Journal of the Science of Food and Agriculture*, *94*(12), 2362–2371.

Guan, Z., Chen, T., Chen, D., Lu, Y., Han, Q., Li, N., … & He, Q. (2023). Leaf litter breakdown and soil microbes in *Catalpa bungei* plantations in response to various fertilization regimes. *Forests*, *14*(4), 699.

Guo, X., Hou, Z., Wu, X., Liu, W., Cai, J., & An, S. (2025). Long-term intercropping shaped soil bacterial microbiome composition and structure of maize fields in a semiarid region. *Soil and Tillage Research*, *247*, 106383.

Gupta, S., Plugge, C. M., Muyzer, G., & Sanchez-Andrea, I. (2024). Harnessing the potential of the microbial sulfur cycle for environmental biotechnology. *Current Opinion in Biotechnology*, *88*, 103164–103172.

Halli, H. M., Govindasamy, P., Choudhary, M., Srinivasan, R., Prasad, M., Wasnik, V. K., … & Pathak, H. (2022). Range grasses to improve soil properties, carbon sustainability, and fodder security in degraded lands of semi-arid regions. *Science of the Total Environment*, *851*, 158211.

Harahap, R. T., Azizah, I. R., Setiawati, M. R., Herdiyantoro, D., & Simarmata, T. (2023). Enhancing upland rice growth and yield with indigenous plant growth-promoting rhizobacteria (PGPR) isolate at N-fertilizers dosage. *Agriculture*, *13*(10), 1987.

Henneron, L., Cros, C., Picon-Cochard, C., Rahimian, V., & Fontaine, S. (2020). Plant economic strategies of grassland species control soil carbon dynamics through rhizodeposition. *Journal of Ecology*, *108*(2), 528–545.

Horwath, W. (2007). Carbon cycling and formation of soil organic matter. In E. A. Paul (Ed.), *Soil Microbiology, Ecology and Biochemistry* (pp. 303–339).

Hu, J., Richwine, J. D., Keyser, P. D., Yao, F., Jagadamma, S., & DeBruyn, J. M. (2022a). Urea fertilization and grass species alter microbial nitrogen cycling capacity and activity in a C4 native grassland. *Peer J*, *10*, 13874–13889.

Hu, Z., Fang, Z., Hu, B., Wen, X., Lou, L., & Cai, Q. (2022b). Profiling of water-use efficiency in switchgrass (*Panicum virgatum* L.) and the relationship with cadmium accumulation. *Agronomy*, *12*(2), 507.

Iqbal, S., Ummara, U., Noreen, S., Akhter, M. S., Jaleel, F., Jabeen, S., …& Al-Qthanin, R. N. (2024). Enhancing systematic tolerance in Bermuda grass (*Cynodon dactylon* L.) through amplified alkB gene expression and bacterial-driven hydrocarbon degradation. *Environmental Science and Pollution Research*, *31*(13), 19871–19885.

Javaid, S., Mushtaq, S., Mumtaz, M. Z., Rasool, G., Naqqash, T., Afzal, M., … & Li, L. (2023). Mineral solubilizing Rhizobacterial strains mediated biostimulation of Rhodes grass seedlings. *Microorganisms*, *11*(10), 2543.

Jia, F., Li, Y., Hu, Q. N., Zhang, L., Mao, L. G., Zhu, L. Z., … & Sun, Y. (2023). Factors impacting the behavior of phytoremediation in pesticide-contaminated environment: A meta-analysis. *Science of the Total Environment*, *892*, 164418.

Jiang, S., Li, J., Wang, Q., Yin, C., Zhan, Y., Yan, Y., … & Ke, X. (2022). Maize growth promotion by inoculation with an engineered ammonium-excreting strain of nitrogen-fixing *Pseudomonas stutzeri*. *Microorganisms*, *10*(10), 1986.

Jin, P., Liu, M., Xu, X., Sun, Y., Kuzyakov, Y., & Gunina, A. (2023). Gross mineralization and nitrification in degraded alpine grassland soil. *Rhizosphere*, *27*, 100778.

Joo, Y. J., Sim, M. S., Elwood Madden, M. E., & Soreghan, G. S. (2022). Significance of the terrestrial sink in the biogeochemical sulfur cycle. *Geophysical Research Letters*, *49*(4), 121–128.

Kang, X., Geng, N., Li, X., Yu, J., Wang, H., Pan, H., … & Lou, Y. (2022). Biochar alleviates phytotoxicity by minimizing bioavailability and oxidative stress in foxtail millet (*Setaria italica* L.) cultivated in Cd-and Zn-contaminated soil. *Frontiers in Plant Science*, *13*, 782963.

Karam, F., Haddad, R., Amacha, N., Charanek, W., & Harmand, J. (2022). Assessment of the impacts of phyto-remediation on water quality of the Litani river by means of two wetland plants (*Sparganium erectum* and *Phragmites australis*). *Water*, *15*(1), 4.

Khaksar, G., Treesubsuntorn, C., & Thiravetyan, P. (2017). Impact of endophytic colonization patterns on *Zamioculcas zamiifolia* stress response and in regulating ROS, tryptophan and IAA levels under airborne formaldehyde and formaldehyde-contaminated soil conditions. *Plant Physiology and Biochemistry*, *114*, 1–9.

Khan, Q., Huang, X., He, Z., Wang, H., Chen, Y., Xia, G., … & Zhang, Y. (2024). An insight into conflict and collaboration between plants and microorganisms. *Chemical and Biological Technologies in Agriculture*, *11*(1), 161.

Kumar, A., Bhattacharya, T., Shaikh, W. A., & Roy, A. (2024). Sustainable soil management under drought stress through biochar application: Immobilizing arsenic, ameliorating soil quality, and augmenting plant growth. *Environmental Research*, *259*, 119531.

Kumar, V., Raghuvanshi, N., Pandey, A. K., Kumar, A., Thoday-Kennedy, E., & Kant, S. (2023). Role of halotolerant plant growth-promoting rhizobacteria in mitigating salinity stress: Recent advances and possibilities. *Agriculture*, *13*(1), 168.

Kurade, M. B., Ha, Y. H., Xiong, J. Q., Govindwar, S. P., Jang, M., & Jeon, B. H. (2021). Phytoremediation as a green biotechnology tool for emerging environmental pollution: A step forward towards sustainable rehabilitation of the environment. *Chemical Engineering Journal*, *415*, 129040.

Laishram, B., Devi, O. R., Dutta, R., Senthilkumar, T., Goyal, G., Paliwal, D. K., … & Rasool, A. (2024). Plant-microbe interactions: PGPM as microbial inoculants/biofertilizers for sustaining crop productivity and soil fertility. *Current Research in Microbial Sciences*, *8*, 100333.

Li, D., Wang, J., Chen, R., Chen, J., Zong, J., Li, L., … & Guo, H. (2024a). Nitrogen acquisition, assimilation, and seasonal cycling in perennial grasses. *Plant Science*, *342*, 112054.

Li, W., Yang, J., Zhan, F., Guo, J., Zhang, Y., Ba, Y., … & He, Y. (2024b). Characteristics of the rhizospheric AMF community and nutrient contents of the dominant grasses in four microhabitats of the subalpine zone in northwestern Yunnan, China. *Forests*, *15*(4), 656.

Lin, Y., Hu, H. W., Gao, G. F., & Cai, Y. (2023). Nitrogen-cycling microorganisms under global change: Response and feedback. *Frontiers in Microbiology*, *14*, 1289–237.

Liu, H., Menchyk, N., Bethea, F., Baldwin, C., Taylor, J., & Patrick, C. (2019). Turfgrass hyper-accumulative characteristics to alleviate heavy and toxic metal stresses. In M. Pessarakli (Ed.), *Handbook of Plant and Crop Stress, Fourth Edition* (pp. 523–538). Taylor and Francis.

Liu, Y., Fan, Y., Kuzyakov, Y., Dai, J., Zhang, C., & Li, C. (2025). Effects of maize/soybean intercropping on nitrogen mineralization and fungal communities in soil. *Plant and Soil*, 2025, 1–15.

Magadlela, A., Lembede, Z., Egbewale, S. O., & Olaniran, A. O. (2023). The metabolic potential of soil microorganisms and enzymes in phosphorus-deficient KwaZulu-Natal grassland ecosystem soils. *Applied Soil Ecology*, *181*, 104647.

Mahala, D. M., Maheshwari, H. S., Yadav, R. K., Prabina, B. J., Bharti, A., Reddy, K. K., … & Ramesh, A. (2021). Microbial transformation of nutrients. *Rhizosphere Microbes: Soil and Plant Functions*, *23*, 175.

Mahmud, K., Panday, D., Mergoum, A., & Missaoui, A. (2021). Nitrogen losses and potential mitigation strategies for a sustainable agroecosystem. *Sustainability*, *13*(4), 2400.

Malik, L., Hussain, S., Shahid, M., Mahmood, F., Ali, H. M., Malik, M., … & Shahzad, T. (2024). Co-applied biochar and drought tolerant PGPRs induced more improvement in soil quality and wheat production than their individual applications under drought conditions. *PeerJ*, *12*, e18171.

Manjunath, M., Khokhar, A., Chary, G. R., Singh, M., Yadav, S. K., Gopinath, K. A., … & Singh, V. K. (2023). Microbial consortia enhance the yield of maize under sub-humid rainfed production system of India. *Frontiers in Sustainable Food Systems*, *7*, 1108492.

Mankar, M. K., Sharma, U. S., & Sahay, S. (2022). Priming effect of native rhizosphere bacteria on little millet (*Panicum sumatrense*). *Die Bodenkultur*, *73*, 55–66.

Marzec, M., Jóźwiakowski, K., Dębska, A., Gizińska-Górna, M., Pytka-Woszczyło, A., Kowalczyk-Juśko, A., & Listosz, A. (2018). The efficiency and reliability of pollutant removal in a hybrid constructed wetland with common reed, manna grass, and Virginia mallow. *Water*, *10*(10), 1445.

Masinire, F., Adenuga, D. O., Tichapondwa, S. M., & Chirwa, E. M. (2021). Phytoremediation of Cr (VI) in wastewater using the vetiver grass (*Chrysopogon zizanioides*). *Minerals Engineering*, *172*, 107141.

Masotla, M. K. L., Melato, F. A., & Mokgalaka-Fleischmann, N. S. (2023). Extraction potential *of Lolium perenne* L. (Perennial Rye Grass) for metals in landfill soil: Its tolerance and defense strategies. *Minerals*, *13*(7), 873.

Mateos, K., Chappell, G., Klos, A., Le, B., Boden, J., Stüeken, E., & Anderson, R. (2023). The evolution and spread of sulfur cycling enzymes reflect the redox state of the early earth. *Science Advances*, *9*(27), 4847–4859.

Mattoo, R., 2023. Microbial roles in the terrestrial and aquatic nitrogen cycle—Implications in climate change. *FEMS Microbiology Letters*, *370*, 1–19.

Murali, M., & Navi, V. (2024). Assessing the impact of elite indigenous *Azospirillum* strains on growth of wheat (*Triticum aestivum* L.) across varied nitrogen regimes. *Mysore Journal of Agricultural Sciences*, *58*(4), 175–180.

Murali, M., Singh, S. B., Gowtham, H. G., Shilpa, N., Prasad, M., Aiyaz, M., & Amruthesh, K. N. (2021). Induction of drought tolerance in *Pennisetum glaucum* by ACC deaminase producing PGPR-*Bacillus amyloliquefaciens* through Antioxidant defense system. *Microbiological Research*, *253*, 126891.

Mushtaq, N. U., Alghamdi, K. M., Saleem, S., Shajar, F., Tahir, I., Bahieldin, A., ... & Hakeem, K. R. (2022). Selenate and selenite transporters in proso millet: Genome extensive detection and expression studies under salt stress and selenium. *Frontiers in Plant Science*, *13*, 1060154.

Mythra, R., Srikanth, G. S., Jagadeesh, U., Bhagyashree, K. B., & Manjunath, A. (2022). Effect of AM fungus on phosphorus nutrition of maize and pigeon pea in alfisols as influenced by different phosphorus amendments of North Carolina Rock Phosphate (NCRP). *Kavaka*, *58*, 28–35.

Nagar, S., Talwar, C., Motelica-Heino, M., Richnow, H. H., Shakarad, M., Lal, R., & Negi, R. K. (2022). Microbial ecology of sulfur biogeochemical cycling at a Mesothermal hot spring atop northern Himalayas, India. *Frontiers in Microbiology*, *13*, 848010–848023.

Nagar, S., Talwar, C., & Negi, R. K. (2024). Integrated multi-platform approaches to gain insights into ecosystem's fundamental ecology and habitat specific alterations. *bioRxiv*, 2024-06.

Nascimento, D. B. D., Lopes, M. L. S., Izidro, J. L. P. S., Bezerra, R. C. A., Gois, G. C., Amaral, T. N. E. D., … & Coêlho, J. J. (2024). Nitrogen, phosphorus, and potassium cycling in pasture ecosystems. *Ciência Animal Brasileira*, *25*, e–76743.

Ogwu, M. C., Lori, T., Aliu, O. O., Febnteh, E. B., Izah, S. C., & Abdelkhalek, S. T. (2024). Agricultural air pollution: Impacts, sources, and mitigation strategies. In S. T. Abdelkhalek (Ed.), *Air Pollutants in the Context of One Health: Fundamentals, Sources, and Impacts* (pp. 395–423). Springer Nature Switzerland.

Pan, B., Huang, Y., Xia, L., Liang, J., Liu, R., Luo, Y., … & Lam, S. K. (2025). Estimating fractions of N_2O emissions from nitrification and denitrification using data assimilation. *Biogeochemistry*, *168*(5), 71.

Patidar, P., Sannagoudar, M. S., Ghosh, A., Singh, A. K., Misra, S., Khandibagur, V., … & Chandra, A. (2023). Tropical range grasses can sustain soil functions despite nutrient depletion in semiarid degraded land. *Frontiers in Sustainable Food Systems*, *7*, 1230156.

Permana, B. H., Thiravetyan, P., & Treesubsuntorn, C. (2022). Effect of airflow pattern and distance on removal of particulate matters and volatile organic compounds from cigarette smoke using Sansevieria trifasciata botanical biofilter. *Chemosphere*, *295*, 133919.

Plunge, S., Gudas, M., & Povilaitis, A. (2022). Effectiveness of best management practices for non-point source agricultural water pollution control with changing climate–Lithuania's case. *Agricultural Water Management*, *267*, 107635.

Prasad, R., Shivay, Y. S., & Mandi, S. (2022). Phytosiderophores and absorption of iron and other cations by plants. In S. K. Upadhyay (Ed.), *Cation Transporters in Plants* (pp. 385–399). Academic Press.

Qin, L., Xiao, Z., Ming, A., Teng, J., Zhu, H., Qin, J., & Liang, Z. (2024). Soil phosphorus cycling microbial functional genes of monoculture and mixed plantations of native tree species in subtropical China. *Frontiers in Microbiology*, *15*, 1419645–1419658.

Rajkumar, M., Narayanasamy, S., & Uthandi, S. (2024). A root-associated Bacillus albus LRS2 and its metabolites for plant growth promotion and drought stress tolerance in little millet (*Panicum sumatrense* L.). *Plant Stress*, *12*, 100446.

Rakkammal, K., Pandian, S., & Ramesh, M. (2024). Physiological and biochemical response of finger millet plants exposed to arsenic and nickel stress. *Plant Stress*, *11*, 100389.

Rasool, A., Hafiz Shah, W., Padder, S. A., Tahir, I., Alharby, H. F., Hakeem, K. R., & ul Rehman, R. (2023). Exogenous selenium treatment alleviates salinity stress in Proso Millet (*Panicum miliaceum* L.) by enhancing the antioxidant defence system and regulation of ionic channels. *Plant Growth Regulation*, *100*(2), 479–494.

Rich, J. J., Arevalo, P., Chang, B. X., Devol, A. H., & Ward, B. B. (2020). Anaerobic ammonium oxidation (anammox) and denitrification in Peru margin sediments. *Journal of Marine Systems*, *207*, 103122.

Rodrigues, M. Â., Correia, C. M., & Arrobas, M. (2024). The application of a foliar spray containing *Methylobacterium symbioticum* had a limited effect on crop yield and nitrogen recovery in field and pot-grown maize. *Plants*, *13*(20), 2909.

Rousk, J., & Bengtson, P. (2014). Microbial regulation of global biogeochemical cycles. *Frontiers in Microbiology*, *5*, 103–105.

Satognon, F. (2024). Advancements in tall fescue (Festuca arundinacea) in the US: Origin, American introduction, development, and improvement. *Journal of Crop Improvement*, *38*(6), 665–687.

Shapiro, A. M. (2022). Understanding morphological responses of understanding morphological responses of Typha (cattail)(cattail) species to nutrient pollution. MSc thesis, University of Northern Iowa.

Singh, P., Halli, H. M., Tomar, M., Wasnik, V. K., Saini, R. P., Mahesha, H. S., … & Yadav, V. K. (2022). Role of range grasses in conservation and restoration of biodiversity. In P. Kumar, R. S. Tomar, J. A. Bhat, M. Dobriyal, & M. Rani (Eds.), *Agro-Biodiversity and Agri-Ecosystem Management* (pp. 53–69). Springer Nature Singapore.

Sladkovska, T., Wolski, K., Bujak, H., Radkowski, A., & Sobol, Ł. (2022). A review of research on the use of selected grass species in removal of heavy metals. *Agronomy*, *12*(10), 2587.

Smilauer, P., Košnar, J., Kotilínek, M., Pecháčková, S., & Šmilauerová, M. (2021). Host age and surrounding vegetation affect the community and colonization rates of arbuscular mycorrhizal fungi in a temperate grassland. *New Phytologist*, *232*(1), 290–302.

Sokol, N. W., Slessarev, E., Marschmann, G. L., Nicolas, A., Blazewicz, S. J., Brodie, E. L., ... & Pett-Ridge, J. (2022). Life and death in the soil microbiome: How ecological processes influence biogeochemistry. *Nature Reviews Microbiology*, *20*(7), 415–430.

Sorkhoh, N. A., Al-Mailem, D. M., Ali, N., Al-Awadhi, H., Salamah, S., Eliyas, M., & Radwan, S. S. (2011). Bioremediation of volatile oil hydrocarbons by epiphytic bacteria associated with American grass (*Cynodon* sp.) and broad bean (*Vicia faba*) leaves. *International Biodeterioration & Biodegradation*, *65*(6), 797–802.

Sosa, O. A. (2018). Phosphorus redox reactions as pinch hitters in microbial metabolism. *Proceedings of the National Academy of Sciences*, *115*(1), 7–8.

Su, Y., Ma, X., Le, J., Li, K., Han, W., & Liu, X. (2021). Decoupling of nitrogen and phosphorus in dominant grass species in response to long-term nitrogen addition in an Alpine Grassland in Central Asia. *Plant Ecology*, *222*, 261–274.

Sullins, K. N., Dillard, S. L., Held, D. W., & Carroll, E. P. (2023). Utility of plant growth-promoting Rhizobacteria for sustainable production of Bermudagrass forage. *Microorganisms*, *11*(4), 863.

Talaie, A., & Ponraj, M. (2024). Impact of Houseplants on Reducing Indoor Air Pollution. *Journal of Environmental Treatment Techniques*, *12*(3), 1–13.

Tao, F., Huang, Y., Hungate, B. A., Manzoni, S., Frey, S. D., Schmidt, M. W., ... & Luo, Y. (2023). Microbial carbon use efficiency promotes global soil carbon storage. *Nature*, *618*(7967), 981–985.

Thakur, T. K., Barya, M. P., Dutta, J., Mukherjee, P., Thakur, A., Swamy, S. L., & Anderson, J. T. (2023). Integrated phytobial remediation of dissolved pollutants from domestic wastewater through constructed wetlands: An interactive macrophyte-microbe-based green and low-cost decontamination technology with prospective resource recovery. *Water*, *15*(22), 3877.

Tian, J., Ge, F., Zhang, D., Deng, S., & Liu, X. (2021). Roles of Phosphate Solubilizing Microorganisms from managing soil phosphorus deficiency to mediating biogeochemical P cycle. *Biology*, *10*, 158–167.

Torres-Cuesta, D., Mora-Motta, D., Chavarro-Bermeo, J. P., Olaya-Montes, A., Vargas-Garcia, C., Bonilla, R., & Estrada-Bonilla, G. (2023). Phosphate-solubilizing bacteria with low-solubility fertilizer improve soil P availability and yield of kikuyu grass. *Microorganisms*, *11*(7), 1748.

Twible, L. E., Whaley-Martin, K., Chen, L. X., Colenbrander Nelson, T., Arrey, J. L., Jarolimek, C. V., … & Warren, L. A. (2024). pH and thiosulfate dependent microbial sulfur oxidation strategies across diverse environments. *Frontiers in Microbiology*, *15*, 1426584–1426593.

Ujang, F. A., Roslan, A. M., Osman, N. A., Norman, A., Idris, J., Farid, M. A. A., … & Hassan, M. A. (2021). Removal behaviour of residual pollutants from biologically treated palm oil mill effluent by *Pennisetum purpureum* in constructed wetland. *Scientific Reports*, *11*(1), 18257.

Upadhayay, V. K., Chitara, M. K., Mishra, D., Jha, M. N., Jaiswal, A., Kumari, G., … & Sharma, A. (2023). Synergistic impact of nanomaterials and plant probiotics in agriculture: A tale of two-way strategy for long-term sustainability. *Frontiers in Microbiology*, *14*, 1133968.

Vitaliano, S., Cascone, S., & D'Urso, P. R. (2024). Mitigating built environment air pollution by green systems: An in-depth review. *Applied Sciences*, *14*(15), 6487.

Wang, C., Liang, Q., Liu, J., Zhou, R., Lang, X., Xu, S., … & Yang, K. Q. (2023). Impact of intercropping grass on the soil rhizosphere microbial community and soil ecosystem function in a walnut orchard. *Frontiers in Microbiology, 14*, 1137590.

Wang, W., Li, M., Chen, P., Yuan, S., Wang, K., Wang, S., & Jiang, X. (2025). Role of nitrogen cycling functional genes and their key influencing factors in eutrophic aquatic ecosystems. *Environmental Reviews, 33*, 1–10.

Wang, Y., Wang, J., Wang, Y., Wang, X., Jin, B., Chen, C., & Zhao, X. (2024). Contribution of litter and root to soil nutrients in different rocky desertification grasslands in a karst area. *Plants, 13*(16), 2329.

Wang, Y., Xiao, J., Ma, Y., Luo, Y., Hu, Z., Li, F., … & Yuan, L. (2021). Carbon fluxes and environmental controls across different alpine grassland types on the Tibetan Plateau. *Agricultural and Forest Meteorology, 311*, 108694.

Wei, Z., Maxwell, T., Robinson, B., & Dickinson, N. (2022). Grasses procure key soil nutrients for clovers. *Nature Plants, 8*(8), 923–929.

Wu, H., Cui, H., Fu, C., Li, R., Qi, F., Liu, Z., ... & Qiao, M. (2024). Unveiling the crucial role of soil microorganisms in carbon cycling: A review. *Science of the Total Environment, 909*, 168627.

Xiao, X., Han, L., Chen, H., Wang, J., Zhang, Y., & Hu, A. (2023). Intercropping enhances microbial community diversity and ecosystem functioning in maize fields. *Frontiers in Microbiology, 13*, 1084452.

Yadav, S., Ambastha, S., Pipil, H., Haritash, A. K., & Reddy, K. R. (2025). Deciphering the sustainable stormwater management strategies for urban areas: A review. *Water Resources Management, 39*(7), 2971–2991.

Yang, J., Liang, X., Jiang, N., Li, Z., & Zu, Y. (2022). Three amendments reduced the bioavailability of heavily contaminated soil with arsenic and cadmium and increased the relative feeding value of *Lolium perenne* L. *Science of the Total Environment, 847*, 157572.

Yasin, M., Tauseef, M., Zafar, Z., Rahman, M., Islam, E., Iqbal, S., & Afzal, M. (2021). Plant-microbe synergism in floating treatment wetlands for the enhanced removal of sodium dodecyl sulphate from water. *Sustainability, 13*(5), 2883.

Zaheer, M. S., Raza, M. A. S., Saleem, M. F., Khan, I. H., Ahmad, S., Iqbal, R., & Manevski, K. (2019). Investigating the effect of *Azospirillum brasilense* and *Rhizobium pisi* on agronomic traits of wheat (*Triticum aestivum* L.). *Archives of Agronomy and Soil Science, 65*(11), 1554–1564.

Zhang, J. H., Huang, J., Hussain, S., & Zhang, H. (2021). Increased ammonification, nitrogenase, soil respiration and microbial biomass N in the rhizosphere of rice plants inoculated with rhizobacteria. *Journal of Integrative Agriculture, 20*(10), 2781–2796.

Zhao, Z., Zhao, Y., Marotta, F., Xamxidin, M., Li, H., Xu, J., … & Wu, M. (2024). The microbial community structure and nitrogen cycle of high-altitude pristine saline lakes on the Qinghai-Tibetan plateau. *Frontiers in Microbiology, 15*, 1424368–1424375.

Zheng, G., Wei, K., Kang, X., Fan, W., Ma, N. L., Verma, M., … & Ge, S. (2023). A new attempt to control volatile organic compounds (VOCs) pollution-modification technology of biomass for adsorption of VOCs gas. *Environmental Pollution, 336*, 122451.

Zhu, W. N., Zhang, G. L., Zhang, P. J., Zhang, Q. Q., Ren, J. T., Xu, B. Y., & Qing, H. (2021). Decomposition characteristics of leaf litters and roots of six main plant species and their relationships with functional traits in Stipa grandis steppe. *Chinese Journal of Plant Ecology, 45*(6), 606.

Zhuang, X., Wang, S., & Wu, S. (2024). Electron transfer in the biogeochemical sulfur cycle. *Life, 14*(5), 591.

5 Endophytic Bacteria as Natural Allies in Mitigating Abiotic Stress

Mechanisms, Applications, and Prospects for Sustainable Agriculture

Rachana Singh, Aditya Sinha Roy, Bhumika Singh, and Shivanshi Tyagi

5.1 INTRODUCTION

The term "endophyte" is derived from the Greek words *endo* (inside) and *phyte* (plant). De Bary (1866) first used it to describe microorganisms residing within healthy plants. Earlier, in 1846, Leveille had already reported fungi inhabiting wheat leaves, which he referred to as endophytic fungi. Wilson (1995) later defined endophytes as fungi or bacteria that internally colonize living plant tissues, either for part or all of their life cycle, without producing visible disease symptoms. Essentially, endophytes include all microbes living within plant tissues without causing harm to their hosts (Okoroafor, 2022). Plants require light, water, carbon, and essential nutrients for proper growth, development, and reproduction. Due to their sessile nature, however, they are constantly exposed to a variety of environmental and biotic stresses. Key abiotic factors such as drought, extreme temperatures, salinity, heavy metals, and nutrient deficiencies significantly limit growth and productivity (Kumar et al., 2022). Endophytes, which inhabit roots, stems, leaves, and seeds, coexist with plants without interfering with their normal physiological processes. They contribute to plant health by improving nutrient acquisition, producing beneficial secondary metabolites, and alleviating disease symptoms. Bacteria, actinomycetes, and fungi are the predominant endophytes, forming complex interactions with their host plants while receiving protection against environmental fluctuations (Passari et al., 2017). Abiotic stressors, originating from nonliving sources, represent major constraints on agricultural productivity. Plants often encounter stress conditions including ultraviolet (UV) radiation, osmotic imbalance, oxidative stress, drought, salinity, extreme temperatures, and heavy metal exposure. To withstand such conditions, plants trigger extensive molecular, cellular, and physiological adaptations (Rawat et al., 2021).

DOI: 10.1201/9781003664451-5

In response to salt and drought stress, rhizobacteria and endophytic bacteria enhance plant survival by producing osmoprotectants, stabilizing subcellular structures, and scavenging reactive oxygen species (ROS; Sharma et al., 2025). Beneficial microbes, such as *Burkholderia* sp., *Arthrobacter* sp., and *Bacillus* sp., stimulate proline synthesis under stress. For example, inoculation of *Zea mays* with *Rhizobium* sp. and *Pseudomonas* sp. improves proline accumulation, decreases electrolyte leakage, enhances leaf water retention, and promotes selective potassium (K^+) uptake, collectively enhancing salinity tolerance (Getahun et al., 2019). Certain rhizobacteria, including *Pseudomonas* sp., can also endure drought and nutrient limitations by secreting exopolysaccharides (EPS). Most abiotic stressors are interconnected and elicit similar cellular damages, such as protein denaturation, ROS accumulation, membrane instability, and altered enzymatic activities (Hasanuzzaman et al., 2021). Based on their lifestyle, endophytic bacteria are categorized as facultative, obligate, opportunistic, or passenger endophytes (Pandey et al., 2018). Plants lacking endophytes in nature are more vulnerable to environmental stress and pathogen invasion (Eid et al., 2019). Severe abiotic stresses can reduce crop survival, biomass, and yield by up to 70%, threatening global food security. Drought, salinity, and temperature extremes are the most critical stressors affecting plant productivity (Oshunsanya et al., 2019).

This chapter will provide a comprehensive overview of endophytic bacteria, including their classification, ecology, and diversity. It will explore the mechanisms of plant responses to abiotic stresses, highlight the functional traits of endophytes in stress alleviation, and examine the underlying molecular and genetic pathways. Additionally, the chapter will present specific case studies on endophyte-mediated stress mitigation, discuss applications in sustainable agriculture and biotechnology, and review recent biotechnological tools and advancements, along with the current challenges and limitations in this field.

5.2 CLASSIFICATION AND CHARACTERISTICS

Endophytic bacteria can be categorized based on their life strategies into obligate, facultative, opportunistic, and passenger types (Pandey et al., 2018). In modern agriculture, crop productivity is increasingly constrained by both biotic factors, such as insects and phytopathogens, and abiotic stresses, including extreme temperatures, nutrient imbalances, salinity, drought, flooding, and organic or inorganic contaminants, many of which are exacerbated by climatc change (Zörb et al., 2019). Endophytes, which primarily comprise bacteria and fungi, inhabit internal plant tissues without causing disease or visible symptoms. They can colonize the apoplast, intercellular spaces, and xylem vessels and have been isolated from fruits, roots, flowers, stems, leaves, and seeds (Compant et al., 2021). These microbes maintain close interactions with the rhizosphere and secrete a variety of metabolic compounds that support plant health (Meena et al., 2025; Sahoo et al., 2025). By mitigating the effects of pathogens and enhancing tolerance to contaminated soils, endophytes perform roles analogous to plant growth–promoting rhizobacteria (PGPR), including the degradation or transformation of pollutants. Classification based on lifestyle reflects how endophytes enter the host, their mode of interaction with plant tissues, and the mechanisms they employ to sustain their populations. In this context, they are generally classified as obligate, facultative, or opportunistic endophytes.

5.2.1 Facultative Endophytes

Facultative endophytes are the most extensively studied and commonly isolated group of bacterial endophytes due to their remarkable ecological flexibility. They can exist as free-living organisms in diverse habitats, including soil, the rhizosphere, and aquatic environments (Kuźniar et al., 2025) or transition into an endophytic lifestyle by colonizing internal plant tissues. This adaptability reflects their broad metabolic capacity, enabling them to utilize varied nutrient sources and tolerate varied environmental conditions (Singh & Dubey, 2018). Their shift between lifestyles is influenced by environmental cues and plant signals that favor internal colonization. Facultative endophytes typically enter plants through wounds caused by mechanical damage or insect activity, as well as through natural openings such as lenticels, stomata, and hydathodes. They may also actively penetrate epidermal cells using hydrolytic enzymes like cellulases and pectinases. Once inside, they colonize intercellular spaces and vascular tissues, sometimes systemically, while maintaining asymptomatic associations (Kandel et al., 2017). A major advantage of facultative endophytes is their ease of cultivation on standard laboratory media, which distinguishes them from many obligate endophytes and has facilitated extensive research into their beneficial traits. Common examples include *Pseudomonas* species, known for nitrogen fixation, phosphate solubilization, and biocontrol activity; *Bacillus* species, which produce antibiotics and lipopeptides with strong antagonistic effects against pathogens (Salazar et al., 2023); *Enterobacter* species, which contribute to nitrogen fixation and phytohormone production; and certain *Azotobacter* strains that can exhibit endophytic nitrogen-fixing activity. Together, these facultative endophytes play vital roles in stimulating plant growth, boosting stress tolerance, and offering natural alternatives to chemical inputs in sustainable agriculture.

5.2.2 Obligate Endophytes

Obligate endophytes represent the highest level of co-evolution between bacteria and their plant hosts, as they are entirely dependent on living within plant tissues for survival, growth, and reproduction (Malik et al., 2025). Their inability to persist outside the host or grow on standard artificial media reflects extensive genetic and metabolic specialization, often involving genome reduction and highly tailored biochemical pathways. These endophytes exhibit extreme host dependence, relying on the plant for essential nutrients, protection from environmental stresses, and dispersal, which contributes to their non-culturability or limited growth in vitro. A key characteristic of many obligate endophytes is vertical transmission, in which the microbes are transferred from parent plants to their offspring, most often via seeds, ensuring stable, long-term associations across generations (Frank et al., 2017). They frequently inhabit highly specialized plant structures such as leaf nodules in *Psychotria* and *Pavetta* species or root nodules in legumes, where the host may evolve dedicated cellular mechanisms to accommodate and nourish these symbionts. Due to their intense specialization and reliance on vertical transmission, obligate endophytes typically display very narrow host specificity, often associating with only one or a few closely related plant species via complex molecular recognition processes. Classic examples include rhizobia such as *Bradyrhizobium*, *Sinorhizobium*, and *Rhizobium* spp., which develop nitrogen-fixing root nodules in legumes (Dudeja et al., 2012), and *Burkholderia* species that inhabit leaf nodules of genera like *Psychotria* and *Pavetta*, where they are vertically transmitted and contribute to nitrogen fixation and other essential metabolic functions. Additionally, numerous endophytes detected solely through molecular approaches (e.g., 16S rRNA sequencing) are considered obligate due to their persistent unculturability, suggesting long-term evolutionary integration with their hosts, although their ecological functions are still being uncovered.

5.2.3 Opportunistic Endophytes

Opportunistic endophytes are environmental bacteria that temporarily colonize internal plant tissues when conditions allow, even though their primary ecological niche lies outside the plant in habitats such as soil, water, or air (Frank et al., 2017). These microbes enter plants mainly to exploit accessible resources such as sugars, amino acids, and other carbon-rich compounds or to seek temporary refuge from environmental stresses. Their presence inside plants is typically short-lived, as they do not form the stable or systemic associations characteristic of well-adapted facultative endophytes. Opportunistic endophytes usually gain entry through passive means, including wounds caused by insect feeding or mechanical injury, cracks in roots, or the movement of bacteria through the transpiration stream from the soil into the xylem ("Internalization and Infiltration," 2020). Once inside, they may proliferate briefly if conditions are favorable, but they do not engage in strong co-evolutionary or symbiotic interactions with the host. They rarely induce specific plant responses or cause notable metabolic changes, and any benefits they provide are incidental rather than adaptive. The presence and abundance of these microbes can vary widely, influenced by environmental conditions, the health of the host plant, and the surrounding microbial community. For instance, common soil bacteria, such as nonspecific *Enterobacteriaceae* or *Arthrobacter* species, may occasionally be recovered from surface-sterilized plant tissues. However, they typically do not establish stable or long-term endophytic associations, particularly when opportunities for entry arise through root damage or other openings.

5.2.4 Passenger Endophytes

Passenger endophytes represent the most incidental and least functionally relevant category of microorganisms found within plant tissues. These bacteria enter the plant purely by chance, lacking any specialized mechanisms for colonization or meaningful interaction with the host (Frank et al., 2017). Their presence is typically the result of random environmental events rather than an adaptive strategy, with accidental entry occurring through processes such as water infiltration into small cracks, dust particles settling on wounds, or bacteria becoming trapped as roots mechanically push through soil aggregates. Once inside, passenger endophytes exhibit minimal to no physiological interaction with the plant; they neither proliferate significantly nor induce detectable host responses, and they offer no clear benefits or harms. Their survival within tissues is generally short-lived, as they may persist only briefly by utilizing residual nutrients without active metabolism or colonization. Detection usually occurs through culture-dependent or molecular analyses following rigorous surface sterilization, though identifying them often raises the challenge of distinguishing true internal passengers from external contaminants. Because of their transient and noninteractive nature, studying their ecological relevance is difficult. Examples include ubiquitous environmental bacteria such as *Sphingomonas*, *Acinetobacter*, or various *Streptomyces* strains that may inadvertently enter plant tissues, as well as microbes present in irrigation water that are taken up into roots but fail to establish replicating populations.

5.3 HABITAT DIVERSITY OF ENDOPHYTIC BACTERIA

Endophytic bacteria occupy diverse plant-associated niches, including the rhizosphere, root endosphere, stems, leaves (phyllosphere), seeds, and even extreme habitats such as mangroves and saline soils. Many originate from soil or the phyllosphere and gain entry into plants through natural openings or disruptions, such as root emergence sites, wounds, or stomata. Once inside, they establish themselves within intercellular spaces and vascular tissues, potentially spreading systemically and interacting closely with host physiological processes (Ryan et al., 2008; Brader et al., 2017). Certain endophytes are vertically transmitted via seeds, creating a heritable seed microbiome that supports early seedling growth and influences plant performance and competitive dynamics in the emerging rhizosphere (Xun et al., 2021). In coastal and saline ecosystems, such as mangrove forests, endophytic communities are often enriched with halotolerant and stress-adapted lineages that enhance host salt tolerance, contribute to nutrient cycling, and improve resilience. Mangrove propagules and roots, in particular, act as reservoirs of beneficial microbes capable of promoting crop growth under salinity stress (Soldan et al., 2019). The composition of endophytic communities is shaped by environmental factors, host-specific traits, including species, organ type, and genotype, and the mode of microbial transmission. This results in organ-specific microbial assemblages, such as distinct communities in roots versus leaves, and context-dependent functional traits, including plant growth promotion, biocontrol activity, and alleviation of abiotic stresses (Bhattacharjee et al., 2023).

5.4 HOST RANGE AND SPECIFICITY OF ENDOPHYTIC BACTERIA

Endophytic bacteria exhibit a wide range of host–microbe interactions, from generalist strains that can colonize multiple plant species to specialist strains with strong host or tissue specificity. Generalist endophytes, including various *Pseudomonas*, *Bacillus*, and *Microbacterium* species, are commonly recovered from seeds, roots, and aerial tissues across diverse plant taxa, reflecting their ability to establish themselves in multiple hosts and tissue types. Their frequent presence in seed microbiomes further highlights their ecological adaptability and versatility (Ryan et al., 2008).

In contrast, several endophytes–plant associations exhibit strong host or tissue specificity. Certain bacteria are vertically transmitted through seeds and appear adapted to specific host genotypes; for example, *Microbacterium* and *Bacillus* species have been documented as seed-transmitted endophytes in switchgrass and wheatgrass, suggesting host-specific functional roles (Fadiji et al., 2025). Obligately specialized vertical transmission, however, remains uncommon. A notable exception is the symbiosis between *Burkholderia* species and leaf-nodulating plants in genera such as *Ardisia*, *Pavetta*, and *Psychotria*, where bacteria exhibit exceptionally high fidelity to their hosts (Frank et al., 2017). Host specificity is shaped by a combination of microbial traits such as motility, adhesion, biofilm formation, and tolerance to host immune responses and plant traits, including tissue structure, developmental stage, and genotype (Barret et al., 2016). Transmission mode also influences host specificity, with vertically transmitted endophytes generally exhibiting higher host fidelity, while horizontally acquired taxa from soil, air, pollen, or insects tend to behave as generalists (Fadiji et al., 2025; Frank et al., 2017). Seed microbiome research further highlights specificity, showing that bacterial communities differ not only among species but even among plant lines, shaped by both inherited microbial taxa and environmentally acquired members (Abdelfattah et al., 2023; Hu et al., 2025; Bintarti et al., 2022). Collectively, these patterns indicate that while a core set of endophytes is consistently transmitted, the broader seed microbiome is dynamic and strongly influenced by host identity and environmental context. Understanding the determinants of endophyte host range and specificity is essential for leveraging these microbes in agriculture and biotechnology: generalist strains may serve as broad-spectrum bioinoculants, whereas specialist strains can provide targeted benefits to particular crops or cultivars, especially when vertically inherited (Fadiji et al., 2025; Frank et al., 2017). Ongoing integration of genomics, microbiome profiling, and functional assays is essential for uncovering the molecular mechanisms driving these plant–microbe interactions and for designing targeted microbial strategies to support sustainable plant management. Endophytic microorganisms have been extensively shown to improve plant tolerance to drought, salinity, extreme temperatures, and other abiotic stresses across a wide range of host species (Table 5.1).

5.5 PLANT ABIOTIC STRESSES AND THEIR MECHANISMS

Plants are constantly exposed to a variety of environmental challenges that can significantly affect their growth, development, and productivity. Among these, abiotic stresses are considered the most critical threats to global agriculture (Bintarti et al., 2022). These stresses arise from nonliving factors such as drought, salinity, extreme temperatures (heat and cold), flooding, heavy metal toxicity, nutrient deficiencies or excesses, and UV radiation (Figure 5.1). Such stresses not only reduce crop yields but also jeopardize global food security, particularly in the context of a growing population and changing climate. Being sessile, plants cannot escape adverse conditions and have therefore evolved complex morphological, physiological, biochemical, and molecular strategies to survive and maintain growth under stress. The severity of these stresses is influenced by factors such as intensity, duration, plant developmental stage, and inherent tolerance of the species or cultivar. Common consequences of

TABLE 5.1
Representative Endophytic Organisms Reported to Confer Tolerance against Various Abiotic Stresses in Different Host Plants

Scientific Name	Endophytic Organism	Abiotic Stress	Reference
Arabidopsis thaliana	*Burkholderia phytofirmans*	Cold and salinity	Ledger et al. (2016)
Capsicum annuum	*Bacillus* sp.	Osmotic stress	Bello et al. (2022)
Zea mays	*Pantoea agglomerans*	Salinity	Gond et al. (2015)
Nicotina tabacum	*Pseudomonas* sp.	Osmotic stress	Begum et al. (2022)
Scutellaria baicalensis (Baikal skullcap)	*Brachybacterium endophyticum*	Salt stress	Tuo et al. (2018)
Festuca arizona	*Neotyphodium*	Drought	Morse et al. (2002)
Oryza sativa	*Curvularia protuberata*	Drought	Singh et al. (2025)
Cucumis sativus	*Thermomyces* sp.	Heat stress	Khan et al. (2024)

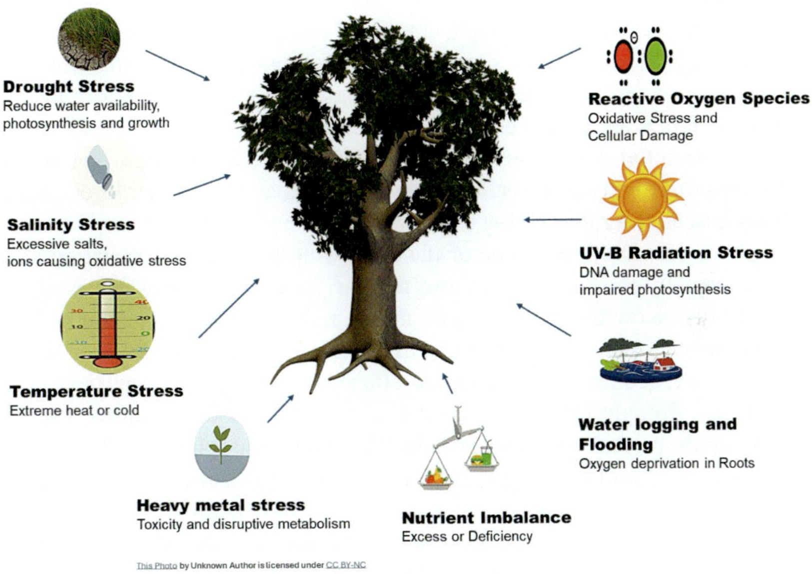

FIGURE 5.1 Abiotic stressors and their primary physiological impacts on plants.

abiotic stress include reduced photosynthesis, impaired nutrient and water uptake, accumulation of ROS, metabolic disruption, and, in extreme cases, plant death (Sachdev et al., 2021). Climate change has intensified these stresses, increasing the frequency and severity of extreme events such as prolonged droughts, elevated soil salinity, heatwaves, and erratic rainfall, thereby threatening global food security. To cope, plants adopt a range of physiological adaptations, such as reducing leaf surface area, modifying root architecture, and closing stomata to minimize water loss. They accumulate osmoprotectants like proline and soluble carbohydrates and activate antioxidant enzymes, including superoxide dismutase (SOD), catalase (CAT), and peroxidases, to neutralize ROS. Hormonal regulation, particularly involving abscisic acid (ABA) and other stress-related hormones, coordinates these responses. At the molecular level, transcription factors such as DREB (dehydration-responsive element binding), NAC, WRKY, MYB, and bZIP orchestrate the expression of stress-responsive genes, while signal transduction pathways involving nitric oxide (NO), ROS, MAPKs (mitogen-activated protein kinases), and calcium mediate stress perception and response (Thilakarathne et al., 2025). Epigenetic modifications, including DNA methylation, histone remodeling, microRNAs, and other noncoding RNAs, also contribute to stress memory and the regulation of adaptive responses (Sintaha, 2025). Advances in omics technologies—such as proteomics, metabolomics, transcriptomics, and genomics—have deepened our understanding of plant stress responses, and applications of transgenic technology, marker-assisted breeding, and CRISPR/Cas9 gene editing are being used to improve abiotic stress tolerance in crops (Villalobos-López et al., 2022). Beneficial microorganisms, including biostimulants and PGPR, provide environmentally sustainable strategies for enhancing stress tolerance. Despite these advancements, challenges remain. Most studies focus on single stress factors, whereas plants in natural environments often face multiple, simultaneous stresses, such as drought combined with heat. Understanding the interactions between different stress signaling pathways and translating this knowledge into crop improvement strategies remains critical. Abiotic stresses, acting individually or in combination, negatively impact plant growth, productivity, and survival, leading to substantial crop losses worldwide (Gull et al., 2019). Each type of stress affects plants differently at the cellular, physiological, and molecular levels, making their study essential for developing effective tolerance strategies.

Drought Stress: Drought is one of the most widespread and devastating abiotic stresses, especially in arid and semiarid regions. It occurs when water availability in the soil falls below the plant's requirements, reducing water potential and causing cellular dehydration. Physiologically, drought leads to decreased turgor pressure, stomatal closure, reduced photosynthetic efficiency, and impaired nutrient uptake. Biochemically, drought induces accumulation of ROS, causing oxidative stress and damage to lipids, proteins, and nucleic acids. To counter this, plants synthesize osmoprotectants such as proline, glycine betaine, soluble sugars, and polyols, which help maintain cellular water balance and protect cellular structures. Morphologically, plants reduce leaf area, increase root depth and density, and adjust root-to-shoot ratios to optimize water absorption. At the molecular level, drought activates ABA-mediated signaling pathways, triggering transcription factors like DREB, NAC, and MYB, which regulate the expression of stress-responsive proteins, including late embryogenesis abundant proteins, dehydrins, aquaporins, and protective enzymes. Signal transduction networks involving calcium ions, MAPKs, nitric oxide (NO), and ROS coordinate stress perception and downstream responses. Epigenetic regulation and small RNAs also contribute to stress memory and adaptive responses.

Salinity Stress: Salinity stress arises primarily from the accumulation of soluble salts, particularly sodium chloride, in the soil. High salt concentrations impose both osmotic stress, by reducing water uptake, and ionic stress, due to the accumulation of toxic Na^+ and Cl^- ions. Physiologically, salinity stress results in osmotic imbalance, reduced photosynthesis, leaf necrosis, and impaired nutrient acquisition. Plants mitigate ionic toxicity by compartmentalizing Na^+ into vacuoles via NHX antiporters, maintaining a high potassium-to-sodium ratio, and activating the salt overly sensitive (SOS) pathway to extrude excess sodium. Osmolytes such as proline, glycine betaine, and soluble sugars accumulate to maintain osmotic homeostasis. Antioxidant enzymes, including SOD, CAT, and ascorbate peroxidase (APX), reduce ROS-mediated oxidative damage. At the molecular level, salinity induces the expression of stress-responsive genes through ABA-dependent and independent pathways, activating transcription factors such as DREB, NAC, WRKY, and bZIP. Cross-talk with other signaling molecules, including calcium, ROS, and MAPKs, ensures coordinated responses under saline conditions.

Temperature Stress: Temperature extremes, both heat and cold, disrupt plant metabolism, growth, and reproduction. Heat stress causes protein denaturation, membrane fluidization, enzyme inactivation, increased transpiration, reduced pollen viability, and ROS accumulation. Plants respond by upregulating heat shock proteins (HSPs), which act as molecular chaperones to stabilize and refold proteins, and activating heat shock transcription factors that regulate downstream protective genes. Cold stress, on the other hand, leads to ice crystal formation, decreased membrane fluidity, impaired enzyme activity, and ROS generation. Plants enhance cold tolerance through accumulation of soluble sugars and compatible solutes, production of antifreeze proteins, modification of membrane lipid composition, and activation of the CBF (C-repeat binding factor) pathway, which induces cold-responsive genes. Both heat and cold stress responses involve cross-talk with hormonal pathways, particularly ABA, and ROS-mediated signaling for stress adaptation.

Waterlogging and Flooding: Excess water around roots causes oxygen depletion (hypoxia or anoxia), limiting aerobic respiration and nutrient uptake. Plants shift to anaerobic metabolism, producing energy through fermentation pathways mediated by enzymes such as pyruvate decarboxylase and alcohol dehydrogenase. Morphological adaptations include the development of aerenchyma (air-filled tissues) to facilitate internal oxygen transport and adventitious roots to maintain nutrient and water uptake. Ethylene signaling plays a crucial role by triggering programmed cell death in submerged tissues and promoting the growth of new roots. Physiological adjustments also include increased carbohydrate mobilization and anaerobic metabolic enzyme regulation to sustain energy production during prolonged flooding.

Heavy Metal Stress: Exposure to toxic metals such as cadmium, lead, mercury, and arsenic impairs photosynthesis, respiration, and nutrient uptake. Metals disrupt membrane integrity, induce ROS generation, and inactivate key enzymes. Plants counter heavy metal stress by synthesizing metal-chelating peptides, including metallothioneins and phytochelatins, which sequester metals into vacuoles, reducing cytoplasmic toxicity. Antioxidant systems, including SOD, CAT, and peroxidases, detoxify ROS, while transporters such as HMA (heavy metal ATPases) regulate metal movement and accumulation. Stress signaling pathways involving calcium, ROS, and MAPKs coordinate gene expression and defense mechanisms under heavy metal stress.

UV-B Radiation Stress: UV-B radiation causes DNA damage, protein degradation, and disruption of photosynthetic pigments. Plants sense UV-B through the UVR8 photoreceptor, which activates protective genes. Flavonoids and phenolics accumulate in epidermal tissues, serving as natural sunscreens, while antioxidant enzymes mitigate ROS damage. DNA repair mechanisms, including photolyases and nucleotide excision repair pathways, restore UV-induced genetic damage. Cross-talk with hormonal and ROS-mediated signaling ensures coordinated protective responses.

Nutrient Imbalance: Deficiency or excess of essential nutrients disrupts metabolic processes and limits growth. Nutrient deficiency can cause chlorosis, necrosis, stunted growth, and reduced enzymatic activity, whereas excess nutrients can lead to oxidative stress and ionic imbalance. Plants adapt through root architecture modifications, upregulation of nutrient transporters, and activation of stress-responsive genes regulated by transcription factors and signaling networks. Osmotic adjustments, antioxidant defenses, and metabolic reprogramming help maintain cellular homeostasis under nutrient stress.

5.6 FUNCTIONAL TRAITS OF ENDOPHYTIC BACTERIA IN MITIGATING ABIOTIC STRESS

Abiotic stresses, such as drought, salinity, nutrient deficiencies, extreme temperatures, and heavy metal toxicity, can induce significant physiological and morphological changes in plants. These effects include stunted growth, premature senescence, purpling of leaves due to phosphorus deficiency, altered root exudation patterns including the release of organic acids, protons, and phenolics to enhance nutrient availability and chlorosis resulting from nitrogen, magnesium, or iron shortages. Endophytic bacteria residing within plant tissues have emerged as vital allies in mitigating these stresses by producing growth-promoting compounds, enhancing antioxidant defenses, and regulating complex signaling pathways that support plant adaptation and resilience.

5.6.1 Production of Plant Growth–Promoting Substances

A wide range of compounds that promote plant growth are produced by endophytic bacteria, including phytohormones, which are essential for boosting plant resistance and growth in harsh conditions.

Indole-3-acetic acid (IAA): Many endophytes produce IAA, an auxin that regulates cell elongation, cell division, root initiation, and vascular differentiation. IAA-producing endophytes such as *Enterobacter cloacae*, *Azospirillum brasilense*, and *Bacillus subtilis* significantly improve root system architecture by stimulating lateral root formation and increasing root hair density, thereby enhancing water and nutrient absorption under drought, salinity, and heavy metal stress conditions.

Gibberellins: Endophytes that generate gibberellins, such as *Rhizobium* sp. and *Acinetobacter calcoaceticus*, encourage stem elongation, seed germination, and blooming. In difficult conditions, GA-producing endophytes guarantee increased vigor and shoot growth in crops like rice and wheat by preventing stress-induced stunting.

Cytokinins (CKs): CK-producing endophytes, such as *Pseudomonas fluorescens* and *Bacillus amyloliquefaciens*, prolong photosynthetic activity, maintain chlorophyll content, and postpone senescence in stressed plants. In drought and nutrient-deficient soils, CK-producing bacteria help crops maintain more production by preserving nutrient mobilization and postponing leaf ageing.

5.6.2 Nutrient Mobilization

Endophytic bacteria mobilize essential elements that are scarce in the soil, acting as organic biofertilizers.

Phosphate Solubilization: Endophytes like *Burkholderia cepacia*, *Bacillus megaterium*, and *Pseudomonas putida* release phosphatases and organic acids (such citric and gluconic) that break down inorganic phosphate into forms that plants can use. Root development, ATP (adenosine triphosphate) production, and general metabolic activity are all supported by increased phosphorus intake and are essential in saline and phosphorus-deficient soils.

Nitrogen Fixation: Nitrogenase enzyme complexes found in a number of diazotrophic endophytes, such as *Azoarcus* species, *Gluconacetobacter diazotrophicus*, and *Herbaspirillum seropedicae*, transform atmospheric nitrogen into ammonia. This mechanism improves photosynthetic efficiency, boosts chlorophyll synthesis, and lessens reliance on chemical nitrogen fertilizers especially in drought-prone and nutrient-poor soils.

5.6.3 Antioxidative Enzyme Production

Abiotic stresses such as extreme temperatures, salinity, dehydration, and heavy metal toxicity often lead to the excessive accumulation of ROS, including superoxide radicals, hydrogen peroxide, and hydroxyl radicals. High levels of ROS cause oxidative stress, damaging proteins, lipids, and nucleic acids, although at low concentrations, ROS can act as signaling molecules. Endophytic bacteria help plants mitigate oxidative damage by enhancing their antioxidant defense systems. For example, *Bacillus cereus* and *P. fluorescens* produce SOD, which converts superoxide radicals into hydrogen peroxide. CAT produced by strains such as *E. cloacae* further detoxifies hydrogen peroxide into water and oxygen. Additionally, endophytes like *Azospirillum* species and *Burkholderia phytofirmans* PsJN induce the activities of APX and glutathione reductase, maintaining redox homeostasis and protecting key metabolic processes during stress.

5.6.4 Quorum Sensing (QS) and Signaling Molecules

QS and signaling molecules are central to the stress-mitigation strategies of endophytic bacteria. Through QS, these microbes detect their population density and modulate gene expression, ensuring that stress-response pathways are activated at appropriate times. In Gram-negative bacteria, N-acyl homoserine lactones (AHLs) serve as primary QS mediators; for example, *P. putida* and *Burkholderia vietnamiensis* use AHLs to coordinate siderophore production and biofilm formation. In Gram-positive bacteria, small peptides act as signaling molecules, as observed in *B. subtilis*, which regulates sporulation and enhances stress tolerance. These communication systems control multiple beneficial traits, including biofilm development, siderophore-mediated iron acquisition, and the production of protective enzymes. Beyond bacterial self-regulation, QS molecules can interact directly with plants, modulating gene expression, shaping root system architecture to improve water and nutrient uptake, and activating induced systemic resistance (ISR) to strengthen plant defenses against both abiotic and biotic stresses. Another important bacterial adaptation involves the production of exopolysaccharides (EPS), which form a hydrated matrix around bacterial cells. EPS improves bacterial survival under challenging conditions such as drought or high salinity by retaining moisture, stabilizing soil structure, and protecting cells from desiccation. Furthermore, EPS can immobilize toxic heavy metals in the rhizosphere, lowering their bioavailability and mitigating their detrimental effects on both microbial communities and host plant tissues.

Endophytic bacteria also emit volatile organic compounds (VOCs) that act as airborne signals, influencing plant physiology and stress responses even without direct contact. VOCs can induce antioxidant pathways, regulate stomatal behavior to optimize water use during drought, and stimulate the expression of defense-related genes. Collectively, the coordinated actions of QS-regulated mechanisms, EPS-mediated protection, and VOC-triggered plant responses create a synergistic network that enhances plant growth, improves tolerance to environmental stresses, and supports long-term ecological stability.

5.7 APPLICATION METHODS OF ENDOPHYTIC BACTERIA FOR ABIOTIC STRESS MITIGATION IN PLANTS

Endophytic microorganisms inhabit a wide range of plant tissues, including seeds, roots, stems, and leaves. They enhance plant growth and stress resilience by improving nutrient uptake, promoting nitrogen fixation, and synthesizing phytohormones. During colonization, bacterial cells secrete exopolysaccharides, which provide protection against oxidative stress. Arbuscular mycorrhizal fungi (AMF) similarly improve nutrient acquisition and bolster tolerance to abiotic stresses. AMF–plant symbioses are critical for plant development, with extensive mycelial networks surrounding roots to facilitate efficient nutrient absorption (Jajoo & Mathur, 2021). Common mycorrhizal networks further optimize the distribution of nitrogen and phosphorus, supporting plant performance under challenging conditions. These plant–microbe interactions involve intricate biochemical and molecular mechanisms that alleviate the adverse effects of drought, salinity, heavy metal toxicity, and nutrient deficiency, often mediated through phytohormones such as gibberellins (GA), auxins, and cytokinins (CKs).

5.7.1 Anticipated Remedial Tools

Effective microorganisms (EM) and related microbial technologies are increasingly valued for sustainable farming due to their role in improving soil fertility and ecological stability. Cyanobacteria (blue-green algae) produce extracellular polysaccharides (EPS) that improve soil aggregation, enhance microbial colonization, and act as renewable biomass inputs. EPS secretion supports biofilm formation and nutrient retention, while also reducing metal toxicity in contaminated soils (Ameen et al., 2023). Beyond this, cyanobacteria fix atmospheric nitrogen and release bioactive metabolites, thereby stimulating plant growth while suppressing pathogens (Singh et al., 2025). Soil microbial biomass (MB) serves as a key fertility indicator, yet land-use changes often reduce MB, impairing ecosystem resilience (Dey et al., 2025). PGPR counter these losses by producing phytohormones, osmoprotectants, and ACC (acetyl-CoA carboxylase) deaminase, which alleviate stress in saline and nutrient-poor soils. Modern strategies such as PGPR-based seed priming or biofilmed biofertilizers (BFBFs) improve microbial persistence, enhance nutrient-use efficiency, and reduce disease incidence (Fiodor et al., 2023). Reintroducing AMF and beneficial PGPR to degraded soils further restores ecosystem function and promotes plant performance, highlighting the promise of microbial tools in climate-resilient agriculture.

5.7.2 Biofertilizers

Biofertilizers offer an environmentally friendly alternative to chemical fertilizers and are increasingly valued for their role in sustainable agriculture. They consist of living microorganisms, including PGPR, AMF, cyanobacteria, and nitrogen-fixing bacteria, which enhance nutrient availability and plant growth by colonizing the rhizosphere or internal plant tissues. Their use improves soil fertility, boosts crop productivity, and reduces reliance on synthetic fertilizers, contributing to environmental sustainability (Rawat et al., 2025). Biofertilizers act through both direct and indirect mechanisms. Directly, they fix atmospheric nitrogen (*Rhizobium*, *Azotobacter*, *Azospirillum*), solubilize phosphorus (*Bacillus* and *Pseudomonas* spp.), mobilize potassium, and synthesize phytohormones such as gibberellins, cytokinins, and IAA). Cyanobacteria and AMF further improve soil organic matter content and enhance water-use efficiency, thereby increasing plant resilience under drought and salinity. Indirectly, biofertilizers suppress soilborne pathogens through mechanisms such as siderophore production, antibiosis, competition for nutrients and space, and induction of systemic resistance (ISR).

Recent innovations include BFBFs, in which beneficial microbes are embedded within protective biofilms to enhance survival, colonization, and effectiveness under field conditions. Research shows that BFBFs outperform free-living microbial inoculants, particularly in degraded or stress-prone soils. Moreover, combining multiple microbial species into consortia has been reported to increase crop yields by up to 32% compared to single-strain applications. Beyond promoting plant growth, biofertilizers improve soil physicochemical properties, such as aggregate stability, nutrient cycling, and organic matter content, thereby supporting long-term soil fertility. They also contribute to reducing greenhouse gas emissions by lowering the need for nitrogenous fertilizers, which are significant sources of nitrous oxide (Bhardwaj et al., 2023). However, their effectiveness depends on crop type, soil fertility, and environmental conditions, emphasizing the need for tailored microbial consortia and locally adapted strains.

5.7.3 Microbial Plant Biostimulants

Microbial biostimulants have emerged as an effective approach to enhance plant performance, particularly under climate-induced stresses such as salinity, drought, and heat. Derived primarily from PGPR, AMF, and cyanobacteria, these biostimulants improve photosynthesis, nutrient uptake, and overall growth through the production of phytohormones, siderophores, and organic acids (Ameen et al., 2023). They also regulate osmolyte accumulation, enhance ROS scavenging, and stimulate defense pathways, thereby enhancing plant tolerance to abiotic stresses (Fadiji et al., 2022). Cyanobacteria-based biostimulants further contribute to soil fertility by enhancing soil structure through EPS secretion, while producing bioactive compounds that increase crop resilience. Unlike chemical fertilizers and pesticides, biostimulants focus on enhancing plant metabolic efficiency and resistance significantly than direct nutrition, making them highly compatible with integrated crop management (Singh et al., 2025). Advances in strain selection, microbial formulations, and delivery systems are helping overcome variability in field performance, although success depends on crop species, soil microbiome interactions, and application timing. As climate change reduces arable land, microbial biostimulants are presumed to play a critical role in maintaining productivity and food security.

5.7.4 Synthetic Microbial Communities (SynComs)

SynComs represent a frontier in agricultural biotechnology, designed through synthetic biology, metagenomics, and bioinformatics to improve crop productivity. Unlike natural communities, SynComs are simplified, defined consortia with complementary functions, such as nutrient cycling, hormone regulation, and pathogen suppression (Munir et al., 2022). For example, rationally assembled SynComs from beneficial PGPR and AMF have demonstrated improved yield and stress resilience in controlled greenhouse conditions (Martins et al., 2024). Similarly, field trials show SynCom application enhances nutrient-use efficiency and disease endurance in cereals and legumes. The design process typically includes identifying keystone taxa from native microbiomes, testing interactions under controlled conditions, and validating their stability in soils. Improvements in high-throughput sequencing and modeling are enabling the creation of crop-specific SynComs tailored to indigenous soil and climatical conditions. However, challenges remain in ensuring stability and persistence in heterogeneous field environments. Integration with plant breeding for exudate profiles that favor SynCom colonization is a promising avenue. Thus, SynComs offer a reproducible, predictable, and scalable solution for harnessing beneficial microbiomes in sustainable agriculture.

5.7.5 Biochar–Microbe Synergy

Biochar (BC), produced by pyrolysis of organic biomass under low-oxygen conditions, enhances soil fertility by increasing pH, porosity, and nutrient retention. Its porous structure provides ideal habitats for beneficial microbes, promoting microbial colonization and activity. When combined with PGPR, AMF, or EM, BC improves nutrient-use efficiency, enhances soil fertility, and increases crop yield more effectively than individual applications (Kizito et al., 2019). BC–microbe consortia have been shown to improve soil aggregation, water retention, and reduce greenhouse gas emissions, particularly in degraded soils (Majumder & Biswas, 2024). Methylotrophic bacteria in BC–microbe systems produce plant hormones such as auxins and cytokinins while suppressing pathogens through antagonistic metabolites. Additionally, co-composting BC with organic fertilizers further enhances nutrient availability and microbial stability (Phares et al., 2022). Recent field studies indicate that BC application with microbial inoculants can significantly increase yield and quality of cereals and vegetables. Overall, the BC–microbe synergy represents a multifunctional, eco-friendly strategy to enhance soil fertility, carbon sequestration, and sustainable crop production.

5.7.6 Industrial Applications

Beyond agriculture, microbial biotechnology has vast industrial applications that contribute to environmental sustainability. PGPR-based products improve plant resilience to abiotic stress, while microbial bioremediation reduces the bioavailability of toxic metals, preventing their accumulation in food crops (Khatoon et al., 2020). In renewable energy, thermophilic microbes are increasingly employed in lignocellulosic bioethanol production through consolidated bioprocessing, integrating enzymatic hydrolysis and fermentation. Engineered strains of yeast and bacteria capable of fermenting pentoses and hexoses are also improving bioethanol yields. Additionally, microbial processes are central to waste-to-energy technologies, biogas production, and recovery of industrial by-products. Phytoremediation assisted by microbes further supports soil detoxification and reclamation of degraded lands, thereby linking industrial and environmental sustainability. Collectively, microbial applications extend from crop productivity to renewable energy and environmental remediation, offering integrated solutions to global challenges such as climate change, energy demand, and food security.

5.7.7 Customized Plant Microbiome Engineering

Customized plant microbiome engineering focuses on manipulating root-associated microbial communities to improve yield, resilience, and sustainability. Strategies include altering root exudates via plant breeding or genetic engineering to recruit beneficial microbes, as well as direct inoculation with tailored microbial consortia (Du et al., 2021). For instance, inoculation of tomato with a PGPR consortium increased yield by 30%, highlighting the potential of multi-strain approaches. Soybean inoculation with *Bradyrhizobium japonicum* and *B. subtilis* improved nodulation, biomass, and stress tolerance. Such microbial consortia enhance nutrient uptake, hormone production, and defense signaling, thereby supporting plant growth under stress (Bartucca et al., 2022). Advances in "microbiome-aware breeding" now focus on selecting crop varieties with root traits that encourage beneficial microbial colonization (Du et al., 2021). Integration of omics approaches, such as metabolomics and metagenomics, further allows precise design of microbiome interventions

(Bartucca et al., 2022). Customized plant microbiome engineering thus represents a frontier in sustainable agriculture, combining host genetics, microbial inoculants, and precision biotechnology to maximize crop productivity in variable environments.

5.8 CHALLENGES AND LIMITATIONS

Endophytic bacteria show great promise in enhancing plant resilience, yet several challenges limit their agricultural application. Host specificity restricts broad use, as compatibility between host and bacterium determines success; for example, *B. phytofirmans's strain* PsJN improves cold tolerance in grapevines but not in cereals or legumes. Environmental variability further complicates outcomes since benefits seen in controlled studies often diminish in heterogeneous soils, fluctuating climates, and diverse native microbiota (Sharma et al., 2025). A major hurdle is inconsistent colonization, and stability colonization success depends on plant stage, environment, and strain mechanisms, yet persistence across the plant life cycle often fails. Moreover, plants host complex microbiomes, and the introduction of exogenous strains may disrupt beneficial microbes, leading to microbial dysbiosis. Another challenge is the limited mechanistic understanding of how endophytes function through hormone modulation, antioxidant enhancement, or nutrient uptake, making strain selection and optimization difficult (Kumar et al., 2022). Additionally, variability in stress alleviation occurs, as strains effective against drought in one host may not confer salinity tolerance in another. Beyond biology, regulatory and safety concerns slow adoption due to risks of horizontal gene transfer, nontarget effects, and environmental persistence. Commercialization hurdles also remain: ensuring viable formulations, affordable delivery, and stable large-scale production is complex and costly. Lastly, the long-term ecological impact of introduced endophytes on biodiversity and microbial balance remains poorly understood, necessitating rigorous risk assessments.

5.9 FUTURE PERSPECTIVE

Research on endophytic bacteria as natural partners in helping plants cope with abiotic stresses is still at an early stage, yet the prospects are highly promising. With climate change driving more frequent droughts, rising salinity, temperature fluctuations, and nutrient deficiencies, these microbes are poised to become vital tools for sustainable agriculture, reducing dependence on chemical inputs while supporting healthier and more resilient crops. Future work should emphasize the use of integrative omics approaches—genomics, transcriptomics, proteomics, and metabolomics—to unravel the molecular mechanisms of endophyte–plant interactions and identify genes, pathways, and metabolites linked to stress tolerance. The development of SynComs, designed from complementary microbial taxa, offers a reliable alternative to single-strain inoculants and may provide consistent benefits across diverse crops and environments. Microbiome engineering also holds significant potential, whether through selective crop breeding for root exudates that recruit beneficial microbes or direct inoculation with engineered strains, enabling plants to establish built-in microbial resilience. Translating laboratory findings to field applications will require scaling up biofertilizer and biostimulant technologies, with advanced formulations such as encapsulation, biofilm-based carriers, or nano-delivery systems improving microbial survival and persistence under real-world conditions. Tailoring microbial consortia to regional soils and cropping systems will further enhance their effectiveness. At the same time, biosafety and ecological risk assessments are essential to prevent unintended consequences such as disruption of native microbiomes or horizontal gene transfer. Beyond agriculture, the multifunctional potential of endophytic bacteria extends to phytoremediation of polluted soils, bioenergy generation, and carbon sequestration, positioning them as valuable allies not only in food security but also in broader environmental and industrial sustainability.

5.10 CONCLUSION

Abiotic stresses, including drought, salinity, extreme temperatures, nutrient deficiencies, oxidative damage, and heavy metal toxicity, pose significant challenges to agriculture worldwide, threatening crop yields and food security. Traditional interventions, such as chemical fertilizers and conventional breeding for stress tolerance, often face limitations in cost, sustainability, and effectiveness. Endophytic bacteria, which reside harmlessly within plant tissues, offer a sustainable alternative by enhancing plant resilience and growth under adverse conditions. These microbes improve drought tolerance by regulating water uptake, stimulating osmolyte synthesis, and optimizing root development. They help plants withstand salinity through ACC deaminase–mediated ethylene reduction and sodium ion sequestration, while mitigating heavy metal toxicity via siderophore production and metal ion immobilization. Thermal stress is alleviated through the induction of HSPs and antioxidant enzymes, whereas cold tolerance is supported by antifreeze proteins and activation of cold-responsive genes. Endophytes also enhance nutrient availability through nitrogen fixation, phosphate solubilization, and mineral mobilization, supporting plant growth in nutrient-limited soils. Beyond stress mitigation, they contribute to sustainable agriculture through applications as biofertilizers, biostimulants, microbial consortia, and microbiome engineering strategies, improving soil health, nutrient efficiency, and crop productivity while minimizing environmental impact. Their use extends to industrial purposes, including phytoremediation, bioenergy generation, and ecological restoration. Despite challenges such as host specificity, inconsistent colonization, environmental variability, and regulatory constraints, advances in omics, synthetic biology, and microbiome engineering offer opportunities to optimize endophyte performance in the field. Overall, endophytic bacteria hold immense potential as eco-friendly bioinoculants, enabling climate-resilient, high-yield, and environmentally sustainable farming systems, thereby supporting global food security and ecological balance.

REFERENCES

Abdelfattah, A., Tack, A. J. M., Lobato, C., Wassermann, B., & Berg, G. (2023). From seed to seed: The role of microbial inheritance in the assembly of the plant microbiome. *Trends in Microbiology*, *31*(4), 346–355. https://doi.org/10.1016/j.tim.2022.10.009

Ameen, F., Mumtaz, S., Ali, B., Hussain, I., Hafeez, A., Gul, A., ... & Ali, S. (2023). The impact of Cu-polluted and organic soil on the fibrous plant; insights into plant growth promotion, antioxidant defences system, and oxidative stress. *Functional Plant Biology*, *50*(8), 623–632.

Barret, M., Guimbaud, J. F., Darrasse, A., & Jacques, M. A. (2016). Plant microbiota affects seed transmission of phytopathogenic microorganisms. *Molecular Plant Pathology*, *17*(6), 791. https://doi.org/10.1111/MPP.12382

Bartucca, M. L., Cerri, M., Del Buono, D., & Forni, C. (2022). Use of biostimulants as a new approach for the improvement of phytoremediation performance—A review. *Plants*, *11*(15), 1946.

Begum, N., Wang, L., Ahmad, H., Akhtar, K., Roy, R., Khan, M. I., & Zhao, T. (2022). Co-inoculation of arbuscular mycorrhizal fungi and the plant growth-promoting rhizobacteria improve growth and photosynthesis in tobacco under drought stress by up-regulating antioxidant and mineral nutrition metabolism. *Microbial Ecology*, *83*(4), 971–988. https://doi.org/10.1007/s00248-021-01815-7

Bello, A. S., Ben-Hamadou, R., Hamdi, H., Saadaoui, I., & Ahmed, T. (2022). Application of cyanobacteria (*Roholtiella* sp.) liquid extract for the alleviation of salt stress in bell pepper (*Capsicum annuum* L.) plants grown in a soilless system. *Plants*, *11*(1), 104. https://doi.org/10.3390/PLANTS11010104

Bhardwaj, A. K., Malik, K., Rani, M., Mandal, U. K., Basak, N., Singh, A., Yadav, R. K., Chaudhari, S. K., & Sharma, D. K. (2023). Residue recycling options and their implications for sustainable nitrogen management in rice–wheat agroecosystems. *Ecological Processes*, *12*(1), 1–19. https://doi.org/10.1186/s13717-023-00464-7

Bhattacharjee, A., Dubey, S., & Sharma, S. (2023). "Next-generation bioformulations" for plant growth promotion and stress mitigation: A promising approach for sustainable agriculture. *Journal of Plant Growth Regulation*, *42*(10), 6741–6759. https://doi.org/10.1007/s00344-023-10996-z

Bintarti, A. F., Sulesky-Grieb, A., Stopnisek, N., & Shade, A. (2022). Endophytic microbiome variation among single plant seeds. *Phytobiomes Journal*, *6*(1), 45–55. https://doi.org/10.1094/PBIOMES-04-21-0030-R

Brader, G., Compant, S., Vescio, K., Mitter, B., Trognitz, F., Ma, L. J., & Sessitsch, A. (2017). Ecology and genomic insights into plant-pathogenic and plant-nonpathogenic endophytes. *Annual Review of Phytopathology*, *55*, 61–83. https://doi.org/10.1146/annurev-phyto-080516-035641

Compant, S., Cambon, M. C., Vacher, C., Mitter, B., Samad, A., & Sessitsch, A. (2021). The plant endosphere world – bacterial life within plants. *Environmental Microbiology*, *23*(4), 1812–1829. https://doi.org/10.1111/1462-2920.15240

Dey, S., Dhar, A., Jana, S., Tripathi, S., Barua, S., Bera, M., Sarkar, S., Garai, S., & Moulick, D. (2025). Biofertilizer and remediation of soil and other environments. In *Sustainable Remediation for Pollution and Climate Resilience*, pp. 535–605. https://doi.org/10.1007/978-981-96-5674-5_20

Du, J. X., Li, Y., Ur-Rehman, S., Mukhtar, I., Yin, Z., Dong, H., Wang, H., Zhang, X., Gao, Z., Zhao, X., Xin, X., & Ding, X. (2021). Synergistically promoting plant health by harnessing synthetic microbial communities and prebiotics. *iScience*, *24*(8). https://doi.org/10.1016/j.isci.2021.102918

Dudeja, S. S., Giri, R., Saini, R., Suneja-Madan, P., & Kothe, E. (2012). Interaction of endophytic microbes with legumes. *Journal of Basic Microbiology*, *52*(3), 248–260. https://doi.org/10.1002/JOBM.201100063

Eid, A. M., Salim, S. S., Hassan, S. E. D., Ismail, M. A., & Fouda, A. (2019). Role of endophytes in plant health and abiotic stress management. In *Microbiome in Plant Health and Disease: Challenges and Opportunities*, pp. 119–144. https://doi.org/10.1007/978-981-13-8495-0_6

Fadiji, A. E., Lanrewaju, A. A., Omomowo, I. O., Parra-Cota, F. I., & Santos-Villalobos, S. de los. (2025). Harnessing seed endophytic microbiomes: A hidden treasure for enhancing sustainable agriculture. *Plants*, *14*(15), 2421. https://doi.org/10.3390/PLANTS14152421

Fadiji, A. E., Orozco-Mosqueda, M. del C., Santos-Villalobos, S. de los, Santoyo, G., & Babalola, O. O. (2022). Recent developments in the application of plant growth-promoting drought adaptive rhizobacteria for drought mitigation. *Plants*, *11*(22), 3090. https://doi.org/10.3390/PLANTS11223090

Fiodor, A., Pandey, A. K., Rana, A., Pranaw, K., Blasi, B., Ajijah, N., Fiodor, A., Kumar Pandey, A., Rana, A., & Pranaw, K. (2023). Plant Growth-Promoting Bacteria (PGPB) with biofilm-forming ability: A multifaceted agent for sustainable agriculture. *Diversity*, *15*(1), 112. https://doi.org/10.3390/D15010112

Frank, A. C., Guzmán, J. P. S., & Shay, J. E. (2017). Transmission of bacterial endophytes. *Microorganisms*, *5*(4), 70. https://doi.org/10.3390/MICROORGANISMS5040070

Getahun, A., Muleta, D., Assefa, F., & Kiros, S. (2019). Field application of rhizobial inoculants in enhancing faba bean production in acidic soils: An innovative strategy to improve crop productivity. In *Salt Stress, Microbes, and Plant Interactions: Causes and Solution: Volume 1*, pp. 147–180. https://doi.org/10.1007/978-981-13-8801-9_7

Gond, S. K., Torres, M. S., Bergen, M. S., Helsel, Z., & White, J. F. (2015). Induction of salt tolerance and up-regulation of aquaporin genes in tropical corn by rhizobacterium *Pantoea agglomerans*. *Letters in Applied Microbiology*, *60*(4), 392–399. https://doi.org/10.1111/LAM.12385

Gull, A., Lone, A. A., Wani, N. U. I., Gull, A., Lone, A. A., & Wani, N. U. I. (2019). Biotic and abiotic stresses in plants. In *Abiotic and Biotic Stress in Plants*. https://doi.org/10.5772/INTECHOPEN.85832

Hasanuzzaman, M., Parvin, K., Bardhan, K., Nahar, K., Anee, T. I., Masud, A. A. C., & Fotopoulos, V. (2021). Biostimulants for the regulation of reactive oxygen species metabolism in plants under abiotic stress. *Cells*, *10*(10), 2537. https://doi.org/10.3390/CELLS10102537

Hu, H., Geng, S., Zhu, Y., He, X., Pan, X., & Yang, M. (2025). Seed-borne endophytes and their host effects. *Microorganisms*, *13*(4), 842. https://doi.org/10.3390/MICROORGANISMS13040842

Jajoo, A., & Mathur, S. (2021). Role of arbuscular mycorrhizal fungi as an underground saviuor for protecting plants from abiotic stresses. *Physiology and Molecular Biology of Plants*, *27*(11), 2589–2603.

Kandel, S. L., Joubert, P. M., & Doty, S. L. (2017). Bacterial endophyte colonization and distribution within plants. *Microorganisms*, *5*(4), 77. https://doi.org/10.3390/MICROORGANISMS5040077

Khalid, S. J., Ain, Q., Khan, S. J., Jalil, A., Siddiqui, M. F., Ahmad, T., Badshah, M., & Adnan, F. (2021). Targeting Acyl Homoserine Lactones (AHLs) by the quorum quenching bacterial strains to control biofilm formation in Pseudomonas aeruginosa. *Saudi Journal of Biological Sciences*, *29*(3), 1673. https://doi.org/10.1016/J.SJBS.2021.10.064

Khan, S., Irshad, S., Mehmood, K., Hasnain, Z., Nawaz, M., Rais, A., Gul, S., Wahid, M. A., Hashem, A., Abd_Allah, E. F., & Ibrar, D. (2024). Biochar production and characteristics, its impacts on soil health, crop production, and yield enhancement: A review. *Plants*, *13*(2), 166. https://doi.org/10.3390/PLANTS13020166

Khatoon, A., Rehman, S. U., Aslam, M. M., Jamil, M., & Komatsu, S. (2020). Plant-derived smoke affects biochemical mechanism on plant growth and seed germination. *International Journal of Molecular Sciences*, *21*(20), 7760.

Kizito, S., Luo, H., Lu, J., Bah, H., Dong, R., & Wu, S. (2019). Role of nutrient-enriched biochar as a soil amendment during maize growth: Exploring practical alternatives to recycle agricultural residuals and to reduce chemical fertilizer demand. *Sustainability*, *11*(11), 3211.

Kumar, A., Bhattacharya, T., Mukherjee, S., & Sarkar, B. (2022). A perspective on biochar for repairing damages in the soil–plant system caused by climate change-driven extreme weather events. *Biochar*, *4*(1), 1–23. https://doi.org/10.1007/s42773-022-00148-zKuźniar, A., Kruczyńska, A., Włodarczyk, K., Vangronsveld, J., & Wolińska, A. (2025). Endophytes as permanent or temporal inhabitants of different ecological niches in sustainable agriculture. *Applied Sciences*, *15*(3), 1253. https://doi.org/10.3390/APP15031253

Ledger, T., Rojas, S., Timmermann, T., Pinedo, I., Poupin, M. J., Garrido, T., Richter, P., Tamayo, J., & Donoso, R. (2016). Volatile-mediated effects predominate in *Paraburkholderia phytofirmans* growth promotion and salt stress tolerance of Arabidopsis thaliana. *Frontiers in Microbiology*, *7*(NOV), 196962. https://doi.org/10.3389/fmicb.2016.01838Malik, A., Suthar, M., Jain, K., Sharma, N., Dhabhai, A., Meena, M., & Marwal, A. (2025). Evolutionary and ecological factors influencing endophyte interactions. *Microorganisms for Sustainability*, *53*, 145–160. https://doi.org/10.1007/978-981-96-4100-0_7

Martins, S., Pereira, S., Dinis, L. T., & Brito, C. (2024). Enhancing olive cultivation resilience: Sustainable long-term and short-term adaptation strategies to alleviate climate change impacts. *Horticulturae*, *10*(10), 1066. https://doi.org/10.3390/HORTICULTURAE10101066

Meena, M., Zehra, A., Swapnil, P., & Seth, C. S. (Eds.). (2025). *Metabolic Sustainability of Endophytes: Volume 52*. https://doi.org/10.1007/978-981-96-4004-1

Morse, L. J., Day, T. A., & Faeth, S. H. (2002). Effect of *Neotyphodium* endophyte infection on growth and leaf gas exchange of Arizona fescue under contrasting water availability regimes. *Environmental and Experimental Botany*, *48*(3), 257–268. https://doi.org/10.1016/S0098-8472(02)00042-4

Munir, S., Li, Y., He, P., He, P., He, P., Cui, W., ... & He, Y. (2022). Defeating Huanglongbing pathogen Candidatus Liberibacter asiaticus with indigenous citrus endophyte *Bacillus subtilis* L1-21. *Frontiers in Plant Science*, *12*, 789065.

Okoroafor, U. E. (2022). Endophytic microbes: Potential tools for sustainable crop production. *Nigeria Agricultural Journal*, *53*(3), 109–114. https://www.ajol.info/index.php/naj/article/view/243438

Oshunsanya, S. O., Nwosu, N. J., & Li, Y. (2019). Abiotic stress in agricultural crops under climatic conditions. In *Sustainable Agriculture, Forest and Environmental Management*, pp. 71–100. https://doi.org/10.1007/978-981-13-6830-1_3.

Pandey, P. K., Singh, S., Singh, M. C., Singh, A. K., Yadav, S. K., Pandey, A. K., & Heisnam, P. (2018). *Diversity, Ecology, and Conservation of Fungal and Bacterial Endophytes*, pp. 393–430. https://doi.org/10.1007/978-3-319-96971-8_15

Passari, A. K., Mishra, V. K., Singh, G., Singh, P., Kumar, B., Gupta, V. K., Sharma, R. K., Saikia, R., Donovan, A. O., & Singh, B. P. (2017). Insights into the functionality of endophytic actinobacteria with a focus on their biosynthetic potential and secondary metabolites production. *Scientific Reports*, *7*(1), 1–17https://doi.org/10.1038/s41598-017-12235-4Phares, C. A., Amoakwah, E., Danquah, A., Afrifa, A., Beyaw, L. R., & Frimpong, K. A. (2022). Biochar and NPK fertilizer co-applied with plant growth promoting bacteria (PGPB) enhanced maize grain yield and nutrient use efficiency of inorganic fertilizer. *Journal of Agriculture and Food Research*, *10*, 100434.

Rawat, A., Negi, A., Riyal, I., Sharma, H., & Dwivedi, C. (2025). Comprehensive review of slow-release fertilizers utilizing metal–organic frameworks: Innovations, applications, and future directions. *Environmental Geochemistry and Health*, *47*(10), 1–21. https://doi.org/10.1007/s10653-025-02745-w

Rawat, P., Shankhdhar, D., & Shankhdhar, S. C. (2021). *Abiotic Stress: Its Outcome and Tolerance in Plants*, pp. 79–106. https://doi.org/10.1007/978-981-15-7094-0_5

Ryan, R. P., Germaine, K., Franks, A., Ryan, D. J., & Dowling, D. N. (2008). Bacterial endophytes: Recent developments and applications. *FEMS Microbiology Letters*, *278*(1), 1–9. https://doi.org/10.1111/J.1574-6968.2007.00918.X

Sachdev, S., Ansari, S. A., Ansari, M. I., Fujita, M., & Hasanuzzaman, M. (2021). Abiotic stress and reactive oxygen species: Generation, signaling, and defense mechanisms. *Antioxidants*, *10*(2), 277. https://doi.org/10.3390/ANTIOX10020277

Sahoo, A., Meher, I., Chhotaray, S. P., Meena, M., Swapnil, P., Arora, J., & Zehra, A. (2025). Fungal endophytes: A resource for producing extracellular enzymes. *Microorganisms for Sustainability*, *52*, 105–145. https://doi.org/10.1007/978-981-96-4004-1_6

Salazar, B., Ortiz, A., Keswani, C., Minkina, T., Mandzhieva, S., Pratap Singh, S., Rekadwad, B., Borriss, R., Jain, A., Singh, H. B., & Sansinenea, E. (2023). Bacillus spp. as bio-factories for antifungal secondary metabolites: Innovation beyond whole organism formulations. *Microbial Ecology*, *86*(1), 1–24. https://doi.org/10.1007/s00248-022-02044-2

Sharma, P., Kapoor, D., Kumar, R., & Husen, A. (2025). Rhizosphere and osmolytes coordination in plant systems under adverse situations. In *Roles of Osmolytes in Changing Environment*, pp. 101–117. https://doi.org/10.1016/B978-0-443-26549-5.00020-5

Sharma, U. C., Datta, M., & Sharma, V. (2025). *Recent Advances and Future Challenges*, pp. 523–580. https://doi.org/10.1007/978-3-031-76357-1_6

Singh, M., Jha, S., Pathak, D., & Maisnam, G. (2025). Advancing biofertilizers: The evolution from single-strain formulations to synthetic microbial communities (SynCom) for sustainable agriculture. *Discover Plants*, *2*(1), 1–21. https://doi.org/10.1007/S44372-025-00318-W

Singh, R., & Dubey, A. K. (2018). Diversity and applications of endophytic actinobacteria of plants in special and other ecological niches. *Frontiers in Microbiology*, *9*(AUG). https://doi.org/10.3389/FMICB.2018.01767

Sintaha, M. (2025). Molecular mechanisms of plant stress memory: Roles of non-coding RNAs and alternative splicing. *Plants*, *14*(13), 2021. https://doi.org/10.3390/PLANTS14132021

Soldan, R., Mapelli, F., Crotti, E., Schnell, S., Daffonchio, D., Marasco, R., Fusi, M., Borin, S., & Cardinale, M. (2019). Bacterial endophytes of mangrove propagules elicit early establishment of the natural host and promote growth of cereal crops under salt stress. *Microbiological Research*, *223–225*, 33–43. https://doi.org/10.1016/J.MICRES.2019.03.008

Thilakarathne, A. S., Liu, F., & Zou, Z. (2025). Plant signaling hormones and transcription factors: Key regulators of plant responses to growth, development, and stress. *Plants*, *14*(7), 1070. https://doi.org/10.3390/PLANTS14071070

Tuo, L., Yan, X. R., Li, F. N., Bao, Y. X., Shi, H. C., Li, H. Y., & Sun, C. H. (2018). *Brachybacterium endophyticum* sp. nov., a novel endophytic actinobacterium isolated from bark of *Scutellaria baicalensis georgi*. *International Journal of Systematic and Evolutionary Microbiology*, *68*(11), 3563–3568. https://doi.org/10.1099/ijsem.0.003032

Villalobos-López, M. A., Arroyo-Becerra, A., Quintero-Jiménez, A., & Iturriaga, G. (2022). Biotechnological advances to improve abiotic stress tolerance in crops. *International Journal of Molecular Sciences*, *23*(19), 12053. https://doi.org/10.3390/IJMS231912053

Xun, W., Shao, J., Shen, Q., & Zhang, R. (2021). Rhizosphere microbiome: Functional compensatory assembly for plant fitness. *Computational and Structural Biotechnology Journal*, *19*, 5487–5493. https://doi.org/10.1016/J.CSBJ.2021.09.035

Zörb, C., Geilfus, C. M., & Dietz, K. J. (2019). Salinity and crop yield. *Plant Biology*, *21*, 31–38. https://doi.org/10.1111/PLB.12884

6 Harnessing Ethnobotany and Traditional Knowledge for Sustainable Agri-Productivity

The Role of Beneficial Microbes in Hunger Alleviation

Japneet Kaur Gandhi and Anuradha Kumari

6.1 INTRODUCTION

Millions of people face persistent hunger and food insecurity in the world, especially in developing countries, despite the tremendous improvement in agricultural technology. The existing conventional farming, which relies on the extensive use of chemicals, has contributed to the degradation of the environment, reduced soil fertility, and biodiversity loss, among others, which also contribute to the vulnerability of agroecosystems and rural livelihoods. The need for sustainable and environmentally sound approaches that improve agricultural productivity, mitigate the ecological impacts of the agricultural sector, and support local food systems cannot be overstated (Pretty et al., 2018).

Ethnobotany and traditional knowledge are an incredible storehouse of experience, which the indigenous and local communities have developed over generations concerning food, medicine, and agriculture using plants and natural resources. These traditional practices are highly based on the local ecosystems, and focus on sustaining soil health, diversity of crops, and symbiotic associations of plants with useful microorganisms. These beneficial microorganisms include plant growth–promoting rhizobacteria, mycorrhizal fungi, and nitrogen-fixing bacteria, which significantly contribute to increasing soil fertility, nutrient cycling processes, and plant resistance to biotic or abiotic stresses (Glick, 2012; Smith and Read, 2010). Their implementation in farming systems can lead to a lesser reliance on the presence of synthetic

 DOI: 10.1201/9781003664451-6

fertilizers and pesticides, sustainable crop production, and food security (Adesemoye and Kloepper, 2009). The combined use of ethnobotanical and useful microbes as part of a long-term agricultural productivity source has not been properly harnessed in spite of their potential. Traditional practices are not systematically documented and scientifically validated, and there are limited efforts in aligning these practices with current agricultural science (Suman et al., 2022). Furthermore, globalization and cultural changes that are threatening the disappearance of traditional knowledge endanger the sustainability of such resources of great value. The proposed interdisciplinary studies and cooperation of scientists, communities, and policymakers are required to fill these gaps and create a context-specific, sustainable solution (Gómez-Baggethun et al., 2013). This chapter outlines the potential of utilizing ethnobotanical knowledge and helpful microbes to realize sustainable agriculture and help solve hunger issues. It attempts to discuss scalable, environmentally friendly agricultural technologies that can empower the local community, protect biodiversity, and guarantee food security despite climate change and resource limitations by recording and certifying traditional methods and reconciling them with scientific knowledge of useful microbes (Pretty et al., 2018; Adesemoye & Kloepper, 2009).

6.1.1 Importance of Sustainable Agri-Productivity

Sustainability is a term that can be used when referring to the practice of upkeep or conservation of resources to sustain their continued utilization in the future (Hamdane et al., 2023). Sustainability in this context is defined in terms of natural resources, especially those used in agriculture. Currently, the overexploitation of these resources poses a major issue, and agricultural practices exert a high level of pressure and direct impact on the availability and quality of natural resources (Hamdane et al., 2023).

Many of the key global challenges faced by the world today can be directly or indirectly correlated to nutrition, or lack thereof (Kumar et al., 2021). Achieving zero hunger, ending poverty, and ensuring access to adequate nutrition, which would go on to positively affect the health status of the society, are some of the current pressing issues (Arora, 2018). These issues are interrelated and can be addressed simultaneously. The agricultural sector is currently under threat, owing to the excessive use of chemical fertilizers, which has already increased the soil salinity and rendered the soil unfit for cultivation (Clemensen et al., 2020). Apart from this, since the onset and spread of industrialization, there has been an increase in water and soil contamination due to heavy metal pollution (Jiang et al., 2022).

In the mid-20th century, the 'green revolution', which boosted agricultural productivity, also led to an increase in global pollution and loss of biodiversity (Ahirwar et al., 2020). South Asian countries, home to 25% of the world's 1.75 billion people, account for 3.4% of the world's total land area, making them vulnerable to soil and land degradation because any negative effects directly hinder the agricultural output (Saleem et al., 2024). As a result, there is a growing pursuit in the use of sustainable approaches such as microbial biotechnologies (Arora, 2018; Ahirwar et al., 2020).

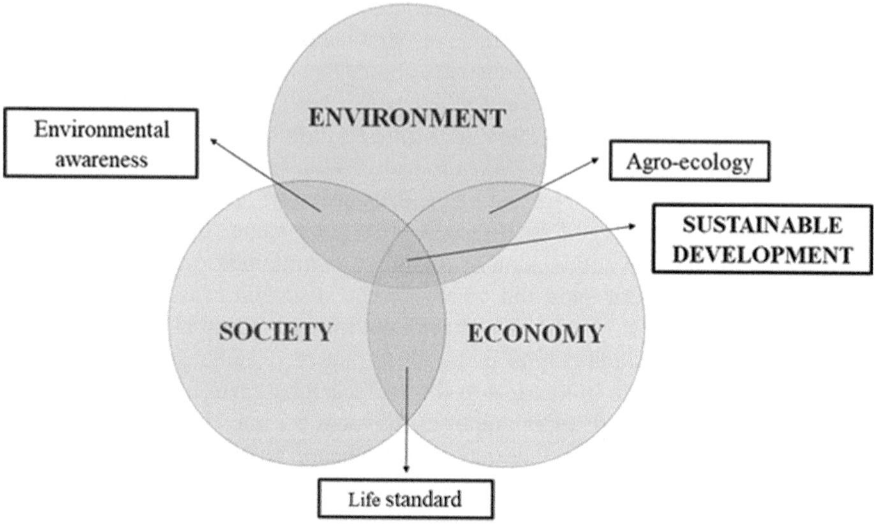

FIGURE 6.1 Three interrelated factors that determine sustainable development.

Sustainable agriculture embraces the natural potential of living organisms, ensuring higher productivity and ecological stability (Ahirwar et al., 2020; Jiang et al., 2022). One more benefit of opting for sustainable practices is its positive correlation with biological richness (Clemensen et al., 2020). The conservative practices utilize ecological management and stabilization as a means to promote phytodiversity and maintain soil quality by improving the water-holding capacity of the soil and making it resilient toward drought conditions (Arora, 2018; Clemensen et al., 2020; Jiang et al., 2022). Implementing diversification can also be exercised via crop rotation using multiple crop sequences (Arora, 2018; Hamdane et al., 2023; Saleem et al., 2024).

When aiming to implement sustainable development as a solution, three key factors need to be addressed: environment, society, and economy; the intersection between each of them is better explained through Figure 6.1, as agro-ecology, environmental awareness, and life standards, all three being determinants of sustainable development (Ahirwar et al., 2020; Hamdane et al., 2023). Integrating traditional approaches with modern innovations can help achieve the end goal of sustainable development (Arora, 2018; Hamdane et al., 2023).

6.1.2 Scope of Ethnobotany and Traditional Knowledge

Our ancestors have instinctively explored Earth's biodiversity for thousands of years, passing down their discoveries along generations, resulting in the current extensive knowledge that we have gained about plants and their potential (Beyene et al., 2016). To date, 30,000 edible plant species have been identified globally, and yet only 150 of them have been commercialized (Li et al., 2020). On a global level, more than half of the nutrition supply to humans comes from three staple crops: wheat, rice, and maize.

And across ages, only 7,000 of the identified plant species have been used as a source of nutrition (Li et al., 2020).

A research study carried out in 2024 by Pironon et al. is based on the distribution of 35,687 plant species, categorizing them based on their food, medicinal, and material uses. The study concluded that the highest utilization, as well as species richness, was observed in indigenous and autochthonous areas. The underutilized plants exhibit more nutritional potential and more resilience toward climate change (Ray et al., 2020). Incorporating them into our diet can transform the face of the agricultural and nutritional systems, a practice already in place in economically disadvantaged parts of the world (Ray et al., 2020).

This traditional knowledge about phytodiversity has been passed down through generations and conserved by the local communities. They utilize indigenous plants for a variety of purposes, including nutrition, folk medicine, etc. (Hamdane et al., 2023). Thousands of plants, which have not yet been discovered scientifically, have been recognized by the native people for their medicinal purposes (Kumar et al., 2021). These plants are addressed by their vernacular names, and their therapeutic potential is well understood by the tribes residing in that area (Kumar et al., 2021). Reflecting this, Agenda 21 of the 1992 Rio Earth Summit also advocated for the involvement of indigenous people in sustainable development practices due to their invaluable traditional knowledge (Kumar et al., 2021).

Since the plants and understanding about their medicinal properties are not accessible, and there is a lack of awareness about their properties, they remain scientifically unexplored, there is no competition for the consumption and utilization of these natural resources, making them a part of the communities' cultural inheritance, along with ensuring their cultivation by conforming them to their native habitats (Jiang et al., 2022; Hamdane et al., 2023). These locally grown plants are also well adapted to the agro-climatic conditions and are resilient to change, further emphasizing the need to introduce diversification of edible and medicinal plants in areas with an unpredictable atmosphere (Kumar et al., 2021; Hamdane et al., 2023). These conservative or organic agricultural practices help restore favorable growth conditions as compared to modernized agriculture, resulting in higher species richness in areas where organic agriculture is exercised (Jiang et al., 2022).

Ethnobotany serves as a bridge between the indigenous people–plant interactions, which we refer to as traditional knowledge, and modern science (Hamdane et al., 2023). There have been several issues in expanding this knowledge and making it accessible to everyone, including the loss of indigenous languages and increased modernization, although this dormant knowledge can be revived as a way to change the world's agricultural scenario (Kumar et al., 2021).

6.2 ETHNOBOTANICAL PLANTS WITH NUTRITIONAL AND THERAPEUTIC VALUE

As discussed in the previous section, medicinal plants have been used as a treatment and alleviation strategy for various disorders, as depicted in Figure 6.2. The presence of various metabolites such as phenolics, flavonoids, and alkaloids contributes to the medicinal uses of these plants (Verma et al., 2018). The plants that this section

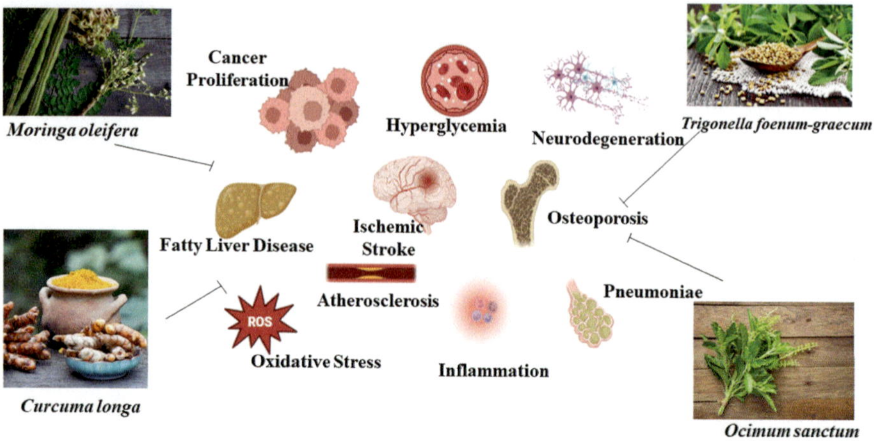

FIGURE 6.2 Medicinal plants and their effect on human health.

focuses on include *Moringa oleifera* (drumstick tree), *Curcuma longa* (turmeric), *Ocimum sanctum* (Holy Basil), and *Trigonella foenum-graecum* (fenugreek), as examples of several indigenous species that possess anticancer, anti-inflammatory, neurodegenerative, antioxidant, and antihyperglycemic properties, marking their roles in the treatment of various ailments, also discussed in this section.

6.2.1 *Moringa oleifera*: A Superfood with Multifunctional Benefits

M. oleifera, also referred to as 'drumstick tree' or 'horseradish tree', is a perennial herb belonging to the family *Moringaceae*. The herb has minimal growth requirements, owing to its fast-growing and drought-resistant characteristics (Mazumder et al., 2024). The appearance of the plant leaves is cylindrical and tri-pinnate, and they are attached to delicate, drooping branches (Mazumder et al., 2024). The plant has its origins in India but has now spread to every tropical and subtropical region of the world, ranging from Africa to Central America to the Caribbean Islands (Mahaveerchand & Abdul Salam, 2024). The plant serves as a valuable source of nutrition, particularly in underdeveloped countries facing challenges such as poverty and malnutrition, due to its economic value and affordability (Singh & Chaturvedi, 2021).

M. oleifera has been used as a food source for ages, and all parts of this plant are fit for human consumption (Singh & Chaturvedi, 2021). The plant is referred to as a 'Superfood', due to its high nutritional content. According to Mazumder et al. (2024), a superfood is a low-calorie food that is rich in vitamins, minerals, and antioxidants. *Moringa* leaves are an abundant storehouse of proteins, as well as nine essential amino acids. The plant also contains copious amounts of minerals, such as calcium, potassium, and zinc, and vitamins like beta-carotene, pyridoxine, and nicotinic acid (Singh & Chaturvedi, 2021; Tiwari et al., 2023). Along with that, the bioactive substances present in the plant constitute flavonoids, phenols, alkaloids, terpenoids, etc., which contribute to its medicinal potential (Tiwari et al., 2023). Research conducted by Zayed M. S. in 2012 recorded that mixed cultures of microbes like *Azotobacter chroococcum* and *Saccharomyces cerevisiae* positively impact the growth parameters of *M. oleifera*, and the same consortium, along with *Bacillus megatherium* and *Bacillus circulans*, increases the protein and mineral content in the plant.

In India and Africa, *M. oleifera* is used in herbal therapies and is considered a panacea. It can cure about 300 disorders (Srivastava et al., 2023). *M. oleifera* has the potential to prevent malnutrition in children via various beneficial approaches, such as elevated hemoglobin levels, weight gain, and reduction in anemia. Along with the physical development of children, it also aids in fulfilling their micronutrient deficiencies (Mahaveerchand & Abdul Salam, 2024). The bioactive elements present in *M. oleifera* leaves can reverse the DNA damage occurring during cancer and other degenerative diseases, being effective in enhancing the chemotherapeutic interventions (Srivastava et al., 2023). The anticancer activity is aided by the antiproliferative potential of the plant, along with the overexpression of caspases 3 and 9, which induce the apoptotic cascade. Besides this, its antioxidant potential helps neutralize free radicals and inhibit the progression of cancer (Mahaveerchand & Abdul Salam, 2024; Mazumder et al., 2024). Particularly, niazimicin, a glucosinolate, has been explored enough to establish antitumor effects exerted by it (Mahaveerchand & Abdul Salam, 2024).

Studies conducted *in vivo* on guinea pigs have confirmed that *M. oleifera* leaves have hepatoprotective activity, which is validated through their ability to reduce accumulated cholesterol and 256 triglycerides in nonalcoholic fatty liver disease (Srivastava et al., 2023). The plant has a flavonoid called quercetin, which helps alleviate liver lipid levels and regulates plasma creatinine levels, and enzymes such as aspartate alkaline phosphatase and aminotransferase (Srivastava et al., 2023). *Moringa* can reverse both Type 1 and Type 2 diabetes (Mazumder et al., 2024). In cases of high glucose concentration, mitochondria generate reactive oxygen species (ROS), which are scavenged by quercetin; as a result, beta cells are preserved, which would otherwise undergo apoptosis in the presence of free radicals, and thus hyperglycemia can be avoided (Tiwari et al., 2023; Mazumder et al., 2024). The plant also exhibits neuroprotective abilities by lowering ROS, ensuring blood supply to the brain, shielding neurons from harm, and preventing ischemia, which is also due to the presence of quercetin and other bioactive compounds like chlorogenic acid (Mahaveerchand & Abdul Salam, 2024).

Apart from this, preclinical and clinical trials as discussed in Table 6.1 have shown that various other therapeutic properties, such as antihypertensive, antipyretic, antifungal, antibacterial, bone health maintenance, and eye protection, are exhibited readily by *M. oleifera* (Srivastava et al., 2023; Tiwari et al., 2023; Mazumder et al., 2024). Due to its high nutritional profile and exceptional health benefits, *Moringa* is referred to as the 'Tree of Life' or 'Miracle Tree' (Srivastava et al., 2023).

6.2.2 *Curcuma longa*: Anti-inflammatory and Antioxidant Properties

C. longa, commonly known as 'turmeric', is a perennial herb belonging to the family *Zingiberaceae* and has been recognized for its therapeutic potential in herbal medicine. The herb originated in India and spread to China, East Africa, and West

TABLE 6.1
***In Vivo* Studies Demonstrating Medicinal Effects on Indigenous Plant Species**

Plant Extract	Experimental Model	Disorder	Effect	Dosage	References
Curcuma longa (curcumin)	Rat	Epilepsy	Inhibited activation of NLRP3	100 mg/kg for 7 days	Memarzia et al. (2021)
	Mice	Acute respiratory distress syndrome	Reduced inflammation and fibrosis in lung tissues	50 mg/kg, intranasal for 5 days	Memarzia et al. (2021)
	Clinical trials	Rheumatoid arthritis	Reduced swelling of joints	500 mg capsule for 8 weeks	Razavi et al. (2021)
	Clinical trials	Alzheimer's disease	Increased A-β plaques in brain	100 mM	Razavi et al. (2021)
	Clinical trials	Cancer	Enhanced IL-4 production	10–60 μM	Memarzia et al. (2021)
Trigonella foenum-graecum (fenugreek seed)	*In vivo*	Diabetes-induced neurological consequences	Reduced calcium levels and neuroprotection	5% w/w in diet for 3 weeks	Zameer et al. (2017)
	Rat	Type 2 diabetes	Reduction in serum glucose	-	Visuvanathan et al. (2022)
	Clinical trials	Coronary heart disease	Lowering cholesterol	-	Visuvanathan et al. (2022)
	Rat	Fat-induced obesity	Hunger suppression	-	Visuvanathan et al. (2022)

(Contiuned)

TABLE 6.1 (*Continued*)
***In Vivo* Studies Demonstrating Medicinal Effects on Indigenous Plant Species**

Plant Extract	Experimental Model	Disorder	Effect	Dosage	References
Moringa oleifera (leaf extract)	HeLa cells	Cervical cancer	Anticancer activity	10–100 μg/mL	Mazumder et al. (2024)
	Swiss Albino mice		Bone marrow chromosomes' radiation resistance	150 mg/kg	Mazumder et al. (2024)
Ocimum sanctum (leaf extract)	Swiss Albino mice	Immunomodulatory activity	Delayed hypersensitivity stimulation and humoral immunity improvement	50–200 mg/kg/day for 14 days	Kumar et al. (2013)
	Diabetic animal models	Hypercholesterolemia	Lipid-lowering activity	-	Siva et al. (2016)
	Rat	Diabetes	Lowering blood sugar	-	Siva et al. (2016)
	Rat	Cancer	Reduced squamous cell carcinoma and hematoma	10 weeks	Siva et al. (2016)

Africa by 700 AD, 800 AD, and 1200 AD, respectively (Krup et al., 2013). It is very well known in India, China, Bangladesh, and other Southeast Asian countries as a condiment in their food. The Indian name for turmeric is *Haldi* or *'Haridra'* (Jyotirmayee & Mahalik, 2022; Labban, 2014). People in India are well aware of the wound healing properties in turmeric, and other than that, also consider it pious, utilizing it in their religious rituals. Seventy-eight percent of the world's turmeric output comes from India, since it is the plant's topmost producer, user, and exporter (Jyotirmayee & Mahalik, 2022).

C. longa has been associated with a lot of medicinal undertakings, such as antidiabetic, anti-inflammatory, hepatoprotective, and anticancerous (Krup et al., 2013). Its cosmetological properties are also well known (Labban, 2014). In *Ayurveda*, turmeric has been used for the treatment of hemorrhage, menstrual discomfort, jaundice, etc. (Labban, 2014). These medicinal effects of turmeric possibly arise from its medicinal constituents, which are called curcuminoids. The curcuminoids present in turmeric are curcumin, desmethoxycurcumin, and bisdemethoxycurcumin (Jyotirmayee & Mahalik, 2022). The preclinical studies on curcuminoids, both *in vitro* and *in vivo*, have been conducted for centuries to investigate their biological actions (Boroumand et al., 2018). These studies exhibited the presence of anti-inflammatory, antioxidant, antidiabetic, cardio- and neuro-protective properties in curcuminoids, which is also highlighted in Table 6.1 (Krup et al., 2013; Boroumand et al., 2018; Memarzia et al., 2021; Jyotirmayee & Mahalik, 2022). Curcumin is the main active phytochemical in turmeric, and it is also the component that imparts yellow color to the spice (Razavi et al., 2021).

Curcumin has the potential to act at the cellular level via various signaling pathways, which leads to its various medicinal roles (Razavi et al., 2021). The property of turmeric to reduce inflammation comes from its ability to inhibit the synthesis of inflammatory prostaglandins from arachidonic acid and neutrophil functions during inflammation (Labban, 2014). Curcumin downregulates signaling molecules like cycloxygenase-2, Janus kinases, tumor necrosis factor (TNF), leukotrienes, phospholipases, etc. (Boroumand et al., 2018). These molecules are involved in inflammation. It inhibits the generation of TNF-α, interleukin (IL)-1, 2, 6, 8, 12, etc. (Boroumand et al., 2018; Krup et al., 2013). These inflammatory mediators are regulated by NF-κB, which has a tendency for DNA-binding in instances of oxidative stress. Curcumin, possessing antioxidant activities, reduces the binding of NF-κB with DNA, thus inhibiting its activation, and in turn, suppresses inflammation (Boroumand et al., 2018). There have been instances where oral ingestion of turmeric has proved as effective against inflammation as cortisone or phenylbutazone (Labban, 2014). They also inhibit leukotriene synthesis via the lipoxygenase pathway (Labban, 2014).

Vitamin C and E possess strong antioxidant properties, and the water- and fat-soluble extracts of turmeric and the curcumin component have a comparable property (Krup et al., 2013). These properties can be linked to their ability to combat oxidative damage (Labban, 2014). The antioxidant properties in curcumin come from the phenolic group in its structure, because of which it can donate an H-atom (Boroumand et al., 2018). In peroxidation processes, curcumin can scavenge free radicals and hence can inhibit lipid peroxidation. Curcumin positions itself within the cell membrane and protects the cell from oxidative damage (Boroumand et al.,

2018). It also acts as an antioxidant against hydrogen peroxide and superoxide radical generation (Boroumand et al., 2018). Antioxidant properties of curcumin lead to its effects as a potential anticancer agent on various biological pathways involved during cancer, such as metastasis, mutagenesis, oncogene expression, apoptosis, and tumorigenesis (Krup et al., 2013). Curcumin also shows antiproliferative effects in multiple cancers by inhibiting NF-κB and its downstream mediators mentioned above (Krup et al., 2013). A study was conducted where mice, after being exposed to human prostate cancer cells, were then treated with curcumin. The treated animals showed a decrease in cell proliferation and an increase in apoptosis as opposed to the control mice (Labban, 2014). Curcumin exerts its effects on the cell cycle and its signal transduction pathways, thereby asserting antitumor activity via cyclin-dependent, p53-dependent, and p53-independent pathways (Krup et al., 2013). Due to its properties and tolerance in humans, curcumin can be used as an adjuvant in conjunction with chemotherapeutic drugs (Krup et al., 2013).

6.2.3 *Trigonella foenum-graecum* (Fenugreek): Managing Metabolic Disorders

T. foenum-graecum, commonly known as 'fenugreek', is an angiosperm belonging to the *Fabaceae* family and Fabales order (Syed et al., 2020). The scientific name of this plant has been derived from the Greek word *trignonou*, which means triangle, due to the leaflet shape of this plant, and the origins of this plant belong to Eastern Europe, from where it has spread worldwide as a vegetable and condiment (Bahmani et al., 2016; Zameer et al., 2017; Syed et al., 2020). The species name '*foenum-graecum*' also has a Greek origin, meaning 'Greek hay' (Syed et al., 2020). The plant has a 50 cm long curved stem, and trifoliated leaves. The seeds of the plant have a pebble-like appearance and a yellowish-brown color spectrum. The seeds, when dried, have a bitter taste and can be ground to form a powder; fenugreek powder or paste is often used as a condiment in India (Bahmani et al., 2016; Ghosh et al., 2015; Nagulapalli Venkata et al., 2017). *Trigonella foenum-graecum* has a profound medical history in many cultures around the planet. The Chinese use it to treat kidney stones as well as edema in the legs, whereas Indians have identified its medicinal values in Ayurveda, using it for its digestive properties (Nagulapalli Venkata et al., 2017; Visuvanathan et al., 2022). It is also recognized for easing childbirth, as well as menstrual pains in Egyptian cultures, along with serving spiritual purposes such as incense in religious ceremonies and for mummifying bodies (Bahmani et al., 2016).

Fenugreek has been phytochemically analyzed to reveal a surfeit of secondary metabolites stored in the plant, such as saponins, alkaloids, flavonoids, phenols, amino acids, fatty acids, and vitamins (Zameer et al., 2017). Nitrogen and sulfur-containing amino acids are present in the highest abundance amongst the total amino acid content in fenugreek (Nagulapalli Venkata et al., 2017). Trigonelline is a quaternary ammonium compound; an alkaloid of medicinal importance found in fenugreek, which adds to the hypoglycemic and neuroprotective properties of the plant, along with being a phytoestrogen (Nagulapalli Venkata et al., 2017). Amongst the dietary fibers, galactomannan possesses medicinal properties and exhibits antihyperlipidemic effect (Zameer et al., 2017).

It has been recognized through *in vivo* studies reported in Bahmnai et al. (2016), Zameer et al. (2017), and Syed et al. (2020) that hydroxyisoleucine, which constitutes about 80% content of fenugreek seeds, can increase the glucose uptake through PI3 kinase/Akt-dependent mechanism. This increases glucose utilization and reduces hepatic synthesis of glucose (Nagulapalli Venkata et al., 2017). There have been further studies correlating hyperglycemia with neurological consequences via lipid peroxidation in the synaptosomal membrane, effects on neurotransmitter uptake, etc. (Nagulapalli Venkata et al., 2017). Fenugreek seeds possess the potential to reverse or alleviate hyperglycemia by reducing lipid peroxidation, as well as restoring the synaptosomal membrane fluidity in the brain. This was confirmed by a study on rats demonstrating neuroprotection by the administration of fenugreek and its constituents (Zameer et al., 2017).

Another study by Nagulapalli Venkata et al. (2017) validates the anti-inflammatory potential of fenugreek by means of hydroxyisoleucine found in its extracts. Hydroxyisoleucine lowers the NF-κB activation, mitogen-activated protein kinase, c-jun N-terminal kinase, etc., thereby reducing ROS-induced inflammation. The saponins present in fenugreek also possess the ability to inhibit proinflammatory enzymes (cyclooxygenase and lipoxygenase) and cytokines (IL-1β and TNF-α) (Nagulapalli Venkata et al., 2017). These health-promoting effects of fenugreek and its constituents have been widely studied, supported by *in vitro*, *in vivo*, and clinical experiments as mentioned in Table 6.1 (Syed et al., 2020). An extensive understanding can be developed into cellular and molecular mechanisms of hypocholesterolemia, hyperinsulinemia, apoptosis, autophagy, free radical scavenging, etc., resulting from the disease-preventive effects of fenugreek (Bahmani et al., 2016; Ghosh et al., 2015; Nagulapalli Venkata et al., 2017; Visuvanathan et al., 2022).

6.2.4 *Ocimum sanctum* (Holy Basil): Respiratory and Immune Health

O. sanctum, commonly known as 'Tulsi' or 'Holy Basil', is a culinary herb utilized in Indian cuisine, and a source of medication recognized by Ayurveda for about 3,000 years (Bhadra & Sethi, 2020). It belongs to the family *Lamiaceae* and order Lamiales, and is distributed around various parts of the world, such as Asia, Africa, and Central and Southern America (Joseph & Nair, 2013). The plant comprises a 30–60 cm bush with inverse green leaves, and has a characteristic scent (Bhadra & Sethi, 2020). Flowers of Tulsi are of racemose inflorescence with purple pigmentation, whereas the seeds are yellow. The plant is considered pious and is worshipped in temples and houses of India (Savita et al., 2021). The phytochemical composition of the plant is rich, including compounds like oleic acid, linoleic acid, rosmarinic acid, gallic acid, benzene, galuteolin, clorogenic acid, vitamin C, calcium, phosphorus, and other micronutrients (Joseph & Nair, 2013). Tulsi exhibits a lot of pharmacological effects, such as immunomodulatory, neuroprotective, analgesic, antidiabetic, anti-inflammatory, hepatoprotective, cardioprotective, and anticancerous effects (Joseph & Nair, 2013).

O. sanctum plays a major role in building immunity by actively participating against a diverse range of ailments, starting from cold, cough, to severe infections like COVID-19. During COVID, a particular commercial product by the name 'Tulsi drop' came into demand, due to its ability to foster immunity (Satyendra, 2021). In studies including human volunteers, Tulsi has exhibited its immunomodulatory properties by causing an increase in interferon-γ and IL-4 (Savita et al., 2021). Along with that, there is experimental evidence regarding the antiviral, antifungal, and antibacterial activity of *O. sanctum* (Siva et al., 2016; Satyendra, 2021; Savita et al., 2021). These include infections of the urinary tract, typhoid, cholera, leishmaniasis, etc. (Savita et al., 2021). The function of liver detoxifying enzymes, such as P450, is said to be augmented by the plant, facilitating the body's removal of various toxins, thus contributing to the detoxification of the body (Satyendra, 2021).

A study conducted by Suresh et al. (2022) aimed at understanding the role of basil against Gram-negative bacterial pneumonia caused by *Klebsiella pneumoniae* in lung epithelial cells (A549). The results showed that basil treatment reduced cell death along with a reduced expression of proapoptotic caspases following pneumonia. Upon treatment with basil extract, it was observed that the expression of pro-inflammatory cytokines like IL-1β, IL-6, and TNF-α had diminished. Another study on A549 cells was conducted by Kustiati et al. (2025), where the phenols and flavonoids in the ethanolic extract of *O. sanctum* caused the lung cancer cells to separate from the extracellular matrix, causing a decrease in their cell viability, along with raising intracellular ROS production.

Owing to its wide plethora of medicinal properties as discussed in Table 6.1, and because of its traditional recognition, it is referred to as the 'Elixir of Life' and 'Queen of Herbs' in Ayurveda (Satyendra, 2021).

6.2.5 Other Region-Specific Ethnobotanical Crops for Hunger Alleviation

Ethnobotanical crops that are region-specific play a significant role in alleviating hunger through food security, nutritional diversity, and economic well-being, especially in vulnerable and rural societies. These are usually the indigenous or wild and semi-domesticated plants that have traditionally been used as food, medicine, or other necessities of the indigenous and local populations. Their worth is due to their ease of adaptation to local environmental scenes, the culture, and their use as a supplementary food in times of scarcity or famine. Their potential importance in sustainable agriculture and alleviation of hunger has also been reinforced in scientific evidence and ethnobotanical studies, more so when combined with the presence of beneficial microbes.

These fruits, leafy vegetables, tubers, flowers, and seeds constitute wild edible food plants, which are well documented in many parts, including India, Ethiopia (Fassil et al., 2024), and Africa. In Odisha, India, more than 190 species of wild edible plants, more than 70 families of wild edible plants, are used by the tribal communities. Such plants are very important in the case of food shortage, serving as a back-stop to filling nutrient and energy gaps when the complementary foods grown are unavailable or out of reach in the market (Mallick et al., 2020). A variety of these wild crops, including *M. oleifera*, *Amaranthus* spp., and *Dioscorea* spp., are high in protein, vitamins, minerals, and dietary fiber, and have nutritional advantages over certain staple crops (Tadesse et al., 2024).

The sustainable use and harvesting of such plants can be made through ethnobotanical knowledge that spans generations. This indigenous knowledge encompasses ways of collecting, processing, storing, and conserving, such that it preserves plant diversity and ensures existence through famine or drought. As an example, local communities in tribal communities can often freeze and preserve leaves, flowers, or tubers in case of an emergency, thus increasing food resilience.

In addition to providing food, ethnobotanical crops assist in sustainable agricultural productivity through symbiosis between the crops and beneficial soil microbes. Nutrient-absorbing microorganisms like nitrogen-fixing bacteria and mycorrhizal fungi improve plant growth and nutrient availability, making the crops better suited to low fertility or drought-affected conditions. The use of these interactions between microbe and plants can be harnessed to practice eco-friendly agriculture due to the minimized use of synthetic fertilizers and pesticides. This is especially important in marginal lands where cultivated crops and wild vegetables are bountiful because of their natural ability to adapt with positive microbiomes. Studies show that inoculation of microbial inoculants enhances germination rate, biomass, and yield of crop together, which in turn leads to food security in sensitive agro-ecologies using ethnobotanical crops (Qureshi et al., 2016).

Moreover, ethnobotanical crops also help in ensuring agro-biodiversity and ecosystem health. The various cropping systems, which include these plants and beneficial microorganisms, lead to conservation of soil, erosion control, and increased soil fertility, leading to a sustainable farming system, which are resistant to environmental stressors. This is the balance that allows the sustainability of food production and the livelihood of the community essential to alleviating hunger. There are economic advantages as well that are developed by these crops because they offer income-generating activities such as the collection, processing, and selling of wild edible plants and related products in the local market. This income can be used to enhance access to food and the resilience of households when disasters strike.

6.3 TRADITIONAL KNOWLEDGE IN CROP SELECTION AND MANAGEMENT

Traditional knowledge in crop selection and management is an adaptive knowledge system that was developed with practice over the centuries through empirical evidence, experimentation, and cultural dispersion. Based on its ethnobotanical background, it helps the farming community find, reproduce, and preserve local crops that best fit their regions' agro-ecological niches. Examples of these include choice of seeds on the basis of phenotype, preservation of landraces as a source of genetic diversity, land and crop-specific planting calendar calendars, mixed cropping and inter-cropping as means of achieving ecological resilience, and crop-livestock-soil systems integration.

Scientifically, these conventional practices tend to relatively maximize plant–soil–microbe relations, sometimes inadvertently, in improving nutrient circularity, organic matter preservation, and the diversity of soil microorganisms. Among the techniques that interfere with pest development life cycles are crop rotations and polycultures, which also promote beneficial microbial consortia, like mycorrhizal fungi and nitrogen-fixing bacteria, that confer stress tolerance and increase yields (Puranik et al., 2023; Sruthy et al., 2023). The deliberate combining of this traditional local knowledge with contemporary microbial biotechnology offers a synergistic route to sustainable agri-productivity, climate security, and hunger lifting, especially in resource-constrained ecosystems.

6.3.1 Indigenous Farming Systems and Practices

Utilization of ethnobotany and traditional knowledge on sustaining agri-productivity and alleviating hunger relates to the identification of how local farming systems maximize the combination of locally found plants, ecological management strategies, and beneficial soil microbes to create resilient, nutrient-rich agroecosystems.

6.3.1.1 Integration of Indigenous Plant Species

Indigenous communities have developed sophisticated ways of selecting and nurturing plant species that are most suitable for the climate, soils, and food habits of their particular area. Through ethnobotanical knowledge – the results of generations of experiential learning accumulated over time – farmers know the different crops that provide a range of nutrients and can be grown through rotation to survive the local stressors. In many parts of the world, landraces (e.g., the traditional millets in India, teff in Ethiopia, amaranths in South America) still exist because of their high nutritional value and suitability for low-input farming, and can often yield more than modern hybrids in marginal growing conditions (Garnatje et al., 2017).

6.3.1.2 Ecological Strategies on Agri-Productivity

Local agricultural systems are more productive through time-tested ecological strategies: Intercropping and Polycultures. The native farmers typically cultivate a variety of crops on the same field, which helps to increase pest resistance, optimize the utilization of facilities, and provide microhabitats that assist other helpful organisms. In Odisha, India, intercrop legumes and cereals are grown in tribes to restore natural nitrogen in the soil (Ojeda-Linares et al., 2024).

Use of organic amendments and mulching: application of livestock manures, crop residues, and green manures enriches soil organic matter, fosters carbon sequestration, and maintains soil fertility and productivity (Porras et al., 2020).

Agroforestry and wild plant integrated system: Khasi hill agroforestry, in India, and African home gardens cultivate woody perennials with food crops, supporting biodiversity and long-term productivity of the soil.

6.3.2 Preservation and Documentation of Local Knowledge

Knowledge about local ethnobotanical varieties is important in preservation and documentation of such information to maintain agri-productivity and hunger alleviation, particularly in combination with innovations in microbials. Scientific evidence supports structured and participatory methods with respect to community rights, accuracy, and scalability with modern agronomy (Suminguit, 2005).

6.3.2.1 Community-Driven Documentation Frameworks

Participatory approaches – such as shared field walks, shared naming of plants, and knowledge documentation by showing authorship by local communities – contribute to the increase of accuracy, as well as cultural valence. They allow people to close the epistemic gap between the researcher and the community and enhance the data validity in the future breeding, conservation, and extension programs (Pullaiah et al., 2017). The recording of local language knowledge with community approval maintains linguistic diversity and plant use semantics and enables subsequent teaching and transfer of such knowledge on the family and village level (Sauini et al., 2020).

6.3.2.2 Plant Conservation Techniques and Practices Conducive to Microbes

Ex in vivo and *in situ* conservation of landraces forms the basis of genetic reserves of drought resistance, pest resistance, and nutrient use efficiency. Seed exchange networks and on-farm preservation are regularly accompanied by documentation to uphold ecological relations with soil microbes, which maintain soil health. A compilation of traditional agronomic activities, such as crop rotations, intercropping, and mulching, composting, and plant-based amendments, demystifies the identification of how communities develop desirable microbiomes in the soil that enhance the cycling of nutrients and disease resistance.

6.4 CONCLUSION AND PROSPECTS

Building resilient agroecosystems: By utilizing ethnobotany and indigenous knowledge, particularly through the incorporation of useful microbes, a way forward can be developed to build resilient agroecosystems with the ability to overcome climatic stress, pests, and market volatility. Agriculture practices that benefit the soil fertility, botanical diversity, and the increase of useful soil microorganisms are inherently agronomical practices like diversified polyculture, crop rotation, and localized selection of plants. These microbe–plant interactions improve the cycling of nutrients and the maintenance of water within the soil and also on the crops, leading to predictable yields and sustainable production of foods even in harsh climatic conditions. Communities end up strengthening themselves by self-reliance, becoming less prone to hunger by idolizing these systems.

Fostering collaboration between science and tradition: A significant opportunity is to have two-way participatory partnerships between the local holders of knowledge and scientific researchers. Ethnobotanical research indicates that innovation can be expedited in sustainable agriculture by validating, recording, and testing scientifically the authentic traditional practices, including the utilization of bioactive plants to eliminate pests, microbial inoculants to improve soil health, etc. Such partnerships are the way to make sure that local knowledge is honored, multiplied, and adjusted to contemporary life, and scientific research can rely on ages-old ecological knowledge. The local innovations can be converted to scalable solutions, which can be used to sustain agro-productivity and conservation through joint pilot projects, co-designed extension programs, and knowledge exchanges.

6.4.1 Recommendations for Policy and Practice

In order to achieve the complete utilization of the advantages of combining ethnobotany, traditional knowledge, and useful microbes: Support methodical recordings of community knowledge: Infrastructure local knowledge Invest into community-based mapping, electronic databases, oral histories, and participatory studies to document, maintain, and extend useful traditions and biocultural inventions.

Grow regionally appropriate solutions by extension and capacity development: Educate train agents on scientific and traditional methods of entrepreneurship, create demonstration plots, and facilitate community-owned enterprises to harness and exploit local biodiversity and innovative plant-microbe-systems. Promote gender equity and youth involvement: Socialize women and older adults as key knowledge custodians, encourage them to lead in knowledge sharing, and motivate young people to participate so as to sustain and achieve continuous innovation.

Overall, the utilization of ethnobotany and traditional knowledge, especially the combination of ethnobotanical crops of particular regions with advantageous microbes, is a feasible and scientifically backed method of hunger mitigation. These crops create a variety of nutritious foodstuffs that are locally resistant and adapted. Combining microbial partners to further improve their productivity and sustainability guarantees food security and socio-economic advantages to helpless populations all over the world. Retention of traditional knowledge on ethnobotany, cultivation of local crops, and scientific studies on their symbiotic microbial partnerships are key to the sustainable solution to hunger to reinforce the livelihood of rural populations across the world.

REFERENCES

Adesemoye, A. O., & Kloepper, J. W. (2009). Plant–microbes interactions in enhanced fertilizer-use efficiency. *Applied Microbiology and Biotechnology*, *85*(1), 1–12.

Ahirwar, N. K., Singh, R., Chaurasia, S., Chandra, R., & Ramana, S. (2020). Effective role of beneficial microbes in achieving the sustainable agriculture and eco-friendly environment development goals: A review. *Frontiers in Microbiology*, *5*, 111–123.

Arora, N. K. (2018). Agricultural sustainability and food security. *Environmental Sustainability*, *1*(3), 217–219.

Bahmani, M., Shirzad, H., Mirhosseini, M., Mesripour, A., & Rafieian-Kopaei, M. (2016). A review on ethnobotanical and therapeutic uses of fenugreek (*Trigonella foenum-graceum* L). *Journal of Evidence-Based Complementary & Alternative Medicine*, *21*(1), 53–62.

Beyene, B., Beyene, B., & Deribe, H. (2016). Review on application and management of medicinal plants for the livelihood of the local community. *Journal of Resources Development and Management*, *22*(1), 33–39.

Bhadra, P., & Sethi, L. (2020). A review paper on the Tulsi plant (*Ocimum sanctum*). *Indian Journal of Natural Sciences*, *10*(60), 20854–20860.

Boroumand, N., Samarghandian, S., & Hashemy, S. I. (2018). Immunomodulatory, anti-inflammatory, and antioxidant effects of curcumin. *Journal of Herbmed Pharmacology*, *7*(4), 211–219.

Clemensen, A. K., Provenza, F. D., Hendrickson, J. R., & Grusak, M. A. (2020). Ecological implications of plant secondary metabolites-phytochemical diversity can enhance agricultural sustainability. *Frontiers in Sustainable Food Systems*, *4*, 547826.

Fassil, A., Mazengia, E., Gebreamanuel, B., Dessie, Y., Kumera, B., Atnkut, B., Mullualem, D., Tsega, A. and Van Damme, P., 2024. Harvesting nature's bounty: Exploring the ethnobotanical landscape of wild edible plants in the Awi Agäw community, Northwestern Ethiopia. *Journal of Ethnobiology and Ethnomedicine*, *20*(1), 59.

Garnatje, T., Peñuelas, J., & Vallès, J. (2017). Ethnobotany, phylogeny, and 'omics' for human health and food security. *Trends in Plant Science*, *22*(3), 187–191.

Ghosh, B., Chandra, I., & Chatterjee, S. (2015). Fenugreek (*Trigonella foenum-graecum* L.) and its necessity. *Fire Journal of Engineering and Technology*, *1*(1), 66–67.

Glick, B. R. (2012). Plant growth-promoting bacteria: Mechanisms and applications. *Scientifica*, *2012*(1), 963401.

Gómez-Baggethun, E., Corbera, E., & Reyes-García, V. (2013). Traditional ecological knowledge and global environmental change: Research findings and policy implications. *Ecology and Society: A Journal of Integrative Science for Resilience and Sustainability*, 18(4), 72.

Hamdane, S., Marcelino, S., Gaspar, P. D., & Paço, A. (2023). Sustainable Agricultural Practices for the Production of Aromatic and Medicinal Plants: Evidence and Recommendations. doi: 10.20944/preprints202307.1380.v1; *Preprints (www.preprints.org)*.

Jiang, L., Chen, Y., Wang, X., Guo, W., Bi, Y., Zhang, C., Wang, J., & Li, M. (2022). New insights explain that organic agriculture as sustainable agriculture enhances the sustainable development of medicinal plants. *Frontiers in Plant Sciences*, *13*, 959810. https://doi.org/10.3389/fpls.2022.959810

Joseph, B. A. B. Y., & Nair, V. M. (2013). *Ocimum sanctum* Linn.(Holy basil): Pharmacology behind its anti-cancerous effect. *International Journal of Pharma and Bio Sciences*, *4*(2), 556.

Jyotirmayee, B., & Mahalik, G. (2022). A review on selected pharmacological activities of *Curcuma longa* L. *International Journal of Food Properties*, *25*(1), 1377–1398. https://doi.org/10.1080/10942912.2022.2082464

Krup, V., Prakash, L. H., & Harini, A. (2013). Pharmacological activities of turmeric (*Curcuma longa* Linn): A review. *Journal of Homeopathy & Ayurvedic Medicine*, *2*(133), ISSN 2167–1206.

kumar Tiwari, A., Jaiswal, S. B. S., & Tiwari, A. (2023). A Comprehensive Review on *Moringa oleifera*: The Green Superfood for Wellness. *Journal of Emerging Technologies and Innovative Research (JETIR)*, *10*(10), (ISSN-2349 516). www.jetir.org.

Kumar, A., Kumar, S., Komal, Ramchiary, N., & Singh, P. (2021). Role of traditional ethnobotanical knowledge and indigenous communities in achieving sustainable development goals. *Sustainability*, *13*(6), 3062.

Kumar, A., Rahal, A., Chakraborty, S., Tiwari, R., Latheef, S. K., & Dhama, K. (2013). *Ocimum sanctum* (Tulsi): A miracle herb and boon to medical science – A review. *International Journal of Agronomy and Plant Production*, *4*(7), 1580–1589.

Kustiati, U., Nugrahaningsih, D. A. A., Kusindarta, D. L., & Wihadmadyatami, H. (2025). Lung cancer: Animal model of lung cancer, molecular carcinogenesis of lung cancer, and antitumor effect of *Ocimum sanctum* against lung cancer. *Open Veterinary Journal*, *15*(2), 482.

Labban, L. (2014). Medicinal and pharmacological properties of Turmeric (*Curcuma longa*): A review. *International Journal of Pharmaceutical and Bio-Medical Science*, *5*(1), 17–23.

Li, X., Yadav, R., & Siddique, K. H. M. (2020). Neglected and underutilized crop species: The key to improving dietary diversity and fighting hunger and malnutrition in Asia and the Pacific. *Frontiers in Nutrition*, *7*, 593711. https://doi.org/10.3389/fnut.2020.593711. PMID: 33330593; PMCID: PMC7710905.

Mahaveerchand, H., & Abdul Salam, A. A. (2024). Environmental, industrial, and health benefits of Moringa oleifera. *Phytochemistry Reviews*, *23*(5), 1497–1556.

Mallick, S. N., Sahoo, T., Naik, S. K., & Panda, P. C. (2020). Ethnobotanical study of Wild Edible Food Plants used by the tribals and rural populations of Odisha, India for food and livelihood security. *Plant Archives (09725210)*, *20*(1), 661–669.

Mazumder, S., Ghosh, S., Bose, S., Chakraborty, A., & Dinda, S. C. (2024). Archaeobotanical elucidation of Moringa oleifera: It's biological, ethnopharmacological and phytochemical activities. *Journal of Natural Remedies*, *24*(4), 703–711.

Memarzia, A., Khazdair, M. R., Behrouz, S., Gholamnezhad, Z., Jafarnezhad, M., Saadat, S., & Boskabady, M. H. (2021). Experimental and clinical reports on anti-inflammatory, antioxidant, and immunomodulatory effects of *Curcuma longa* and curcumin, an updated and comprehensive review. *BioFactors*, *47*(3), 311–350.

Nagulapalli Venkata, K. C., Swaroop, A., Bagchi, D., & Bishayee, A. (2017). A small plant with big benefits: Fenugreek (*Trigonella foenum-graecum* Linn.) for disease prevention and health promotion. *Molecular Nutrition & Food Research*, *61*(6), 1600950.

Ojeda-Linares, C., Casas, A., González-Rivadeneira, T., & Nabhan, G. P. (2024). The dawn of ethnomicrobiology: An interdisciplinary research field on interactions between humans and microorganisms. *Journal of Ethnobiology and Ethnomedicine*, *20*(1), 86.

Pironon, S., Ondo, I., Diazgranados, M., Allkin, R., Baquero, A. C., Cámara-Leret, R., Canteiro, C., Dennehy-Carr, Z., Govaerts, R., Hargreaves, S., & Hudson, A. J. (2024). The global distribution of plants used by humans. *Science*, *383*,293–297. https://doi.org/10.1126/science.adg8028

Porras, G., Chassagne, F., Lyles, J. T., Marquez, L., Dettweiler, M., Salam, A. M., Samarakoon, T., Shabih, S., Farrokhi, D. R., & Quave, C. L. (2020). Ethnobotany and the role of plant natural products in antibiotic drug discovery. *Chemical Reviews*, *121*(6), 3495–3560.

Pretty, J., Benton, T. G., Bharucha, Z. P., Dicks, L. V., Flora, C. B., Godfray, H. C. J., Goulson, D., Hartley, S., Lampkin, N., Morris, C. and Pierzynski, G., Vara Prasad, P. V., Reganold, J., Rockström, J., Smith, P., Thorne, P. & Wratten, S. (2018). Global assessment of agricultural system redesign for sustainable intensification. *Nature Sustainability*, *1*(8), 441–446.

Pullaiah, T., Krishnamurthy, K. V., & Bahadur, B. (2017). *Ethnobotany of India, Volume 3: North-East India and the Andaman and Nicobar Islands*. Apple Academic Press.

Puranik, S., Shukla, L., Kundu, A., Kamil, D., Paul, S., Venkadasamy, G., Salim, R., Singh, S. K., Kumar, D., & Kumar, A. (2023). Exploring potent fungal isolates from sanitary landfill soil for in vitro degradation of dibutyl phthalate. *Journal of Fungi*, *9*(1), 125.

Qureshi, R., Ghazanfar, S. A., Obied, H., Vasileva, V., & Tariq, M. A. (2016). Ethnobotany: A living science for alleviating human suffering. *Evidence-based Complementary and Alternative Medicine: eCAM*, *2016*, 9641692.

Ray, A., Ray, R., & Sreevidya, E. A. (2020). How many wild edible plants do we eat—Their diversity, use, and implications for sustainable food system: An exploratory analysis in India. *Frontiers in Sustainable Food Systems*, *4*, 56. https://doi.org/10.3389/fsufs.2020.00056

Razavi, B. M., Ghasemzadeh Rahbardar, M., & Hosseinzadeh, H. (2021). A review of therapeutic potentials of turmeric (*Curcuma longa*) and its active constituent, curcumin, on inflammatory disorders, pain, and their related patents. *Phytotherapy Research*, *35*(12), 6489–6513.

Saleem, A., Anwar, S., Nawaz, T., Fahad, S., Saud, S., Ur Rahman, T., Khan, M. N. R., & Nawaz, T. (2024). Securing a sustainable future: The climate change threat to agriculture, food security, and sustainable development goals. *Journal of Umm Al-Qura University for Applied Sciences*, *11*(3), 595–611.

Satyendra, M. (2021). Phytochemical efficiency of *Ocimum sanctum* (Tulsi) in health enhancement and disease prevention: A review. *IJARESM*, *9*, 1346–1351.

Sauini, T., Stern da Fonseca-Kruel, V., Baptistela Yazbek, P., Matta, P., Cassas, F., da Cruz, C., Barretto, E. H. P., Santos, M. A., Gomes, M. A. S., Garcia, R. J. F., Honda, S., Passero, L. F. D., Conde, B. E., & Rodrigues, E. (2020). Participatory methods on the recording of traditional knowledge about medicinal plants in Atlantic forest, Ubatuba, São Paulo, Brazil. *PLoS One*, *15*(5), e0232288.

Savita, K. K., Irchhiaya, R., & Tiwari, P. (2021). *Ocimum sanctum* ameliorate anti-oxidant, anti-microbial and immune system for the control of diabetes against coronavirus: An episode. *European Journal of Pharmaceutical and Medical Research*, *8*(7), 308–319.

Singh, A., & Chaturvedi, N. (2021). Medicinal value, herbal gold to combat malnutrition and health benefit of moringa leaf and multipurpose uses of all parts–a review. *Plant Arch*, *21*(1), 1327–1332.

Siva, M., Shanmugam, K. R., Shanmugam, B., Venkata, S. G., Ravi, S., Sathyavelu, R. K., & Mallikarjuna, K. (2016). *Ocimum sanctum*: A review on the pharmacological properties. *International Journal of Basic and Clinical Pharmacology*, *5*(3), 558–565.

Smith, S. E., & Read, D. J. (2010). *Mycorrhizal Symbiosis*. Academic Press.

Srivastava, S., Pandey, V. K., Dash, K. K., Dayal, D., Wal, P., Debnath, B., Singh, R., & Dar, A. H. (2023). Dynamic bioactive properties of nutritional superfood Moringa oleifera: A comprehensive review. *Journal of Agriculture and Food Research*, *14*, 100860.

Sruthy, K. S., Shukla, L., Kundu, A., Singh, S. K., Abdulrahman Alodaini, H., Hatamleh, A. A., Santoyo, G., & Kumar, A. (2023). Effect of microbial consortium constructed with lignolytic ascomycetes fungi on degradation of rice stubble. *Journal of Fungi*, 9(5), 567.

Suman, J., Rakshit, A., Ogireddy, S. D., Singh, S., Gupta, C., & Chandrakala, J. (2022). Microbiome as a key player in sustainable agriculture and human health. *Frontiers in Soil Science*, *2*, 821589.

Suminguit, V. J. (2005). Ethnobotanical documentation: A user's guide. *Asia-Pacific Database on Intangible Cultural Heritage (ICH) by Asia-Pacific Cultural Centre for UNESCO (ACCU),* Paris.

Suresh, A., Rao, T. C., Solanki, S., Suresh, M. V., Menon, B., & Raghavendran, K. (2022). The holy basil administration diminishes the NF-kB expression and protects alveolar epithelial cells from pneumonia infection through interferon gamma. *Phytotherapy Research*, *36*(4), 1822–1835.

Syed, Q. A., Rashid, Z., Ahmad, M. H., Shukat, R., Ishaq, A., Muhammad, N., & Ur Rahman, H. U. (2020). Nutritional and therapeutic properties of fenugreek (*Trigonella foenum-graecum*): A review. *International Journal of Food Properties*, *23*(1), 1777–1791. https://doi.org/10.1080/10942912.2020.1825482

Tadesse, D., Masresha, G., Lulekal, E., & Wondafrash, M. (2024). A systematic review exploring the diversity and food security potential of wild edible plants in Ethiopia. *Scientific Reports*, *14*(1), 17821.

Verma, R. K., Kumari, P., Maurya, R. K., Kumar, V., Verma, R. B., & Singh, R. K. (2018). Medicinal properties of turmeric (*Curcuma longa* L.): A review. *International Journal of Chemistry Studies*, *6*(4), 1354–1357.

Visuvanathan, T., Than, L. T. L., Stanslas, J., Chew, S. Y., & Vellasamy, S. (2022). Revisiting *Trigonella foenum-graecum* L.: Pharmacology and therapeutic potentialities. *Plants*, *11*(11), 1450.

Zameer, S., Najmi, A. K., Vohora, D., & Akhtar, M. (2017): A review on therapeutic potentials of *Trigonella foenum graecum* (fenugreek) and its chemical constituents in neurological disorders: Complementary roles to its hypolipidemic, hypoglycemic, and antioxidant potential. *Nutritional Neuroscience*. https://doi.org/10.1080/1028415X.2017.1327200

Zayed, M. S. (2012). Improvement of growth and nutritional quality of Moringa oleifera using different biofertilizers. *Annals of Agricultural Sciences*, *57*(1), 53–62.

7 Battling Salinity

Microbial Soil Fertility Programs for Arid Lands

Neha Singh, Nisha Yadav, and Shaili Srivastava

7.1 INTRODUCTION

Currently, 46.2% of the surface of the earth is made up of semiarid and dry regions, home to 3 billion people. Global warming is driving the expansion of arid and semi-arid areas, particularly in regions where agriculture relies heavily on crop cultivation as the primary source of food production (Mirzabaev et al., 2019). Abiotic factors such as salinity, aridity, and elevated temperatures exert substantial influence on plant development, growth, and yield. High-temperature stress, in particular, negatively affects crop productivity and soil fertility on a global scale. These adverse conditions are increasingly driven by anthropogenic environmental degradation and accelerated climate change, which collectively exacerbate the vulnerability of agricultural systems (Ma & Yang, 2017).

Drought stress is caused by high temperatures and prolonged precipitation deficits, which reduce and aggravate soil moisture and impede plant growth (Singh et al., 2022). As a result, there is an increase in mineral buildup, particularly salt ions, and water evaporation, and the soil becomes unstructured and infertile (Yadav et al., 2020). Soil is classified as salinized when its electrical conductivity (EC) surpasses 4 dS/m and the sodium chloride (NaCl) concentration exceeds 40 mM. Such conditions impose substantial osmotic and ionic stress on plants and soil microbial communities, thereby adversely affecting their physiological functions, metabolic activities, and overall ecosystem productivity. This tension may be made worse by the use of synthetic fertilizers, saline watering, ocean salt deposition caused by rain or wind, and a lack of precipitation (Seleiman et al., 2021). Salinity impairs plant growth by causing oxidative and osmotic stress, disrupting soil microbial communities, reducing photosynthesis, and degrading soil quality and crop productivity. The "plant microbiota" is a varied population of microorganisms connected to plants that contain bacteria, fungi, protists, and archaea (Wang et al., 2017). Microorganisms are fundamental regulators of plant growth and development, colonizing a wide range of plant-associated niches, including the phyllosphere, endosphere, rhizosphere, and root tissues. Within these habitats, they interact with their host plants through diverse ecological relationships, mutualistic, commensal, or parasitic, thereby influencing physiological processes, nutrient acquisition, stress tolerance, and overall plant health (Etesami & Beattie, 2017). Plant microbiota are mostly found in the soil, while they can also be found in seeds, water, air, insects, and mammals. Environmental stresses

DOI: 10.1201/9781003664451-7

like salinity and drought, along with soil properties regarding the genetic makeup of plants, have an important effect on microbiological diversity and function in the plant microbiome (Corwin, 2021). For instance, in addition to enhancing genetic and metabolic capacities, the plant microbiome offers defensive mechanisms, metabolites, and supportive activities, all of which significantly boost plant fitness.

Additionally, it intricates microbiological communities. Plant growth–promoting bacteria (PGPB) are unique in their capacity to boost plant durability by increased stress tolerance, phytohormone production, minerals solubility, and nitrogen fixing (Orozco-Mosqueda et al., 2021). Plants may require more water to thrive where water supplies are scarce because of nitrogen-fixing bacteria. By lowering the demand using artificial fertilizers, which is essential in delicate dry areas where soil damage might result from overuse of chemical inputs, nitrogen-fixing symbioses promote sustainable farming practices. By increasing resistance and promoting plant survival and productivity, these interactions may assist plants survive environmental stresses like high temperatures and salinity that are common in arid areas. Certain techniques are needed for the management and reclamation of such soils in order to attain sustained productivity. In the past, both chemical and physical techniques have been used to restore saline soil (Chialva et al., 2022). In chemical processes, soluble salts are eliminated by physical techniques like flushing, leaching, and scraping as well as neutralizing substances in alkaline and acidic soils. Even though they were produced, crops that endure salt were unable to compete globally and did badly in areas with high salt concentrations.

7.2 ARID AND SALINE SOILS

The two abiotic factors that most significantly limit plant growth and productivity are drought and salinity, soil salinization and aridity, are impacted by numerous environmental and human impacts, including drainage, irrigation water quality, shallow aquifers, sea floods, and saline groundwater intrusion that encourage buildup of salt, low evapotranspiration-related vegetation cover, and agronomic techniques like fertilization and the use of pesticides (Müller et al., 2016). Generally speaking, “primary” is the classification for soil salinity when salinity accumulates naturally and “secondary” when resource management is to blame. Dry and semiarid areas with naturally saline soils can result from prolonged exposure to weather elements including winds and precipitation. In some regions, there is insufficient precipitation to remove soluble salts from the soil (Abd-Alla et al., 2023). Salinity can be classified into three major categories: transient salinity, groundwater associated (commonly knows as irrigation salinity) and dryland salinity. They are thought to be the planet’s largest biome, occupying 41% of the planet’s surface. According to recent estimations, the size of drylands has reached up to about 45% of the Earth’s total terrestrial surface area, which reflects the considerable growth of arid and semi-arid zones, including Siberia. The expansion of arid lands has been observed not only in higher latitude zones, for example, in Alaska, but also in lower latitudes, especially central and south Asia. Notably, China contributes approximately 55% of the newly emerging arid lands within Asia, representing about 31% of the global total. Drought has become more prevalent throughout North and Southeast Africa, especially in Morocco, Algeria, and Libya.

Soil salinity and sodicity can be effectively mitigated through an integrated application intervention. Chemical procedures such as calcium chloride dihydrate ($CaCl_2{\cdot}2H_2O$) and gypsum ($CaSO_4{\cdot}2H_2O$) promote the displacement of exchangeable sodium (Na^+) from soil colloids by supplying readily soluble calcium ions (Ca^{2+}), thereby improving soil structure and permeability. Concurrently, hydrological practices, particularly controlled leaching and improved irrigation management, facilitate the downward movement of soluble salts from the root zone. Biological approaches, including the use of plant growth–promoting biofertilizers and genetically engineered microbial consortia, enhance sodium mobilization and uptake, contributing to the restoration of soil health and improved crop performance in arid and semiarid agroecosystems (Wang et al., 2014a). Globally, more than 800 million hectares of agricultural land are affected by soil salinity, representing a major constraint to crop productivity and soil health. A soil is classified as *saline* when it contains $\geq$0.25% soluble salts, exhibits an electrical conductivity of the saturation extract (ECe) greater than 4 dS/m at 25°C, and maintains a pH $\leq$ 8.5. Sodium chloride (NaCl) is the predominant salt contributing to salinity; however, considerable accumulation of calcium, magnesium, and potassium carbonates and sulfates may also occur. When excessive sodium is adsorbed onto soil clay particles, the soil may transition from saline to *sodic* conditions, typically characterized by an ECe < 4 dS/m coupled with an elevated pH (> 8.5). Sodic soils exhibit pronounced physicochemical deterioration, including clay dispersion, reduced aggregate stability, impaired infiltration, and diminished hydraulic conductivity, ultimately restricting plant growth and ecosystem functioning.

In addition to having negative impacts on human nutrition and health, overaccumulation of soluble salt in crops has detrimental effects on farming, leading to low crop yields and a substantial financial burden. They are believed to be the biggest biome on Earth, occupying 41% of the planet's surface (Wang et al., 2019).

7.2.1 Saline Soil Microorganism Ecology

Various microbial communities have a big influence on the nutrient cycle. Environmental stress in the soil has a detrimental effect on microorganisms, decreasing their capacity to survive. When stress-tolerant mechanisms are started and activated, surviving cells suffer a metabolic cost. The primary pressures on soil microbe diversity are low moisture and saltiness, while hot, dry weather reduces microorganism activity and metabolism. Because transpiration increases the amount of water that plants need, the negative effects have a greater impact on the rhizosphere of the plant. Because microflora is crucial for maintaining balance between the bacteria' cytoplasm and their surroundings, strong bioenergetics are necessary for living things. The microorganisms must have substantial amount sufficient energy that is adequate for oxygen regulation, because they survive via keeping sodium ions out of the cell (Mdlambuzi et al., 2022).

Plants with high sodium (Na^+) ion concentrations develop more slowly and show necrotic signs. High concentrations of Cl^- degrade chlorophyll. In contrast to the negative soil water potential, the high salt content restricts how much water plant roots can take. High salt concentrations further hinder the rhizosphere's capacity to absorb plant nutrients. Salt stress exposure modulates the expression and functional dynamics of the RuBisCO enzyme in microorganisms, thereby enhancing carbon substrate availability and facilitating more efficient CO_2 fixation to support energy metabolism and related cellular processes. A range of osmo-tolerant protein types are produced in challenging circumstances to aid plants in retaining water and surviving prolonged exposure to extreme salinity stress. Salt-tolerant bacteria can be categorized into four distinct groups based on their sodium chloride requirements for optimal growth. Non-halophilic bacteria proliferate at NaCl concentrations below 0.2 M, whereas slightly halophilic species exhibit preferential growth between 0.2 and 0.5 M. Moderately halophilic organisms demonstrate optimal proliferation at 0.5–2.5 M NaCl, while extreme halophiles require >2.5 M NaCl to sustain growth and metabolic activity. Halotolerant bacteria, however, do not depend on high salinity for growth and perform well even at salt levels below approximately 0.2 M (around 1%). Genera *Bacillus*, *Pseudomonas*, *Micrococcus*, and *Alcaligenes* are the most common halophilic bacteria found in saline soil. In saline soil in various parts of Spain, Gram-positive, rod-shaped *Bacillus*, halotolerant bacteria that produce endospores, were found. Plant growth and development depend on nitrogen (N) recycling processes, such as adsorption and conversion through microbial processes. The opposite of beneficiation is immobility, which is the process that converts transforming biological nitrogen into an inorganic form (Li & Wu, 2008). NaCl considerably slowed down both nitrification and mineralization. Soil treated with 4,000 mg NaCl kg^{-1} exhibited highest level inhibition. Sodium chloride had a comparatively greater inhibitory effect on N mineralization in NH_4^+-treated soils. According to research findings, bacteria that have consumed NO_3^- are more vulnerable to NaCl. Furthermore, the immobilization of the N is postponed by the addition of NaCl. Because salinity affects the diversity of microorganisms, different microbial genotypes have different levels of resistance. Fungi appear to be more susceptible than other microbes to osmotic stress and the salt environment. When soil with varying NaCl concentrations was salinized, the overall fungus count was significantly decreased. In a similar vein, bacteria and actinobacteria are significantly reduced when the salinity level is greater than 5%. Halotolerant microorganisms accumulate the ions required for cell metabolism. The microbial community experiences increased physiological activity as a result of this adaptive mechanism, which has repercussions. The cell uses the substrate less frequently. Considering soil water potential including both osmotic and matric components provides a better understanding of how microbial biomass and metabolic activity respond to salt stress, especially when salinity levels increase in affected soils (Rengasamy, 2006).

In situations of salt stress, EC serves as a marker of microbiological stress. An essential labile component of soil organic matter (OM), microbial biomass serves a variety of purposes, including recycling and transforming soil nutrients and OM and providing plant nutrients (Polash et al., 2019). The enzyme responsible for regulating multiple biogeochemical processes in soil is also abundantly present in microbial secretions. Microbial activities within the rhizosphere play a pivotal role in modulating nutrient availability and facilitating their uptake by plants. Numerous factors affect the microbial population and its function, which in turn affects plant development and supply of nutrients.

7.3 HALOPHILIC BACTERIA

The osmotic pressure in their environment can be balanced by these halophilic microorganisms. Furthermore, halophiles are organisms that need salt for proper growth and development and may endure in environments that are extremely salinized. They fall into the three categories of life, bacteria, eukarya, and archaea, and are incredibly varied. As dormant cells, they can be found in rock salt crystals, soda lakes, and salt ponds. There are two kinds of organisms: those that need salt for growth and development and those that can tolerate it (Sparks, 2003). Halophilic bacteria can endure and flourish in environments that are susceptible to high salt concentrations and associated extreme conditions like low water activity, intense UV radiation, and, in some cases, high temperatures or alkaline pH. They are divided into three categories: mild, moderate, and extreme halophiles. The ideal range of concentrations for slight halophile growth is 0.2–0.5 M (1%–3%) NaCl. 2.5–5.2 M (15%–30%) NaCl level is sufficient for extreme halophile growth, while 0.5–2.5 M (3%–15%) NaCl level is sufficient for moderate halophile growth. Depending on the environment, halophiles can be aerobic, anaerobic, heterotrophic, phototrophic, or chemoautotrophic (Zaman et al., 2018). Numerous environmental abiotic stresses, including droughts, cold salinity, nutritional deficiencies, infections, heavy metals, etc., are encountered by plants in agriculture. This stress condition disrupts normal plant growth and development. One main environmental stressor that affects soil is salinity, which is caused by high temperatures and limited rainfall, with inadequate water in dry and semiarid areas. When there is salt stress, halophilic bacteria help plants to survive by adapting to salinity in a new way (Munns & Tester, 2008).

Osmolyte synthesis, antioxidant enzymes, hormones, and ion exclusion are just a few of the physiological and biochemical tactics that plants use to survive in salt-stressed conditions. Additionally, the defense mechanism, microbial community in soil, including halophilic bacteria, is important for boosting the soil's ability to withstand salt.

7.3.1 Classification of Halophilic Bacteria

Microbes that flourish in highly saline conditions are called halophilic microbes. Halophiles belonging to the family of Halobacteriaceae are some of the most commonly recognized halophiles. Halanaerobium praevalens, which is considered to be one of the early discovered halophilic bacteria, was found in the Great Salt Lake (Zeikus et al., 1983). Following these findings, the developments in phylogeny, membrane lipids, and molecular taxonomy have enabled scientists to identify many new genera and species of halophilic bacteria. The list of halophilic organisms is shown in Table 7.1.

TABLE 7.1
Halophilic Bacterial Species with the Salt-Tolerant Range

Halophilic Bacterial Species	Range of Salinity for Development and Growth (%)	References
Aliifodinibius saliphilus sp. nov.	2–24	Cho and Whang (2020)
Alkalibacterium gilvum	0–17.5	Ishikawa et al. (2013)
Amphibacillus cookii	6–26	Pugin et al. (2012)
Arhodomonas recens	2–25	Saralov et al. (2012)
Bacillus daqingensis	0–16	Wang et al. (2014a)
Desulfohalophilus alkaliarsenatis	12.5–33	Blum et al. (2012)
Fodinibius salinus	10–15	Wang et al. (2012)
Halanaerobacter jeridensis	6–30	Mezghani et al. (2012)
Halanaerobium sehlinense	5–30	Abdeljabbar et al. (2013)
Halanaerocella petrolearia	6–26	Gales et al. (2011)
Halobellus salinus	15–30	Cui et al. (2012)
Halomicroarcula pellucida	20–30	Echigo et al. (2013)
Halomonas huangheensis	1–20	Miao et al. (2014)
Halomonas sambharensis	5–8	Kushwaha et al. (2020)
Halomonas urmiana sp.	0.4–19	Khan et al. (2020)
Kangiella spongicola	2–15	Ahn et al. (2011)
Lentibacillus saliphilus sp. nov. (type strain YIM 93176^T)	1–23	Wang et al. (2020)
Limimonas halophila	15–30	Amoozegar et al. (2013b)
Marinobacter halodurans sp. nov.	2–17	Yoo et al. (2020)
Natribacillus halophilus	7–23	Echigo et al. (2013)
Oceanicola flagellatus	0–21	Liu and Yang (2014)
Salifodinibacter halophilus	25	Parada-Pinilla et al. (2020)
Salinibacter iranicus	12–30	Makhdoumi-Kakhki et al. (2012)
Salisediminibacterium halotolerans	3–30	Jiang et al. (2011)
Saliterribacillus persicus	0.5–22.5	Amoozegar et al. (2013a)
Spiribacter salinus	10–25	León et al. (2014)

7.3.2 Mechanisms by Which Halophilic Bacteria Adapt to Saline Environments

Living bacteria can survive in hostile environments due to their capacity for adaptation. Bacteria that are not acclimated to saltwater settings shrink and finally die when they lose water because they lack cell architecture and activity. To preserve their cellular integrity and functionality under high-salt conditions, halophilic bacteria have evolved two distinct adaptive mechanisms that minimize water loss and enable them to tolerate extreme salinity (Acosta-Motos et al., 2017). Figure 7.1 describes mechanisms by which halophilic bacteria adapt to saline environments.

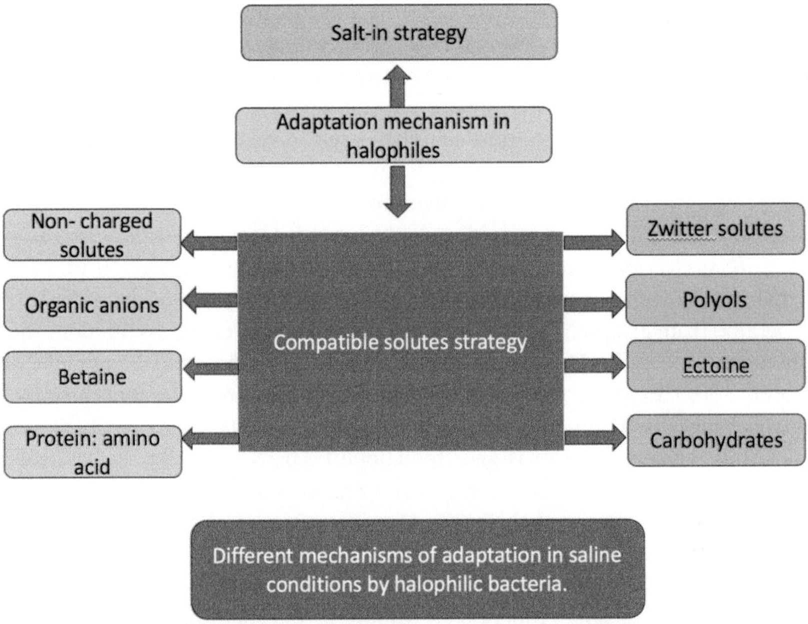

FIGURE 7.1 Different mechanisms of adaptation in saline conditions by halophilic bacteria.

The compatible solute technique is the second tactic, while the salt-in method is the first. The significant KCl accumulation in the bacterial cells keeps the internal and external environments' osmotic equilibrium intact. A cell uses this tactic by altering some physiological metabolisms, including the synthesis of cellular components, enzyme activity, and the functions and configurations of particular organelles. By building up inorganic ions inside their cells, excessive salt-in technique shields halophiles in a salty setting and maintains the equilibrium of salt levels in the surroundings. The cells of bacteria accumulate a large amount of KCl, which keeps the oscillating equal conditions of the internal and external environments. The Cl^- pumps, which are made up of halophiles, move Cl^- from the surroundings into the cytoplasm. At both ends of the channel, arginines and lysines are positioned to improve the absorption and release of Cl^- (Rengasamy, 2010).

The majority of halophilic bacteria shield the cell from elevated salt concentrations by accumulating suitable solutes, including inorganic (K^+, Mg^{2+}, Na^+) and organic (proline, betaine, ectoine, and trehalose) solutes. However, to counteract external osmotic pressure, the majority of bacteria do not have an internal mechanism for the active transfer of water. Consequently, the production or transport of a set of appropriate solutes maintains the internal environment without interfering with the cell's metabolic processes.

According to their chemical composition, the three types of suitable solutes are anionic, zwitterionic, and non-charged. In highly salinized environments, halophilic bacteria employ organic anions to balance their internal environment. Ectoine and L-glutamate are produced by halophilic bacteria, such as *Halomonas* and *Halobacterium*, respectively, in order to live in salinity-stressed environments (Saffriel & Adeel, 2005). Sorbitol, arabitol, glycerol, and mannitol are polyol molecules that certain halotolerant bacteria, such as *Bacillus*, *Pseudomonas*, *Aeromonas*, and *Zymomonas*, use for osmoadaptation in salt-stressed environments. Zwitter solutes generated from neutral amino acids are used as osmolytes by halophilic bacteria under salt stress. As an osmolyte, betaine is a naturally occurring, negatively charged substance that helps cells maintain internal equilibrium by controlling within the cells. This helps cells deal with excessive osmotic stress. *Halomonas*, *Virgibacillus*, *Oceanobacillus*, and *Polaribacter* are among the halophilic bacteria that may methylate the main amine to make a quaternary amine in order to produce betaine from glycine. Through glycine methylation or choline oxidation, several methanogens, including *Methanohalophilus* and *Methanohalobium*, can make betaine (Oren, 2002). Ectoine, a cyclic tetrahydropyrimidine compound that can either be produced internally within the cell or taken up from the outside environment, acts as a compatible solute for salt resistance by halophilic bacteria. *Halorhodospira halochloris*, a bacterium recovered from the hypersaline Mono Lake environment, was found to possess this (Jiang et al., 2007). *Methylophaga*, *Methyllarcula*, *Halomonas*, *Oceanobacillus*, and *Nesterenenkonia* are among the halotolerant and halophilic bacteria that have been identified to contain ectoine osmolytes. A variety of polar, uncharged small molecules function as osmolytes that help microorganisms withstand high-salinity conditions. Among them, glycerol is commonly accumulated by several bacteria and halotolerant yeast species (Garabito et al., 1998).

Several sugars such as trehalose act as protective osmolytes in halotolerant and halophilic microorganisms, enabling them to withstand extreme salinity, temperature fluctuations, and desiccation. In many marine proteobacteria and cyanobacteria, sucrose accumulation is a common response to salt-induced stress. Additionally, certain amino acid–based osmoprotectants, including proline, carboxamides, and acetylated glutamine dipeptides, help safeguard cellular structures against high-salt conditions. These osmolytes are especially prevalent in marine phototrophic microbes and in halophilic purple sulfur bacteria (Llamas et al., 2008).

7.4 HALOPHILIC BACTERIA'S FUNCTION IN PROMOTING PLANT GROWTH UNDER SALT STRESS

Every biotic thing, including plants, is exposed to the harsh circumstances of the earth during its growth and development. They will either fight or come up with a different strategy to get out of the stressful situation. Plants fight back against the unfavorable conditions since they are unable to flee them due to their extreme fragility and sessility (Mandeel, 2006). To withstand the harmful effects of various abiotic stressors, halophilic bacteria utilize diverse adaptive strategies that enhance their growth, survival, and overall tolerance.

7.4.1 The Role of Bacterial Phytohormones

Low-concentration organic substances are called bacterial phytohormones, which affect the physiological and biological functions of plants (Juniper & Abbott 2006). These minuscule amounts of bacterial phytohormones have an impact on the regulation of various processes related to plant development and differentiation. Bacteria associated with plant roots can release developmental signals that influence the parent plant's adaptive responses. PGPB biosynthesize a diverse range of phytohormones including gibberellins, cytokinins, abscisic acid, and indole-3-acetic acid (IAA) which act synergistically to regulate and enhance plant growth, development, and physiological performance. By significantly boosting root length as well as surface area, all these phytohormones prolong roots stimulation, which improves plant health in situations when salt is present (Liang et al., 2007). This increases nutrient absorption.

7.4.2 Aminocyclopropane-1-Carboxylate Deaminase

Ethylene is a gaseous phytohormone that plays a critical role in regulating diverse physiological and developmental processes in plants. Despite its necessity at very low concentrations to elicit biological responses, ethylene is inherently highly flammable, making its handling and application a safety-sensitive aspect in experimental and agricultural settings. It regulates nodulation, adventitious root formation, vegetative growth, and root-zone signaling under salt stress (Herrmann et al., 2005). However, when plants encounter abiotic stresses, they often synthesize excess ethylene, which can negatively impact growth. To address this limitation, chemical inhibitors such as aminoethoxyvinylglycine (AVG) and cobalt ions have been employed to suppress ethylene biosynthesis; however, their high cost and associated environmental risks restrict their practical application. A sustainable and biologically safer alternative is the enzyme ACC deaminase, synthesized by various halotolerant bacterial taxa. This enzyme catalyzes the cleavage of 1-aminocyclopropane-1-carboxylate (ACC) into α-ketobutyrate and ammonia, thereby lowering endogenous ACC levels and mitigating ethylene overproduction in plants exposed to abiotic stress (Larsen, 1986).

7.4.3 Phosphate Solubilization

In plants, ethylene participates in nodulation, adventitious root initiation, vegetative growth regulation, and the transmission of stress-related signals within the rhizosphere (Herrmann et al., 2005). Under abiotic stress conditions, plants frequently overproduce ethylene, resulting in the suppression of growth and disruption of normal developmental processes. Although chemical inhibitors such as AVG and cobalt ions (Co^{2+}) are commonly employed to attenuate ethylene biosynthesis, their elevated cost and associated environmental concerns significantly limit their large-scale or long-term application. An eco-sustainable strategy for improving crop performance in saline environments involves the use of salt-tolerant, plant growth–promoting rhizobacteria (PGPR) that produce aminocyclopropane-1-carboxylate (ACC) deaminase. This enzyme mitigates stress-induced ethylene biosynthesis by catalyzing the conversion of ACC into α-ketobutyrate and ammonia, thereby reducing ethylene-mediated inhibition of plant growth. Larsen (1986) reported that several rhizobacterial isolates can tolerate salinity levels as high as 60 g/L NaCl while maintaining efficient phosphate-solubilizing capabilities. Moreover, *Pseudomonas* spp. has demonstrated significant growth-promoting effects on *Zea mays* L. under saline stress. In addition, inoculation with phosphate-solubilizing bacteria such as

Herbaspirillum seropedicae and *Burkholderia* spp. has been shown to enhance crop biomass by approximately 1.5%–21% in salt-affected soils, underscoring their potential for sustainable agricultural applications.

7.4.4 Antioxidative Activity

Salt stress induces the excessive generation of reactive oxygen species (ROS), which can oxidatively damage essential biomolecules such as proteins, lipids, and nucleic acids, ultimately leading to cellular dysfunction and plant mortality. To counteract this oxidative burden, plants possess intricate antioxidant defense systems that mitigate ROS toxicity. These protective mechanisms include a diverse array of nonenzymatic antioxidants that contribute to ROS scavenging (Edbeib et al., 2016). In addition, several halotolerant PGPR, including *Rhizobium leguminosarum* and *Serratia proteamaculans*, enhance plant tolerance to salt stress by producing key antioxidant enzymes such as superoxide dismutase (SOD), peroxidase (POX), and catalase (CAT), which collectively reduce ROS accumulation and protect cellular integrity. It has been discovered that when soybean plants are exposed to 100 mM NaCl, salinity-tolerant bacteria (*Pseudomonas simiae* AU) boost the exhibition of the CAT and peroxidase genes. PGPR inoculation lessened negative effects of the oxidative stress in saline-stressed environments. They moved for less sunny locations for nonenzymatic operations, which reduced their exposure to ROS (Tanimura et al., 2016). The synthesis of pigments and the packaging of DNA with chromatins associated proteins provides distinct targets for species-specific attacks. Some antioxidants that are not enzymatic also prevent ROS. Conversely, the enzymatic approach generates many enzymes, including glutathione peroxidases, peroxiredoxins, catalases, and superoxide dismutase, without producing more reactive species. Antioxidant enzymes either find and break down the toxic products or change them into less toxic compounds. Additionally, these techniques keep reactive species like ROS at the proper physiological levels.

7.4.5 Siderophore Producers

Iron is a vital micronutrient for plants, functioning as a cofactor in numerous metabolic and redox processes that are critical for optimal growth and development. Despite its abundance in soil, iron is predominantly present as ferric hydroxide, a highly insoluble form that markedly limits its bioavailability. Consequently, plants frequently experience iron deficiency due to the restricted solubility and mobility of this oxidized iron species in the rhizosphere. Under abiotic stress, iron becomes even more limited, creating challenges for both plants and microorganisms (Ciulla et al., 1999). To overcome this limitation, diverse soil microorganisms contribute to mobilizing and solubilizing iron, thereby enhancing its accessibility to plants. By producing siderophores, halotolerant bacteria improve iron availability and help plants survive in salt-stressed environments. Iron is bound and transported into cells by tiny molecules called siderophores. The hydroxamate-based siderophores were generated by the *Chromocurvus halotolerans* strain EG19 and *Halobacillus trueperi* MXM-16.

7.5 ARBUSCULAR MYCORRHIZAL FUNGAL SPECIES

Mycorrhizal fungi establish symbiotic associations with the roots of terrestrial plants and play a pivotal role in enhancing plant tolerance to saline environments. It has been demonstrated that mycorrhizal inoculation modulates ionic homeostasis, increases nutrient solubilization and uptake, enhances photosynthetic efficiency, and improves both morphological and metabolic attributes of plants exposed to salt stress. In particular, arbuscular mycorrhizal (AM) fungi facilitate plant resilience to salinity by extending the functional root absorption zone through their extraradical hyphal networks, thereby improving water and nutrient acquisition. Additionally, AM fungi induce structural and physiological adjustments in host plants, stimulate the production of growth-promoting phytohormones, and interact synergistically with plant metabolic pathways, collectively contributing to improved stress mitigation under saline conditions. Arbuscular mycorrhizal fungi (AMF) not only enhance water and nutrient acquisition but also promote nutrient solubilization, stimulate root system architecture, and improve the overall nutritional status of plants subjected to salt stress. By modulating membrane-associated transport proteins, AMF help maintain ion homeostasis within the host plant, thereby increasing the effective availability of essential nutrients. In several crop species, including lettuce and *Solanum lycopersicum* L., AMF colonization has been reported to substantially elevate chlorophyll content. Under saline conditions, plants typically activate a suite of protective mechanisms, such as the accumulation of compatible osmolytes and the induction of antioxidant defense pathways, to counteract oxidative damage. The presence of AMF further amplifies these protective responses by significantly enhancing the activities of major antioxidant enzymes, including peroxidase, catalase, and superoxide dismutase, thereby improving plant tolerance to salinity.

Mycorrhizal inoculation markedly enhanced the activities of key antioxidant enzymes catalase (CAT), ascorbate peroxidase (AsA-POD), peroxidase (POD), and superoxide dismutase (SOD) during the early phases of salt stress. Consistent with these observations, Hajiboland et al. (2010) and Huang et al. (2010) demonstrated that AMF alleviate salt-induced oxidative damage by upregulating antioxidant defense pathways and promoting the detoxification of ROS. In salinity-stressed peanut plants, AMF colonization further resulted in a significant increase in proline accumulation. Similarly, Sannazzaro et al. (2007) reported that inoculation with *Glomus intraradices* enhanced proline and polyamine levels in two *Lotus glaber* genotypes. Parallel findings have been documented in *Glycine max* and *Cyamopsis tetragonoloba*, where mycorrhizal symbiosis under salt stress led to elevated proline concentrations, suggesting a conserved osmoprotective response across species.

In addition to these physiological responses, AMF substantially improve soil physicochemical properties. The extraradical hyphae secrete glomalin, a recalcitrant glycoprotein that plays a central role in soil aggregation. Although the molecular mechanisms and genetic determinants underpinning glomalin biosynthesis remain poorly resolved, existing evidence indicates that glomalin and related AMF-derived proteins form a stabilizing matrix around hyphal networks, thereby enhancing soil structural integrity and promoting long-term soil health.

7.6 CYANOBACTERIA

Cyanobacteria are prokaryotic, photosynthetic microorganisms capable of fixing both carbon (C) and nitrogen (N). In rice agroecosystems, cyanobacteria commonly referred to as blue-green algae (BGA) can contribute approximately 25%–30% of the total nitrogen input (N ha^{-1} $season^{-1}$) through biological nitrogen fixation (BNF). Beyond nutrient provisioning, BGA also enhance soil health by synthesizing and releasing phytohormones, secondary metabolites, and extracellular polysaccharides. These biomolecules improve soil structure, aggregation, and moisture retention, particularly under conditions of excessive chemical fertilizer use or salt accumulation. In coastal regions of Orissa, eight cyanobacterial strains belonging to the genera *Nostoc*, *Anabaena*, *Calothrix*, and *Aulosira* have been identified and evaluated under field conditions across varying pH, salinity, and drought stress regimes. Due to their inherent tolerance to elevated sodium concentrations during wet seasons, these strains have shown significant potential in the amelioration of sodic soils (Lemanceau et al., 2005). Combining inoculation of gypsum and cyanobacteria induces beneficial alterations in soil physicochemical properties, facilitating the reclamation of salt-affected soils.

Experimental results demonstrate that cyanobacterial application leads to a marked reduction in soil pH, EC, and exchangeable sodium (Na^+), accompanied by an increase in organic carbon (OC) levels. Notably, *Nostoc calcicola* in combination with gypsum exhibits synergistic effects in the remediation of saline–alkaline soils. However, certain cyanobacterial species are known to produce cyanotoxins under environmental stress, and these compounds may exert concentration-dependent phytotoxic effects, influencing seed germination and plant development.

The cyanobacteria grow more readily in alkaline soils with elevated pH and Na^+ concentrations because pH drops noticeably. The various organic compounds that are produced by cyanobacterial activity in soils also contribute to the soil fertility being sustained year after year (Savary et al., 2000). The pH level of the saline/alkaline stress soil decreases when *Nostoc calcicole* is added, demonstrating increased soil fertility. The ability of *N. calcicola* to withstand salt may be the reason for its dominance in saline/alkaline-stressed soils, indicating that it may be a suitable biological strategy for soil reclamation.

7.7 RHIZOSPHERE – A SCENARIO FOR PLANT–MICROBE INTERACTION

The term "rhizosphere," which refers to the region of soil surrounding plant roots and characterized by a higher microbial density than uncultivated soil, was first coined by Lorenz Hiltner (in 1904. Although there have been other definitions of rhizosphere, the Hiltner one is still widely used. Through a secretion mechanism called root exudation or rhizodeposition, the root releases various chemicals in gaseous, liquid, or solid states, which accounts for the greater bacterial population Hiltner observed in the rhizosphere. The exudates secretion activity, including organic acids, lysates, sugars, vitamins, mucilage, and several other compounds, is what causes the movement of carbon from photosynthetic processes from the root to the soil (Bueno Batista & Dixon, 2019).

It is generally estimated that approximately 11% of all fixed carbon, or 27% of carbon allocated to roots, is released into the surrounding soil. However, these amounts, as well as the composition of root exudates, are influenced by factors such as plant species, age, nutritional status, and exposure to biotic and abiotic stresses. There is a strong relationship between the composition of root exudates and the microorganisms associated with them. These exudates can profoundly influence plant growth by acting as chemical signals that attract or repel bacteria, fungi, and viruses. Seeds and fruits have a positive impact on yield, health, and nutritional value. This dynamic process is referred to as the rhizosphere feedback loop. The soil microbiota detects changes in the surroundings of the root caused by rhizodeposition (Athar, 1998). Subsequently, several bacterial populations alter their own physiology in response to the environmental change brought about by the emission of root exudates. Selection favors the bacterial populations that are most suited to this novel situation as a result. This underlies the coevolution between plants and their associated microbiota. Changes in the composition and activity of microbial communities can have significant effects on plant growth. Microorganisms contribute by facilitating nutrient acquisition and enhancing rhizodeposition (Bano et al., 2010).

In addition to generating and consuming trace gases, they may have an impact on the hydrophobicity, acidity, or water presence in the soil. They can also decompose xenobiotic chemicals and chelate metals. But there are also harmful and sometimes fatal parasite interactions with plants. Soil-borne diseases reduce crop product quality and yield on agricultural land, which reduces profits. For example, plant diseases in France result in losses ranging from 5% to 96% of potato production, in contrast to Asia, output losses from bacteria and viruses in rice varied between 24% and 41%. The kind of soil also affects the stability and organization of the community; however, the plant chooses the best species-specific microbial makeup. In actuality, separate microbial communities are hosted by various plant species, and certain microbial taxa are more strongly linked to particular plant species. However, many soil bacterial taxa can be cosmopolitan, meaning they can connect with a variety of plant taxa (Cerezini et al., 2020). Plant development and its associated microbiota have such a close relationship that we have called the plant a holobiont or superorganism, which is the direct result of extremely intricate interactions between microbes and plants.

7.7.1 Rhizobia Use in Arid Legume Cultures

Nitrogen is a crucial element for plant growth and productivity. However, most nitrogen in the soil exists in forms that plants cannot directly assimilate; diazotrophic bacteria are crucial for fixing it there, which plants then absorb as ammonia. One of the best options for preserving the nitrogen supply to appropriate leguminous species planted in arid areas is rhizobia, which, when attached to leguminous plants, are Gram-negative soil bacteria that can fix nitrogen from the atmosphere (Zaidi et al., 2009). An efficient legume rhizobium symbiosis can form in difficult conditions, since rhizobia can withstand the negative impact of osmotic stress and desiccation.

7.7.1.1 Drought Stress

In drylands, drought stress greatly impacts soil quality. In regions with low water potential, the poor, unproductive soil is subject to disaggregation and lacking in important minerals, especially nitrogen (N). Thus, the most effective way to increase N intake is still BNF. Plants have developed a variety of relationships with nitrogen-fixing bacteria in order to survive in arid settings. Rhizobium legume symbiosis has garnered significant interest as a sustainable nitrogen supply in dry natural settings. Rhizobia that tolerate drought have developed to flourish at low soil moisture levels, despite the fact that drought stress impacts bacterial growth and performance.

In reaction to stress, these bacterial communities have experienced various genetic and morphologic alterations (Ben Romdhane et al., 2022). Drought-tolerant rhizobial species have been widely studied for their potential to enhance legume growth in water-limited environments. In southern Ethiopia, researchers assessed selected drought-resistant and highly effective rhizobial strains for their ability to improve growth and yield of soybean (*Glycine max* L.) and common bean (*Phaseolus vulgaris* L.) in low-fertility soils over two consecutive drought years. Soybean plants were inoculated with a consortium of three *Bradyrhizobium strains* (*Bradyrhizobium japonicum* and two *Bradyrhizobium elkanii* isolates), while common bean plants received a mix of *Rhizobium etli* and two *Rhizobium phaseoli strains* (Vardharajula et al., 2011).

In addition to N-fixation, rhizobia possess other plant growth-promoting (PGP) characteristics that could potentially enhance legume development during dry conditions. Phosphorus, the second most limiting nutrient for crop growth, is often deficient in dryland soils due to erosion, water scarcity, and high temperatures. To address this limitation, many studies have focused on identifying rhizobial strains capable of solubilizing phosphorus and utilizing them as biofertilizers. In this context, a diverse range of rhizobia was isolated from the root nodules of chickpea plants grown across Tunisia's arid, semiarid, and Saharan regions. Among these isolates, *Mesorhizobium ciceri* (LL10) was chosen for plant inoculation due to its greater phosphate solubilization capacity. Comparing this chosen isolate to nitrogen-fertilized seedlings, a field trial in a semiarid area showed that it greatly improved nodule formation and proved both pod production and shoot dry matter accumulation by 412%, 256%, and 200%, respectively (Sarapat et al., 2020). Under low water potential, certain rhizobial species synthesize osmolytes that support their own survival and indirectly promote plant growth. Although osmolyte accumulation was first recognized as a plant strategy for maintaining cellular osmotic balance during stress, recent studies reveal that soil bacteria employ comparable mechanisms. Rhizobia synthesize amino acid based osmoprotectants, including glutamine, taurine, and proline, which lower their internal osmotic potential and help shield cells from oxidative damage. Proline, in particular, has been reported to act as both an antioxidant modulator and a free-radical scavenger, and its elevated levels reflect the osmoadaptive response of these bacteria (Chavoshi et al., 2018).

Drought-exposed leguminous plants frequently accumulate elevated amounts of the phytohormone ethylene, which normally regulates many physiological functions. Under stress conditions, elevated ethylene levels can cause premature leaf and pod drop, accelerate aging, limit root growth, and inhibit the formation of nitrogen-fixing nodules. Certain rhizobial strains mitigate these harmful effects by producing the enzyme 1-aminocyclopropane-1-carboxylate deaminase, which converts the ethylene precursor 1-aminocyclopropane-1-carboxylate into ammonia and α-ketobutyrate. By reducing ethylene accumulation in plant tissues, these bacteria help stressed plants sustain growth and development.

The strain *R. leguminosarum biovar viciae* 1066S, which produces 1-aminocyclopropane-1-carboxylate (ACC) deaminase, enhanced drought tolerance in pea plants more effectively than a mutant strain lacking the enzyme. The wild-type strain promoted several critical traits under water-limited conditions, including increased root growth, improved nodule formation, higher shoot biomass, elevated BNF, and greater water-use efficiency. Enhancing ACC deaminase activity through genetic modification strengthens symbiotic interactions under stress. For instance, inoculating mung bean (*Vigna radiata* L.) with a genetically engineered *Bradyrhizobium* sp. SUTN9-2 increased ACC deaminase activity approximately eightfold compared to the wild-type strain, resulting in improved drought resilience, accelerated nitrogen fixation, and increased plant biomass.

7.7.2 Plant Biofertilizers in Arid and Semiarid Areas: Non-Rhizobial Endophytes

Non-rhizobial PGPB candidates have been recognized and classified as viable biofertilizers because of their capacity to form advantageous relationships with legumes and their resistance to different types of abiotic stressors. As an example, Chavoshi et al. (2018) used a combined inoculum of bacteria that solubilizes potassium (*Bacillus circulans* and *Bacillus megaterium*) and phosphates (*Pseudomonas putida* and *Pantoea agglomerans*) to inoculate red beans (*P. vulgaris*) that were under water stress at various vegetative stages.

In comparison to the control, the results showed a notable rise in the biomass of the infected plants as well as a recovery of their physiological functioning. The beneficial effects of the bioinoculant were also demonstrated by examinations of seed protein content and chlorophyll, which showed an upside of 21.73% in protein, which is equivalent to the moderate protein content of plants that receive full irrigation. Also, a number of other methods that non-rhizobial species use to reduce plant stress have been thoroughly investigated. For instance, isolates from the species *Pseudomonas*, *Azotobacter*, and *Azospirillum* are currently among the most researched because of their capacity to generate large amounts of growth regulators and phytohormones, including cytokinins and gibberellins. PGPB produces a phytohormone called auxin, which affects plant growth at different stages either directly or indirectly (Kaushal & Wani, 2016). In plant, IAA promotes bud formation and cell proliferation. Additionally, with increasing number of tips and root hairs, it enhances root structure and proliferation, boosting the plant's water availability and increasing nutrient uptake in water-limited environments (Mohanty et al., 2021). Five *Pseudomonas aeruginosa* species were chosen because they are known to increase plant growth.

The functional analysis of the strains showed that there was high production of indole-3-acetic acid (IAA) up to 116 μg mL^{-1}, phosphate solubilization, and siderophores, which were responsible for the enhanced growth of two mung bean (*Vigna radiata*) cultivars under drought stress.

Field studies demonstrated that the association between *V. radiata* and *P. aeruginosa* strains markedly enhanced plant performance under drought conditions. The inoculated plants exhibited substantial improvements, including up to 125% greater shoot length, 68% higher water content, 293% increase in total yield, 139% enhancement in root length, and a 19% rise in antioxidant activity. These results highlight the crucial role of IAA-producing bacteria in alleviating drought stress by stimulating robust root development, which helps plants cope with limited water availability. Nevertheless, excessively high concentrations of bacterially derived acetic acid are known to inhibit plant growth, particularly root formation. This raises important questions regarding the regulatory mechanisms and signaling pathways used by host plants to modulate bacterial phytohormone production in a manner that aligns with their physiological needs.

7.7.2.1 Saline Soils Application

Extensive microbiological investigations on soil microbial communities have revealed numerous halotolerant strains belonging to genera such as *Pseudomonas*, *Bacillus*, *Enterobacter*, and several others. Introducing these salt-resilient microorganisms into saline agricultural systems has consistently improved crop performance and productivity. Investigation of the native rhizosphere of the wild legume *Sulla carnosa* in northeastern regions led tothe isolation of several salt tolerant bacterial strains possessing diverse plant growthpromoting traits. Among them, three strains capable of tolerating sodium chloride concentrations of up to 10% classified as *Acinetobacter* species and *P. putida* significantly enhanced the salt tolerance and overall vigor of *S. carnosa*. Furthermore, inoculation tests under 200 mM of salt were conducted using *Curtobacterium* sp. endophytic inoculation that had an advantageous effect on soil quality by lowering EC, a measure of the rhizosphere's reduced NaCl ionic toxicity. The inoculated strain's acidification of the soil caused by the secretion of organic acids is primarily responsible for lowering soil EC. Consequently, inoculated plants exhibited increased apical biomass along with enhanced levels of antioxidant enzymes and photosynthetic pigments indicating an increase in the plant's response to salt damage. Dietary legumes including chickpeas, faba beans, and peas are vulnerable to salt stress, just like some wild legumes. High salinity conditions suppress microbial metabolic functions and reduce overall productivity. Disruptions in osmotic balance and ion homeostasis negatively affect growth, ultimately causing cellular injury. In this context, the halotolerant endophytic strain *Kocuria rhizophila* also produces 1-aminocyclopropane-1-carboxylate deaminase and has the ability to enhance salt tolerance in two cultivars of pea. Inoculation with *K. rhizophila* significantly improved water uptake, resulting in a 70%–73% increase in leaf relative water content compared to control. This improvement was associated with a 17%–22% reduction in sodium levels in both plant types, indicating enhanced tolerance to osmotic stress and ion toxicity. *Azospirillum* species, widely recognized as PGPB, exhibit multiple beneficial traits, including ACC deaminase activity, IAA production, and BNF. Previous studies have demonstrated that various *Azospirillum* strains can mitigate the negative effects of abiotic stresses across different crops. El-Esawi et al. (2019) evaluated the salt-tolerant strain *Azospirillum lipoferum* FK1, which can survive at a sodium chloride concentration of 250 mM, to assess its ability to improve the response of chickpea plants under salinity. *Cicer arietinum* adapted more effectively to saline conditions after inoculation. The endophytic strain alleviated oxidative stress by stimulating the production of photosynthetic pigments and promoting the accumulation of protective osmolytes, including proline and gibberellins, within plant tissues (Khan et al., 2021; Kumari et al., 2022).

7.8 EFFECTS OF SALINITY AND ARIDITY ON MICROBIAL COMMUNITIES AND PLANT–MICROBE INTERACTIONS

Salinity negatively influences several important physicochemical characteristics of soil, including pH, nutrient availability, cation exchange capacity, and soil OC content. It also substantially affects the activity and composition of soil microbial communities. Maintaining adequate soil OC is critical for sustaining ecological functions and biological processes. Soil salinization, a progressive degradation process caused by the accumulation of water-soluble salts in the topsoil, reduces agricultural productivity, ecosystem functioning, and ultimately economic stability (Uzma et al., 2022). Crop productivity under such conditions can be preserved and salt or water stress can be mitigated through the cultivation of salt-tolerant varieties, the adoption of alternative farming practices, and management strategies such as drip irrigation and soil drainage. Additionally, incorporating organic amendments like manure, compost, or crop residues can enhance microbial biomass and help alleviate stress. Enzymatic activities contribute to the formation of soil aggregates, the stabilization of OC, and the production of hormones and volatile compounds that trigger plant defense responses and promote growth. Additionally, bacterial synthesis of extracellular polymeric substances enhances plant tolerance to salt and drought stress. Overall, plants benefit from the activities of microbial communities that colonize their root systems (Szoboszlay et al., 2019). Microbial populations are shaped by a combination of biotic and abiotic factors. Studies in both greenhouse and field settings indicate that soil type, plant genotype, developmental stage, and cultivation practices influence the composition and structure of the microbiota. Among these factors, soil properties particularly pH are the primary determinants of microbial community dynamics. Soils with near-neutral pH often support higher bacterial diversity compared to acidic or alkaline soils, such as those found in desert environments. Furthermore, plants can actively modulate their root-associated microbiota, notably through the release of root exudates.

The rhizosphere of cactus plants is dominated by bacterial species *Actinobacteria*, *Proteobacteria*, and *Acidobacteria* in the semiarid region of central Mexico. The microbial communities of Chile's Atacama desert revealed low concentrations of *Acidobacteria*, alpha and beta-*Proteobacteria*, and a major amount of *Actinobacteria* and *Chloroflexi*. *Acidobacteria* live in habitats with different pH, soil types, and temperatures. *Actinobacteria* are typically more abundant than

Acidobacteria in basic soils, including desert soils. The study on Berlanas vineyards supports this finding: the soil in the vineyards is more acidic as compare to the desert, and the *Acidobacteria* (13.9% and 16.7%) come right after the *Actinobacteria and Proteobacteria* (almost 57% of the total). Alpha and beta-*Proteobacteria* are commonly present in soils with high OC content, a characteristic that desert soils generally lack. *Acidobacteria*, *Bacteroidetes*, *Actinobacteria*, *Gemmatimonadetes*, *Proteobacteria*, and *Verrucomicrobia* are among the most prevalent bacterial phyla in soils, remaining relatively abundant across a wide range of environments and dominated microbial communities in diverse biomes such ascold deserts, warm deserts, grasslands, forests, and tundra ecosystems. *Verrucomicrobia* and *Acidobacteria* have demonstrated the reverse trend, generally speaking, desert soils have higher levels of *Actinobacteria* and *Bacteroidetes* than do forests, meadows, and tundra. Although making up less than 5% of the sequences found, their relative amounts vary widely, the *Cyanobacteria*, *Chloroflexi phylum*, *Firmicutes*, and *Gemmatimonadetes* have also been detected in practically all soils. The production of various antimicrobial molecules like phthalates, enzymes, and biosurfactants, which promote plant growth as well as induce systemic resistance in plants, is a characteristic feature of *Firmicutes* and their family *Bacillus* (Lee et al., 2019). The high salinity of desert soils rather than their extremely high pH levels is the cause of their poor diversity. Desert soils are arid, low in nutrients, often have a basic pH, are higher than those of other biomes, and have little to no plant biomass, which lowers the amount of OC inputs that are necessary for bacterial metabolism. The stability and variety of the microbiota are positively correlated with the amount of OM in the soil. The rhizosphere of *Caragana microphylla* included *Actinobacteria*, *Firmicutes*, *Acidobacteria*, and *Proteobacteria*. The primary cause of the increased *Actinobacteria* abundance in the rhizosphere soil was the rhizodeposition-induced reduction in soil pH. Electronic conductivity, pH, total OC, and total phosphorus were the main factors influencing the microbial communities associated with the various dunes, whereas soil total organic C, total N, Na^+, and total P primarily influenced the structure of the rhizospheric bacterial diversity (Neilson et al., 2012).

7.9 PLANT GROWTH–PROMOTING BACTERIA

Salt-affected soils are increasingly widespread due to natural processes and human activities. These soils host diverse halophilic PGPB that enhance nutrient uptake, support plant development, and improve plant tolerance to salt-induced stress. PGPB facilitate plant growth through multiple mechanisms, including the production of plant hormones, mobilization and decomposition of OM, and increasing the availability of essential nutrients such as phosphorus and iron. They also help suppress various plant pathogens. Plant hormones such as auxins and cytokinins, produced by PGPB, stimulate shoot growth and modify root architecture by increasing root branching and root hair formation, which enhances water and nutrient absorption. Certain bacterial species, including *Enterobacter* sp., support plant growth through ACC deaminase activity, hydrogen cyanide production, siderophore secretion, IAA production, and phosphorus solubilization. *Streptomyces sundarbansensis* and *Streptomyces rochei*, for example, produce IAA, contributing to plant development.

Although only a small fraction of soil bacteria (about 1%) is culturable, with typical densities of 10^8–10^9 cells per gram of soil, several genera such as *Streptomyces*, *Thiobacillus*, *Azospirillum*, *Clostridium*, *Serratia*, *Alcaligenes*, *Bacillus*, *Rhizobium*, *Pseudomonas*, and *Klebsiella* have shown effectiveness as PGPB in saline environments. Field studies have demonstrated their potential as inoculants or as components of microbial consortia (Gao et al., 2019). PGPB contribute to maintaining plant turgor, ion homeostasis, and osmotic balance, thereby reducing the impact of salt stress. Beyond enhancing plant growth, salt-tolerant microbial communities support soil health and ecological stability. However, the mechanisms underlying their diversity and functional roles remain largely unexplored and require further investigation.

7.9.1 Production of Phytohormones and ACC Deaminase Activity

PGPB enhance plant development by synthesizing a range of phytohormones, such as auxins, cytokinins, and gibberellins, and by mitigating the adverse effects of ethylene through the activity of 1-aminocyclopropane-1-carboxylate (ACC) deaminase. Ethylene, a gaseous phytohormone, tends to accumulate in plants under various abiotic stresses, which can negatively impact growth. Its levels vary among plant species, organs, and tissues, and ethylene regulates key developmental processes, including seed germination, root hair formation, fruit ripening, leaf abscission, and senescence. Excessive ethylene accumulation under stress conditions can inhibit growth (Bibi, 2017). PGPB produce ACC deaminase, which breaks down the ethylene precursor ACC into ammonia and α-ketobutyrate, effectively controlling ethylene levels in plants. This activity alters root architecture by increasing the number of root tips and expanding root surface area, enhancing nutrient uptake and supporting plant survival under stress conditions. The positive impact of these bacteria under salinity stress is largely attributed to ACC deaminase activity and the resulting reduction in ethylene accumulation. In addition to modulating ethylene, PGPB influence other phytohormones, particularly auxins. Bacterial genera such as *Nocardia*, *Actinobacteria*, *Kitasatospora*, *Frankia*, and *Streptomyces* are capable of producing auxins, including indole-3-butyric acid and IAA, which further promote plant growth and development.

7.9.2 Essential Nutrient Uptake

Plants under salt stress are less able to absorb and accumulate nutrients like P, N, and K as well as water because of their high osmotic potential and ion toxicity. In order to thrive under stress, plants therefore require additional nutrients. Salt inhibits nutrient uptake and transport; crop productivity suffers in salt-affected soils. It is commonly recognized that plant-associated PGPB facilitate plants' uptake of nutrients and water (Shen et al., 1997). Inoculating plants with PGPB promotes nitrogen buildup through both symbiotic and nonsymbiotic interactions with the roots. Soils contain both organic and inorganic fixed forms of phosphorus, most of which plants cannot use. Through the mechanisms of acidification and chelation processes, PGPB can change these inaccessible forms of phosphorus into ones that are available. Another crucial ingredient for plant growth is potassium (K), the majority of which is present in the soil in fixed forms that are not absorbed by plants. Under salt stress, plant access to potassium is often limited. In such conditions, potassium-solubilizing bacteria help fulfill crop potassium requirements by converting mineral potassium into plant-available forms. Additionally, PGPB enhance the availability of other essential nutrients, including iron, copper, manganese, and zinc, through mechanisms such as chelation and soil acidification.

7.10 APPLICATION OF PGPB IN SALT-STRESSED SOIL

Globally, the growing global population, currently at 7.4 billion and projected to reach 9.7 billion by 2050, is driving an increasing demand for food. Chemical fertilizers are used in agriculture despite their detrimental effects on the environment, especially the soil. Additionally, as was already indicated, agricultural productivity is negatively impacted by environmental abiotic stresses including drought, high salt, and temperature extremes. For example, it was recently estimated that the US loses $27.3 billion due to soil salinization (Meena et al., 2016). In a similar vein, drought decreased agricultural output by 17% every year in the tropics, arid regions, and semiarid regions, and by 5% in the main food-producing nations. There is an urgent need for sustainable and eco-friendly agricultural approaches to maintain natural and farming ecosystems while meeting the growing food demand. PGPB offer a viable solution by enhancing soil fertility and supporting crop productivity, even under stress conditions. These bacteria can modify their activity in response to plant stress, influencing the plant's tolerance mechanisms. However, understanding their role in the rhizosphere under adverse conditions is complex, as each plant harbors a unique microbiome shaped by soil type, environmental conditions, and genetic factors, making it difficult to fully define the plant's stress response based on microbial interactions. Both bacteria and fungi are capable of producing ACC deaminase and various phytohormones, which can modulate plant hormone levels and influence hormonal signaling pathways. Under drought, salinity, and pathogen stress, plants often overproduce ethylene, which significantly inhibits root growth. Salinity also reduces potassium availability, but potassium-solubilizing bacteria help by converting insoluble mineral potassium into plant-accessible forms. Additionally, PGPB enhance the availability of other essential nutrients, including iron, copper, manganese, and zinc, through mechanisms such as soil acidification and chelation. The provision of a N (ammonia) source to plants is one of the ensuing consequences of this specific kind of plant–bacterial relationship. Although there is some evidence that PGPB improve plant salt tolerance by changing the status of the endogenous hormone, nothing is known about how these PGPB affect this process. Research conducted on pepper plants under water stress reveals that plants treated with ACC rhizobacteria exhibited increased photosynthetic efficiency and even higher tissue turgidity. When compared to uninoculated plants, these positive benefits result in a 50% increase in biomass and root length. Strains of the genera *Achromobacter*, *Klebsiella*, and *Citrobacter* have shown the strongest resistance to drought. Given its capacity to colonize these species' roots, it is plausible that this protective action may be performed outside of the lab as well (Marasco et al., 2012). PGPB improves soil fertility, enhances plant nutrient uptake, increases the expression of beneficial physiological traist in plants, and lowers soil-borne plant diseases, thereby offering a viable chance to boost crop yields even in stressed soils.

Burkholderia sp. strain P50 is capable of producing ACC deaminase, IAA, extracellular polymeric substances, and the osmolyte proline under salinity stress. Rice seedlings inoculated with strain P50 and exposed to 185 mM NaCl exhibited improvements in multiple morphological and biochemical traits, including enhanced antioxidant production for ROS scavenging. Moreover, seedlings under salt stress treated with *Burkholderia* sp. P50 showed reduced stress-induced ethylene levels compared to uninoculated plants. The reduction in ethylene appears linked to the activity of strain P50, as a mutant lacking ACC deaminase genes failed to lower ethylene under stress conditions. Plants' ability to absorb nutrients can be enhanced by PGPB. When inoculated in *S. lycopersicum* under circumstances of high salinity, *Achromobacter piechaudii* increases the acquisition of P and K and enhances water use efficiency (Mayak et al., 2004). When wheat is grown in a field under extreme saline stress, *Bacillus aquimaris* inoculation dramatically raises the concentrations of N, P, and K in the leaves. PGPB have the capacity to fix atmospheric N in a non-symbiotic manner and solubilize insoluble P. In addition to this, PGPB can release siderophores to get rid of iron, which makes ferri-siderophores easily accessible to plants.

7.10.1 The PGPB: Promising Alternatives for Improving the Dryland Agricultural System

Plants that thrive in challenging settings have a microbiome that is organically evolved to defend them from stress. It is interesting to note that multiple investigations have revealed that the bacterial communities in the soils of extreme ecosystems generally distant, arid, and hostile regions are the same. However, there are some differences in the abundance of each phylum, according to high-throughput sequencing research, which showed that *Proteobacteria*, *Firmicutes*, *Actinobacteria*, and *Bacteroidetes* dominate these communities. All of these regions share difficult environmental circumstances, which may be the reason for the similarities in the diversity profile of bacterial populations. Dry soil microorganisms have evolved a number of defense mechanisms to withstand the extreme circumstances they face, temperatures as high as 50°C, indicating that they could withstand osmotrophic solutes. Halotolerant bacteria, including beneficial genera such as *Halobacillus*, *Rhizobium*, *Klebsiella*, and *Actinoplanes*, dominated the rhizosphere of three halophyte species grown in salt-affected soils in Wujiaqu, China. Likewise, phenotypic analysis of bacteria from the rhizosphere of the native legume *Psoralea corylifolia* L. in salt-affected areas of India indicated strong intrinsic salt tolerance, likely resulting from the selective pressure of the semiarid, high-salinity environment. The rhizospheric microbiome is highly dynamic and adapts to the requirements of the host plant; for instance, the bacterial community associated with *Arabidopsis* shifts over time, suggesting selective recruitment of microbes for specific functions. Similarly, native plants in the harsh environments of Chile seem to associate with particular bacterial groups possessing multiple plant growth–promoting traits, enhancing plant growth and stress tolerance and contributing to the diversity of rhizosphere bacterial communities.

7.10.2 PGPB Consortia for Mitigating Drought and Salinity Stress in Legumes

The application of *Rhizobium* inoculants or other non-rhizobial PGPB individually has demonstrated significant potential for sustainable agriculture. These microorganisms function as biofertilizers, reducing dependency on chemical fertilizers and enhancing plant resilience to biotic and abiotic stresses. Recent research highlights that co-inoculation of *Rhizobium* with other non-rhizobial endophytes can further stimulate plant growth by strengthening symbiotic interactions and increasing the effectiveness of *Rhizobium* within the host. This synergistic approach is especially valuable under adverse conditions such as salinity and drought, which often limit symbiotic efficiency. For instance, the growth of faba bean plants in salt-affected soils was notably improved through co-inoculation with *R. leguminosarum* and *Azotobacter chroococcum*, an extracellular polymeric substance–producing bacterium, illustrating the benefits of bacterial consortia under stress conditions (Abd El-Ghany & Attia, 2020). Under high salinity (100 mM NaCl), co-inoculation enhanced plant yield, proline accumulation, photosynthetic pigments, and relative water content. Furthermore, biochemical analyses revealed improved uptake of essential nutrients such as nitrogen, phosphorus, and potassium, while the concentrations of harmful ions, including sodium and chloride, were reduced.

Co-inoculation has been shown to improve multiple physiological and biochemical traits in plants compared to single-strain inoculation. For example, the symbiotic efficiency of *Mesorhizobium* under saline conditions was enhanced when chickpea plants were co-inoculated with *M. ciceri* and a consortium of non-rhizobial endophytes, including *Phyllobacterium* and *Xanthomonas*, isolated from legumes native to arid regions (Gritli et al., 2022). Co-inoculating *Lathyrus cicera* with *Rhizobium laguerreae* and three *Bacillus* endophytic strains under salt stress significantly improved plant growth and biochemical characteristics. The treatment caused an increase in the expression of salt-inducible genes such as HKT1 and NHX7, which play a role in sodium ions transport and Na^+/H^+ exchange in plant tissues, resulting in salt tolerance. The effectiveness of co-inoculation under drought conditions has also been demonstrated; chickpea plants treated with *M. ciceri* and *Pseudomonas fluorescens* at 40% field water capacity showed synergistic effects, enhancing their resilience to water deficit. Overall, the use of bacterial consortia has been shown to significantly improve crop tolerance to both salinity and drought stress.

7.10.3 Rhizosphere–PGPB Interactions: Cell-to-Cell Communication

Within the intricate and dynamic rhizosphere, plant roots interact with diverse microbial communities, including bacteria, which play a pivotal role in plant health and development. Bacterial communication is critical in this environment for regulating population dynamics and coordinating physiological responses to various environmental stresses. Root exudates act as primary metabolic cues that mediate these plant–microbe interactions. Research indicates that the composition of carbon compounds and other metabolites released by roots can influence whether rhizobacterial populations adopt cooperative or competitive strategies. Network analyses of bacterial communities further reveal that multiple signaling pathways are involved in controlling specific behaviors and population density, particularly under stress conditions. Additionally, bacteria synthesize a broad spectrum of secondary metabolites that facilitate intercellular and interspecies communication, a process broadly described as quorum sensing. The best illustration of bacterial collaboration is the creation of biofilms. This unique structure is one of the various tactics that bacterial endophytes employ to colonize plant surfaces and adapt to salinity and drought stressors (Pathak et al., 2022; Kashyap et al., 2023). According to recent research, this collaboration takes place through the transfer of genetic material between biofilm members and is regulated by Quorum Sensing (QS) systems, specifically acyl-homoserine lactones (AHL) molecules. Studies have discovered additional forms of cell-to-cell communication systems in the rhizosphere, such as extracellular vesicle–based communication and diffusible signal factor (DSF)-mediated signaling. Fatty acids, including cis-2-unsaturates fatty acids, are secreted and detected in DSF-QS. These autoinducers were initially discovered to be regulators of virulence expression in the phytopathogen *Xanthomonas campestris*. A number of genes linked to plant colonization, plant growth promotion, and stress resistance have been found to be stimulated by DFS-mediated signaling in the biocontrol agent *Stenotrophomonas maltophilia* in more recent times. Cultivating all bacterial species in conjunction with plant systems is still a major problem, despite the vast amount of research on the evolution and interactions of the plant microbiome. It is challenging to replicate the intricate natural habitat of plants in a lab setting. As a result, the precise makeup of the phytomicrobiome and how it affects plant development are still not fully understood. Therefore, in order to clarify the relationships among all members of the phytomicrobiome, sophisticated experiments are required. These discoveries will be essential for creating efficient synthetic microbial communities that can improve plant productivity and health in the face of quickly shifting climate circumstances.

7.10.4 Upcoming Difficulties in Using Halophilic Microbes to Reduce Salt Stress

Halophilic PGPB have considerable potential to enhance crop performance in saline soils. Developing bioformulations based on these microbes provides a sustainable approach to maintaining crop yields under salt stress. Recent strategies involve incorporating salt-tolerant bacteria directly into seeds or seedlings to mitigate salinity effects. The success of these approaches depends on the careful design and delivery of microbial formulations, ensuring their functionality under saline conditions. Such bio-products offer an eco-friendly method to boost plant growth and productivity. Additionally, improving soil fertility in salt-affected areas remains crucial, and studies indicate that using multi-strain inoculants is generally more effective than single-strain applications. There is a dearth of research on the specific process underlying mycorrhizal fungal-associated plant development under salt stress, which has to be investigated. Since beneficial microbes like halophilic PGPR offer an inexpensive and environmentally friendly way to reduce salt stress, this potential strategy for addressing the issue of soil salinity will have the biggest impact on the agricultural economy. The use of PGPR to counteract salt stress is one of the most recent areas of study. The variety of choices for preserving the healthy yield of crops in saline environments can be increased by the creation of biological products based on advantageous microorganisms. By applying PGPR to crop seeds and seedlings, a novel method for reducing salt stress in plants has emerged in recent years. Its combination with halophilic PGPR presents a wonderful potential for study on salt tolerance. The success of inoculation technology's use in industrial and agricultural settings is its ultimate goal. The creation of a commercial salt-tolerant bioformulation that works well in salt stress situations depends heavily on the microbial formulation and application technologies. The 21st century is becoming the age of biotechnology in large part because bioformulations provide an environmentally friendly way to improve crop health and productivity. Reclaiming and enhancing the fertility of stressed areas is an additional goal to be concentrated on in addition to bioformulation. If low-cost, user-friendly strain compositions that are stress-tolerant can be created, the agricultural economy will benefit most from the potential strategy of using beneficial microbes, such as PGPR, to address the issue of soil salinity.

7.11 SDG 15: LIFE ON LAND, SDG 14: LIFE UNDERWATER, AND SDG 13: CLIMATE CHANGE

Halophilic bacteria contribute significantly to achieving the United Nations Sustainable Development Goals (SDGs), particularly Goal 13 on Climate Action, Goal 14 on Life Below Water, and Goal 15 on Life on Land. Their exceptional ability to thrive in saline and arid environments positions them as valuable agents for mitigating climate change effects and promoting ecosystem resilience. Improving plant resistance to drought and salinity, halophilic bacteria support agricultural productivity in marginal regions and help mitigate climate change under SDG 13. They help climate-resilient agriculture by reducing stress-induced crop losses through the synthesis of phytohormones, ACC deaminase, and antioxidative enzymes. Additionally, in saline soils, microbial-mediated carbon sequestration reduces greenhouse gas emissions. Halophiles are essential to marine environments for SDG 14, because they control nutrient cycles and break down contaminants. By preserving ecosystem processes in hypersaline and coastal zones, halophilic bacteria help the oceans absorb over 25% of human CO_2. For fisheries to continue and marine life to be preserved, microbial diversity in saline aquatic systems must be preserved. Halophilic microorganisms improve biodiversity and soil fertility in degraded areas, which is relevant to SDG 15 (Arora et al., 2023). They fight desertification, restore vegetation cover in arid areas, and encourage sustainable land use by solubilizing phosphate, generating siderophores, and establishing symbiotic relationships. Their involvement in rhizosphere interactions serves as an example of a natural approach to land rehabilitation.

7.12 CONCLUSIONS

This chapter evaluated the most recent data on salt-affected soils and their bioremediation utilizing different microbial approaches, despite the fact that salt-affected soils have grown in number overall and that managing and reclaiming them presents difficulties. The microbial consortium is crucial to agriculture's long-term development. Halophilic bacteria can also help reduce soil that is stressed by salt. For long-term growth and development, salt-affected soil must be treated with rhizobacteria that promote plant growth and that flourish in salt-stressed soil. The creation and use of halophilic bacteria that promote plant growth could help crops grow over the long term under soil conditions influenced by salt. The development of advantageous halophile microbial formulations can increase the options for maintaining agricultural productivity in soils impacted by salt.

REFERENCES

Abd El-Ghany, M.F.; Attia, M. Effect of exopolysaccharide-producing bacteria and melatonin on faba bean production in saline and non-saline soil. *Agronomy* **2020**, *10*, 316.

Abd-Alla, M.H.; Al-Amri, S.M.; El-Enany, A.-W.E. Enhancing rhizobium–legume symbiosis and reducing nitrogen fertilizer use are potential options for mitigating climate change. *Agriculture* **2023**, *13*, 2092.

Abdeljabbar, H.; Cayol, J.L.; Hania, W.B.; Boudabous, A.; Sadfi, N.; Fardeau, M.L. *Halanaerobium sehlinense* sp. nov. an extremely halophilic, fermentative, strictly anaerobic bacterium from sediments of the hypersaline lake Sehline Sebkha. *Int. J. Syst. Evol. Microbiol.* **2013**, *63*, 2069–2074.

Acosta-Motos, J.; Ortuño, M.; Bernal-Vicente, A.; Diaz-Vivancos, P.; Sanchez-Blanco, M.; Hernandez, J. Plant responses to salt stress: Adaptive mechanisms. *Agronomy* **2017**, *7*, 18.

Ahn, J.; Park, J.W.; Mc Connell, J.A.; Ahn, Y.B.; Häggblom, M.M. *Kangiella spongicola* sp. nov. a halophilic marine bacterium isolated from the sponge *Chondrilla nucula*. *Int. J. Syst. Evol. Microbiol.* **2011**, *61*, 961–964.

Amoozegar, M.A.; Bagheri, M.; Didari, M.; Fazeli, S.A.S.; Schumann, P.; Sanchez-Porro, C.; Ventosa, A. *Saliterribacillus persicus* gen. nov. sp. nov. a moderately halophilic bacterium isolated from a hypersaline lake. *Int. J. Syst. Evol. Microbiol.* **2013a**, *63*, 345–351.

Amoozegar, M.A.; Bagheri, M.; Makhdoumi-Kakhki, A.; Didar, M.; Schumann, P.; Nikou, M.M.; Sánchez-Porro, C.; Ventosa, A. *Aliicoccus persicus* gen. nov. sp. nov. a halophilic member of the Firmicutes isolated from a hypersaline lake. *Int. J. Syst. Evol. Microbiol.* **2014**, *64*, 1964–1969.

Amoozegar, M.A.; Makhdoumi-Kakhki, A.; Ramezani, M.; Nikou, M.M.; Fazeli, S.A.S.; Schumann, P.; Ventosa, A. *Limimonas halophila* gen. nov. sp. nov. an extremely halophilic bacterium in the family *Rhodospirillaceae*. *Int. J. Syst. Evol. Microbiol.* **2013b**, *63*, 1562–1567.

Arora, N.K.; Mishra, I.; Arora, P. SDG 14: Life below water – viable oceans necessary for a sustainable planet. *Environ. Sustain.* **2023**, *6*, 433–439.

Aserse, A.A.; Markos, D.; Getachew, G.; Yli-Halla, M.; Lindström, K. Rhizobial inoculation improves drought tolerance, biomass and grain yields of common bean (*Phaseolus vulgaris* L.) and soybean (*Glycine max* L.) at Halaba and Boricha in Southern Ethiopia. *Arch. Agron. Soil Sci.* **2020**, *66*, 488–501.

Athar, A. Drought tolerance by lentil rhizobia (*Rhizobium leguminosarum*) from arid and semi-arid areas of Pakistan. *Lett. Appl. Microbiol.* **1998**, *26*, 38–42.

Bano, A.; Batool, R.; Dazzo, F. Adaptation of chickpea to desiccation stress is enhanced by symbiotic rhizobia. *Symbiosis* **2010**, *50*, 129–133.

Belimov, A.A.; Zinovkina, N.Y.; Safronova, V.I.; Litvinsky, V.A.; Nosikov, V.V.; Zavalin, A.A.; Tikhonovich, I.A. Rhizobial ACC deaminase contributes to efficient symbiosis with pea (*Pisum sativum* L.) under single and combined cadmium and water deficit stress. *Environ. Exp. Bot.* **2019**, *167*, 103859.

Ben Romdhane, S.; De Lajudie, P.; Fuhrmann, J.J.; Mrabet, M. Potential role of rhizobia to enhance chickpea-growth and yield in low fertility-soils of Tunisia. *Antonie Van Leeuwenhoek* **2022**, *115*, 921–932.

Bibi, F. Diversity of antagonistic bacteria isolated from medicinal plant *Peganum harmala* L. *Saudi J. Biol. Sci.* **2017**, *24*, 1288–1293.

Blum, J.S.; Kulp, T.R.; Han, S.; Lanoil, B.; Saltikov, C.W.; Stolz, J.F.; Miller, L.G.; Oremland, R.S. *Desulfohalophilus alkaliarsenatis* gen. nov. sp. nov. an extremely halophilic sulfate-and arsenate-respiring bacterium from Searles Lake California. *Extremophiles* **2012**, *16*, 727–742.

Bueno Batista, M.; Dixon, R. Manipulating nitrogen regulation in diazotrophic bacteria for agronomic benefit. *Biochem. Soc. Trans.* **2019**, *47*, 603–614.

Cerezini, P.; Kuwano, B.H.; Grunvald, A.K.; Hungria, M.; Nogueira, M.A. Soybean tolerance to drought depends on the associated *Bradyrhizobium* strain. *Braz. J. Microbiol.* **2020**, *51*, 1977–1986.

Chavoshi, S.; Nourmohamadi, G.; Madani, H.; Heidari Sharif Abad, H.; Alavi Fazel, M. The effects of biofertilizers on physiological traits and biomass accumulation of red beans (*Phaseolus vulgaris* cv. Goli) Under Water Stress. *Iran. J. Plant Physiol.* **2018**, *8*, 2555–2562.

Chialva, M.; Lanfranco, L.; Bonfante, P. The plant microbiota: Composition, functions, and engineering. *Curr. Opin. Biotechnol.* **2022**, *73*, 135–142.

Cho, G.Y.; Whang, K.S. *Aliifodinibius saliphilus* sp. nov. a moderately halophilic bacterium isolated from sediment of a crystallizing pond of a saltern. *Int. J. Syst. Evol. Microbiol.* **2020**, *70*, 358–363.

Ciulla, R.A.; Diaz, M.R.; Taylor, B.F.; Roberts, M.F. Organic osmolytes in aerobic bacteria from mono lake, an alkaline, moderately hypersaline environment. *Appl. Environ. Microbiol.* **1999**, *63*, 220–226.

Corwin, D.L. Climate change impacts on soil salinity in agricultural areas. *Eur. J. Soil Sci.* **2021**, *72*, 842–862.

Cui, H.L.; Yang, X.; Zhou, Y.G.; Liu, H.C.; Zhou, P.J.; Dyall-Smith, M.L. *Halobellus limi* sp. nov. and *Halobellus salinus* sp. nov. isolated from two marine solar salterns. *Int. J. Syst. Evol. Microbiol.* **2012**, *62*, 1307–1313.

Daliakopoulos, I.N.; Tsanis, I.K.; Koutroulis, A.; Kourgialas, N.N.; Varouchakis, A.E.; Karatzas, G.P.; Ritsema, C.J. The threat of soil salinity: A European scale review. *Sci. Total Environ.* **2016**, *573*, 727–739.

Datta, P.; Kulkarnni, M.V. Arbuscular mycorrhizal colonization enhances biochemical status and mitigates adverse salt effect on two legumes. *Not. Sci. Biol.* **2014**, *6*, 381–393.

Echigo, A.; Minegishi, H.; Shimane, Y.; Kamekura, M.; Itoh, T.; Usami, R. *Halomicroarcula pellucida gen* nov. sp. nov. a non-pigmented, transparent-colony-forming, halophilic archaeonisolated from solar salt. *Int. J. Syst. Evol. Microbiol.* **2013**, *63*, 3556–3562.

Echigo, A.; Minegishi, H.; Shimane, Y.; Kamekura, M.; Usami, R. *Natribacillus halophilus* gen. nov. sp. nov. a moderately halophilic and alkalitolerant bacterium isolated from soil. *Int. J. Syst. Evol. Microbiol.* **2012**, *62*, 289–294.

Edbeib, M.F.; Wahab, R.A.; Huyop, F. Halophiles: Biology, adaptation, and their role in decontamination of hypersaline environments. *World J. Microbiol. Biotechnol.* **2016**, *32*, 135.

El-Esawi, M.A.; Al-Ghamdi, A.A.; Ali, H.M.; Alayafi, A.A. *Azospirillum lipoferum*FK1 confers improved salt tolerance in chickpea (*Cicer arietinum* L.) by modulating osmolytes, antioxidant machinery and stress-related genes expression. *Environ. Exp. Bot.* **2019**, *159*, 55–65.

Etesami, H.; Beattie, G. *Plant-Microbe Interactions in Adaptation of Agricultural Crops to Abiotic Stress Conditions*; Springer: Berlin/Heidelberg, Germany, **2017**.

Fierer, N.; Leff, J.W.; Adams, B.J.; Nielsen, U.N.; Bates, S.T.; Lauber, C.L.; Owens, S.; Gilbert, J.A.; Wall, D.H.; Caporaso, J.G. Cross-biome metagenomic analyses of soil microbial communities and their functional attributes. *Proc. Natl. Acad. Sci. USA* **2012**, *109*, 21390–21395.

Gales, G.; Chehider, N.; Joulian, C.; Battaglia-Brunet, F.; Cayol, J.L.; Postec, A.; Borgomano, J.; Neria-Gonzalez, I.; Lomans, B.; Ollivier, B. Characterization of *Halanaerocella petrolearia* gen. nov. sp. nov. a new anaerobic moderately halophilic fermentative bacterium isolated from a deep subsurface hypersaline oil reservoir. *Extremophiles* **2011**, *15*, 565–571.

Gao, J.; Luo, Y.; Wei, Y.; Huang, Y.; Zhang, H.; He, W.; Sheng, H.; An, L. Effect of aridity and dune type on rhizosphere soil bacterial communities of *Caragana microphylla* in desert regions of northern China. *PLoS One* **2019**, *14*, e0224195.

Gao, L.; Huang, Y.; Liu, Y.; Mohamed, O.A.A.; Fan, X.; Wang, L.; Li, L.; Ma, J. Bacterial community structure and potential microbial coexistence mechanism associated with three halophytes adapting to the extremely hypersaline environment. *Microorganisms* **2022**, *10*, 1124.

Garabito, M.J.; Marquez, M.C.; Ventosa, A. Halotolerant *Bacillus* diversity in hypersaline environments. *Can. J. Microbiol.* **1998**, *44*, 95–102.

Gritli, T.; Boubakri, H.; Essahibi, A.; Hsouna, J.; Ilahi, H.; Didier, R.; Mnasri, B. Salt stress mitigation in *Lathyrus cicera* by combining different microbial inocula. *Physiol. Mol. Biol. Plants* **2022**, *28*, 1191–1206.

Hajiboland, R.; Aliasgharzadeh, N.; Laiegh, S.F.; Poschenrieder, C. Colonization with arbuscular mycorrhizal fungi improves salinity tolerance of tomato (*Solanum lycopersicum* L.) plants. *Plant Soil* **2010**, *331*, 313–327.

Hayat, R.; Ali, S.; Amara, U.; Khalid, R.; Ahmed, I. Soil beneficial bacteria and their role in plant growth promotion: A review. *Ann. Microbiol.* **2010**, *60*, 579–598.

Herrmann, A.; Witter, E.; Katterer, T. A method to assess whether 'preferential use' occurs after 15N ammonium addition: Implication for the 15N isotope dilution technique. *Soil Biol. Biochem.* **2005**, *37*, 183–186.

Hiltner, L. Über neuere erfahrungen und problem auf dem gebeit der bodenbakteriologie und unter besonderer berucksichtigung der grundungung und brache. *Arb. Dtsch. Landwirschaft Ges.* **1904**, *98*, 59–78.

Hmaeid, N.; Wali, M.; Mahmoud, O.M.-B.; Pueyo, J.J.; Ghnaya, T.; Abdelly, C. Efficient rhizobacteria promote growth and alleviate NaCl-induced stress in the plant species *Sulla carnosa*. *Appl. Soil Ecol.* **2019**, *133*, 104–113.

Huang, Z.; He, C.X.; He, Z.Q.; Zou, Z.R.; Zhang, Z.B. The effects of arbuscular mycorrhizal fungi on reactive oxyradical scavenging system of tomato under salt tolerance. *Agric. Sci. China* **2010**, *9*, 1150–1159.

Ishikawa, M.; Yamasato, K.; Kodama, K.; Yasuda, H.; Matsuyama, M.; Okamoto-Kainuma, A.; Koizumi, Y. *Alkalibacterium gilvum* sp. nov. slightly halophilic and alkaliphilic lactic acid bacterium isolated from soft and semi-hard cheeses. *Int. J. Syst. Evol. Microbiol.* **2013**, *63*, 1471–1478.

Jiang, F.; Cao, S.-J.; Li, Z.H.; Fan, H.; Li, H.F.; Liu, W.J.; Yuan, H.L. *Salisediminibacterium halotolerans* gen. nov. sp. nov. a halophilic bacterium isolated from Xiarinaoer soda lake sediment in Inner Mongolia, China. *Int. J. Syst. Evol. Microbiol.* **2011**, *62*, 2127–2132.

Jiang, H.; Dong, H.; Yu, B.; Liu, X.; Li, Y.; Ji, S.; Zhang, C.L. Microbial response to salinity change in Lake Chaka, a hypersaline lake on Tibetan plateau. *Environ. Microbiol.* **2007**, *9*, 2603–2621.

Juniper, S.; Abbott, L.K. Soil salinity delays germination and limits growth of hyphae from propagules of arbuscular mycorrhizal fungi. *Mycorrhiza* **2006**, *16*, 371–379.

Kashyap, N.; Singh, S.K.; Yadav, N.; Singh, V.K.; Kumari, M.; Kumar, D.; Shukla, L.; Kaushalendra, Bhardwaj, N.; Kumar, A. Biocontrol screening of endophytes: Applications and limitations. *Plants* **2023**, *12*(13), 2480. https://doi.org/10.3390/plants12132480

Kaushal, M.; Wani, S.P. Plant-growth-promoting rhizobacteria: Drought stress alleviators to ameliorate crop production in drylands. *Ann. Microbiol.* **2016**, *66*, 35–42.

Khan, A.A.; Wang, T.; Hussain, T.; Ali, F.; Shi, F.; Latef, A.A.H.A.; Ali, O.M.; Hayat, K.; Mehmood, S.; Zainab, N. Halotolerant-*Koccuria rhizophila* (14asp)-induced amendment of salt stress in pea plants by limiting Na+ uptake and elevating production of antioxidants. *Agronomy* **2021**, *11*, 1907.

Khan, S.A.; Zununi Vahed, S.; Forouhandeh, H.; Tarhriz, V.; Chaparzadeh, N.; Hejazi, M.A.; Hejazi, M.S. *Halomonas urmiana* sp. nov. a moderately halophilic bacterium isolated from Urmia Lake in Iran. *Int. J. Syst. Evol. Microbiol.* **2020**, *70*, 2254–2260.

Kumari, M.; Qureshi, K.A.; Jaremko, M.; White, J.; Singh, S.K.; Sharma, V.K.; Singh, K.K.; Santoyo, G.; Puopolo, G.; Kumar, A. Deciphering the role of endophytic microbiome in postharvest diseases management of fruits: Opportunity areas in commercial up-scale production. *Front. Plant Sci.* **2022,** *13*, 4390. https://doi.org/10.3389/fpls.2022.1026575

Kushwaha, B.; Jadhav, I.; Jadhav, K. *Halomonas sambharensis* sp. nov. a moderately halophilic bacterium isolated from the saltern crystallizer ponds of the Sambhar Salt Lake in India. *Curr. Microbiol.* **2020**, *77*, 1125–1134.

Larsen, H. Halophilic and halotolerant microorganisms—An overview and historical perspective. *FEMS Microbiol. Rev.* **1986**, *39*, 3–7.

Lee, S.A.; Kim, Y.; Kim, J.M.; Chu, B.; Joa, J.-H.; Sang, M.K.; Song, J.; Weon, H.-Y. A preliminary examination of bacterial, archaeal, and fungal communities inhabiting different rhizocompartments of tomato plants under real-world environments. *Sci. Rep.* **2019**, *9*, 9300.

Lemanceau, P.; Offre, P.; Mougel, C.; Gamalero, E.; Dessaux, Y.; Moënne-Loccoz, Y.; Berta, G. Microbial ecology of the rhizosphere. In *Microbiological Methods for Assessing Soil Quality*; Bloem, J., Hopkins, D.W., Benedetti, A., Eds.; CAB International Wallingford: Oxfordshire, UK, **2005**; pp. 228–230.

León, M.J.; Fernández, A.B.; Ghai, R.; Sánchez-Porro, C.; Rodriguez-Valera, F.; Ventosa, A. From metagenomics to pure culture: Isolation and characterization of the moderately Halophilic bacterium *Spiribacter salinus* gen. nov. sp. nov. *Appl. Environ. Microbiol.* **2014**, *80*, 3850–3857.

Li, D.-P.; Wu, Z.-J. Impact of chemical fertilizers application on soil ecological environment. *Ying Yong Sheng Tai Xue Bao J. Appl. Ecol.* **2008**, *19*, 1158–1165.

Liang, Y.; Chen, H.; Tang, M.J.; Shen, S.H. Proteome analysis of an ectomycorrhizal fungus *Boletus edulis* under salt shock. *Mycol. Res.* **2007**, *111*, 939–946.

Liu, W.; Yang, S.S. *Oceanobacillus aidingensis* sp. nov. a moderately halophilic bacterium. *Antonie Van Leeuwenhoek* **2014**, *105*, 801–808.

Llamas, D.P.; Gonzales, M.D.; Gonzales, C.I.; Lopez, G.R.; Marquina, J.C. Effects of water potential on spore germination and viability of *Fusarium* species. *J. Ind. Microbiol. Biotechnol.* **2008**, *35*, 1411–1418.

Ma, Z.; Yang, Q. Global patterns of aridity trends and time regimes in transition. In *Aridity Trend in Northern China*; World Scientific: by Congbin Fu and Huiting Mao, Singapore, **2017**; pp. 67–90.

Makhdoumi-Kakhki, A.; Amoozegar, M.A.; Ventosa, A. *Salinibacter iranicus* sp. nov. and *Salinibacter luteus*sp. nov. isolated from a salt lake, and emended descriptions of the genus *Salinibacter* and of *Salinibacter ruber*. *Int. J. Syst. Evol. Microbiol.* **2012**, *62*, 1521–1527.

Mandeel, Q.A. Biodiversity of the genus *Fusarium* in saline soil habitats. *J. Basic Microbiol.* **2006**, *46*, 480–494.

Marasco, R.; Rolli, E.; Ettoumi, B.; Vigani, G.; Mapelli, F.; Borin, S.; Abou-Hadid, A.F.; El-Behairy, U.A.; Sorlini, C.; Cherif, A.; et al. A drought resistance-promoting microbiome is selected by root system under desert farming. *PLoS One* **2012**, *7*, e48479.

Mayak, S.; Tirosh, T.; Glick, B.R. Plant growth-promoting bacteria confer resistance in tomato plants to salt stress. *Plant Physiol. Biochem.* **2004**, *42*, 565–572.

Mdlambuzi, T.; Tsubo, M.; Muchaonyerwa, P. Maize (*Zea mays* L.) Production from co-application of biogas slurry with chemical fertilizer and effects on soil quality in a semi-arid region of South Africa. *Commun. Soil Sci. Plant Anal.* **2022**, *53*, 2574–2583.

Meena, M.D.; Joshi, P.K.; Jat, H.S.; Chinchmalatpure, A.; Narjari, B. Changes in biological and chemical properties of saline soil amended with municipal solid waste compost and chemical fertilizers in a mustard–pearl millet cropping system. *Catena* **2016**, *140*, 1–8.

Mezghani, M.; Alazard, D.; Karray, F.; Cayol, J.L.; Joseph, M.; Postec, A.; Fardeau, M.L.; Tholozan, J.L.; Sayadi, S. *Halanaerobacter jeridensis* sp. nov. isolated from a hypersaline lake. *Int. J. Syst. Evol. Microbiol.* **2012**, *62*, 1970–1973.

Miao, C.; Jia, F.; Wan, Y.; Zhang, W.; Lin, M.; Jin, W. *Halomonas huangheensis* sp. nov. a moderately halophilic bacterium isolated from a saline–alkali soil. *Int. J. Syst. Evol. Microbiol.* **2014**, *64*, 915–920.

Mirzabaev, A.; Wu, J.; Evans, J.; García-Oliva, F.; Hussein, I.A.; Iqbal, M.H.; Kimutai, J.; Knowles, T.; Meza, F.; Nedjroaoui, D.; et al. Desertification. In *Climate Change and Land: An IPCC Special Report on Climate Change, Desertification, Land Degradation, Sustainable Land Management, Food Security, and Greenhouse Gas Fluxes in Terrestrial Ecosystems*; Shukla, P.R., Skeg, J., Calvo Buendia, E., Masson-Delmotte, V., Pörtner, H.O., Roberts, D.C., Zhai, P., Slade, R., Connors, S., van Diemen, R., et al., Eds.; IPCC: Geneva, Switzerland, **2019**, 249–345.

Mohanty, P.; Singh, P.K.; Chakraborty, D.; Mishra, S.; Pattnaik, R. Insight into the role of PGPR in sustainable agriculture and environment. *Front. Sustain. Food Syst.* **2021**, *5*, 667150.

Müller, D.B.; Vogel, C.; Bai, Y.; Vorholt, J.A. The plant microbiota: Systems-level insights and perspectives. *Annu. Rev. Genet.* **2016**, *50*, 211–234.

Munns, R.; Tester, M. Mechanisms of salinity tolerance. *Annu. Rev. Plant Biol.* **2008**, *59*, 651–681.

Neilson, J.W.; Quade, J.; Ortiz, M.; Nelson, W.M.; Legatzki, A.; Tian, F.; LaComb, M.; Betancourt, J.L.; Wing, R.A.; Soderlund, C.A.; et al. Life at the hyperarid margin: Novel bacterial diversity in arid soils of the Atacama Desert, Chile. *Extremophiles* **2012**, *16*, 553–566.

Oren, A. Molecular ecology of extremely halophilic archaea and bacteria. *FEMS Microbiol. Ecol.* **2002**, *39*, 1–7.

Orozco-Mosqueda, M.; Flores, A.; Rojas-Sánchez, B.; Urtis-Flores, C.A.; Morales-Cedeño, L.R.; Valencia-Marin, M.F.; Chávez-Avila, S.; Rojas-Solis, D.; Santoyo, G. Plant growth-promoting bacteria as bioinoculants: Attributes and challenges for sustainable crop improvement. *Agronomy* **2021**, *11*, 1167.

Parada-Pinilla, M.P.; Díaz-Cárdenas, C.; López, G.; Díaz-Riaño, J.I.; Gonzalez, L.N.; Restrepo, S.; Baena, S. *Salifodinibacter halophilus* gen. nov. sp. nov. a halophilic gammaproteobacterium in the family *Salinisphaeraceae* isolated from a salt mine in the Colombian Andes. *Int. J. Syst. Evol. Microbiol.* **2020**, *70*, 004490.

Pathak, P.; Rai, V.K.; Can, H.; Singh, S.K.; Kumar, D.; Bhardwaj, N.; Roychowdhury, R.; de Azevedo, L.C.B.; Kaushalendra, Verma, H.; Kumar, A. Plant-endophyte interaction during biotic stress management. *Plants* **2022**, *11*(17), 2203. https://doi.org/10.3390/plants11172203

Polash, M.A.; Sakil, M.A.; Hossain, M.A. Plants responses and their physiological and biochemical defense mechanisms against salinity: A review. *Trop. Plant Res.* **2019**, *6*, 250–274.

Pugin, B.; Blamey, J.M.; Baxter, B.K.; Wiegel, J. *Amphibacillus cookii* sp. nov. a facultatively aerobic, spore-forming, moderately halophilic, alkalithermotolerant bacterium. *Int. J. Syst. Evol. Microbiol.* **2012**, *62*, 2090–2096.

Rengasamy, P. World salinization with emphasis on Australia. *J. Exp. Bot.* **2006**, *57*, 1017–1023.

Rengasamy, P. Soil processes affecting crop production in salt-affected soils. *Funct. Plant Biol.* **2010**, *37*, 613.

Saffriel, U.; Adeel, Z. Dryland system. In *Ecosystems and Human Well-Being: Current State and Trends*; Island Press: Washington, DC, USA, **2005**; Volume 53, pp. 623–662.

Sannazzaro, A.I.; Echeverría, M.; Albertó, E.O.; Ruiz, O.A.; Menéndez, A.B. Modulation of polyamine balance in *Lotus glaber* by salinity and arbuscular mycorrhiza. *Plant Physiol. Biochem.* **2007**, *45*, 39–46.

Saralov, A.; Kuznetsov, B.; Reutskikh, E.; Baslerov, R.; Panteleeva, A.; Suzina, N. *Arhodomonasrecens* sp. nov. a halophilic alkane-utilizing hydrogen-oxidizing bacterium from the brines of flotation enrichment of potassium minerals. *Microbiology* **2012**, *81*, 582–588.

Sarapat, S.; Songwattana, P.; Longtonglang, A.; Umnajkitikorn, K.; Girdthai, T.; Tittabutr, P.; Boonkerd, N.; Teaumroong, N. Effects of increased 1-aminocyclopropane-1-carboxylate (ACC) deaminase activity in *Bradyrhizobium* sp. SUTN9-2 on mung bean symbiosis under water deficit conditions. *Microbes Environ.* **2020**, *35*, ME20024.

Savary, S.; Willocquet, L.; Elazegui, F.A.; Castilla, N.P.; Teng, P.S. Rice pest constraints in tropical Asia: Quantification of yield losses due to rice pests in a range of production situations. *Plant Dis.* **2000**, *84*, 357–369.

Seleiman, M.F.; Al-Suhaibani, N.; Ali, N.; Akmal, M.; Alotaibi, M.; Refay, Y.; Dindaroglu, T.; Abdul-Wajid, H.H.; Battaglia, M.L. Drought stress impacts on plants and different approaches to alleviate its adverse effects. *Plants* **2021**, *10*, 259.

Shen, B.; Jensen, R.G.; Bohnert, H.J. Mannitol protects against oxidation by hydroxyl radicals. *Plant Physiol.* **1997**, *115*, 527–532.

Singh, R.; Pandey, K.D.; Singh, M.; Singh, S.K.; Hashem, A.; Al-Arjani, A.B.F.; Abd_Allah, E.F.; Singh, P.K.; Kumar, A. Isolation and characterization of endophytes bacterial strains of *Momordica charantia* L. and their possible approach in stress management. *Microorganisms* **2022**, *10*(2), 290. https://doi.org/10.3390/microorganisms10020290

Sparks, D.L. The chemistry of saline and sodic soil. In *Environmental Soil Chemistry*; Academic Press: Cambridge, MA, USA, **2003**.

Szoboszlay, M.; Näther, A.; Liu, B.; Carrillo, A.; Castellanos, T.; Smalla, K.; Jia, Z.; Tebbe, C.C. Contrasting microbial community responses to salinization and straw amendment in a semiarid bare soil and its wheat rhizosphere. *Sci. Rep.* **2019**, *9*, 9795.

Tanimura, K.; Matsumoto, T.; Nakayama, H.; Tanaka, T.; Kondo, A. Improvement of ectoine productivity by using sugar transporter-overexpressing *Halomonas elongate*. *Enzyme Microb. Technol.* **2016**, *89*, 63–68.

Uzma, M.; Iqbal, A.; Hasnain, S. Drought tolerance induction and growth promotion by indole acetic acid producing *Pseudomonas aeruginosa* in *Vigna radiata*. *PLoS One* **2022**, *17*, e0262932.

Vardharajula, S.; Zulfikar Ali, S.; Grover, M.; Reddy, G.; Bandi, V. Drought-tolerant plant growth promoting *Bacillus* spp.: Effect on growth, osmolytes, and antioxidant status of maize under drought stress. *J. Plant Interact.* **2011**, *6*, 1–14.

Wang, L.; Kaseke, K.F.; Seely, M.K. Effects of non-rainfall water inputs on ecosystem functions. *Wiley Interdiscip. Rev. Water* **2017**, *4*, e1179.

Wang, S.; Sun, L.; Wei, D.; Zhou, B.; Zhang, J.; Gu, X.; Jiang, S. *Bacillus daqingensis* sp. nov. a halophilic, alkaliphilic bacterium isolated from Saline-sodic soil in Daqing, China. *J. Microbiol.* **2014a**, *52*, 548–553.

Wang, W.; Zhuo, L.; Li, M.; Liu, Y.; Wu, P. The effect of development in water-saving irrigation techniques on spatial-temporal variations in crop water footprint and benchmarking. *J. Hydrol* **2019**, *577*, 123916.

Wang, Y.; Hao, Y.; Cui, X.Y.; Zhao, H.; Xu, C.; Zhou, X.; Xu, Z. Responses of soil respiration and its components to drought stress. *J. Soils Sediments* **2014b**, *14*, 99–109.

Wang, Y.; Jiang, G.Q.; Lin, H.P.; Sun, P.; Zhang, H.Y.; Lu, D.M.; Tang, S.K. *Lentibacillus saliphilus*. sp. nov. a moderately halophilic bacterium isolated from a saltern in Korea. *Arch. Microbiol.* **2020**, *203*, 621–627.

Wang, Y.X.; Liu, J.H.; Xiao, W.; Zhang, X.X.; Li, Y.Q.; Lai, Y.H.; Ji, K.Y.; Wen, M.L.; Cui, X.L. *Fodinibius salinus* gen. nov. sp. nov. a moderately halophilic bacterium isolated from a salt mine. *Int. J. Syst. Evol. Microbiol.* **2012**, *62*, 390–396.

Yadav, S.; Modi, P.; Dave, A.; Vijapura, A.; Patel, D.; Patel, M. Effect of abiotic stress on crops. *Sustain. Crop Prod.* **2020**, *3*, 5–16.

Yoo, Y.; Lee, H.; Kwon, B.O.; Khim, J.S.; Baek, S.; Pathiraja, D.; Kim, J.J. *Marinobacter halodurans* sp. nov. a halophilic bacterium isolated from sediment of a salt flat. *Int. J. Syst. Evol. Microbiol.* **2020**, *70*, 6294–6300.

Zaidi, A.; Khan, M.; Ahemad, M.; Oves, M. Plant growth promotion by phosphate solubilizing bacteria. *Acta Microbiol. Immunol. Hung.* **2009**, *56*, 263–284.

Zaman, M.; Shahid, S.A.; Heng, L. *Guideline for Salinity Assessment, Mitigation and Adaptation Using Nuclear and Related Techniques*; Springer: Berlin/Heidelberg, Germany, **2018**; ISBN 978-3-319-96189-7

Zeikus, J.G.; Hegge, P.W.; Thompson, T.E.; Phelps, T.J.; Langworthy, T.A. Isolation and description of *Haloanaerobium praevalens* gen. nov. and sp. nov. *J. Biotechnol.* **1983**, *152*, 114–124.

8 Enriching Tomorrow through Microbes
Unveiling Future Research Demands

Yashpal Singh, Vikram Singh, Monika Singh, Sanjana Bhat, Anushka Das, Arun Bhatt, and Rishendra Kumar

8.1 INTRODUCTION

The essential function and importance of microorganisms in the sustainable advancement of humanity and the ecosystem is the key focus of study. In recent decades, considerable scientific progress across multiple fields has created a possibility to investigate the hidden realm of microbial life. A vast array of invisible and limitless organisms has played a crucial role in shaping our environment and also has a significant impact on human health in multiple ways (Mallik et al., 2024). The field of microbiology emerged ages ago, leading to the development of crucial procedures such as fermentation and the study of various disease patterns (Ray and Bhunia, 2025). Microbiology is an intriguing area that has drawn interest from researchers and scientists worldwide, as microbes have shown significant promise and effect on many biological processes in our environment (Gupta and Verma, 2024). This significant field has substantially enhanced our knowledge of life and the natural environment. It has fundamentally transformed the way scientists across diverse fields, including agriculture, engineering, biotechnology, nutritional research, and environmental science, conduct their work. Early investigations laid the foundation for significant advances in our understanding of microorganisms. During this exploratory phase, several major developments and discoveries shaped the field of microbiology and contributed to its rapid progress. This phase focused on recognizing the framework of microbes, cataloguing their forms and cellular arrangements, and labelling all of them based on taxonomic criteria (Godbole and Sachdev, 2024). During this period, eminent researchers discovered significant findings that explained the crucial functions and vital significance of microbes (Sadasivan et al., 2025). The field of microbiology plays a vital role in tackling significant global issues, such as multidrug resistance, public health challenges, environmental sustainability, and ensuring food and nutritional security (Sharma et al., 2024). Microbiology has achieved remarkable milestones in contemporary science and technology, including the creation of novel

 DOI: 10.1201/9781003664451-8

drugs in pharmacy, advancements in food science and technology, and initiatives aimed at conservation of environment (Eren and Banfield, 2024).

This chapter showcases the immense potential of microorganisms across several emerging fields that are crucial part in promoting sustainable growth in the near future (Guerrieri and Libert, 2024). Akinsemolu et al. (2024) asserted that microbes are important for several uses, such as production of vital pharmaceuticals, helping in the restoration of ecosystems and enhancing crops resilience so they can survive adverse climatic conditions. These unseen organisms can operate at their highest capacity to handle worldwide issues when employed appropriately. Using friendly microorganisms on purpose can strengthen people's immune systems, potentially requiring fewer medications. The linkages increase food safety and security, lower the requirement for chemical fertilizers, and encourage carbon sequestration (Xing et al., 2025).

Microorganisms play a significant role in upholding the environmental integrity, ensuring ecological balance, and contributing to the nutritional security. Furthermore, progress in various domains like Nanobiotechnology, microbial biotechnology and production, and industrial engineering, offers hope for a more sustainable and improved future ahead. Microbes have dramatically improved our understanding of the processes that function within ecosystems and serve a vital role in building a world defined by sustainable growth and development. This chapter mainly focuses on the several roles that the microbes play in promoting a safe, secure, and sustainable future ahead.

8.2 ROLE OF MICROBES IN SUSTAINABLE ENVIRONMENTAL MANAGEMENT

Many serious global concerns such as pollution caused due to irrational use of plastics have affected the human as well as animal health. In the present era, microbes act as an effective option to tackle such prevailing environmental issues, as these bioagents offer a potential approach to minimize such environmental challenges (Devgan et al., 2024). At present, environmental pollution stands as one of the challenging problems encountered worldwide. Various studies have reported the engagement of microbes for waste management, and removal of harmful contaminants from ecosystem through different approaches like environmental bioremediation and plastic degradation (Thirumalaivasan et al., 2024). The capabilities of bacteria can further be utilized for making a clean, contamination-free, and beneficial ecosystem (Velusamy et al., 2024). In this journey to make a cleaner and better place to reside, microbes may act as powerful and potent players to mitigate and reduce the harmful impacts of pollutants and offer a sustainable ecosystem (Figure 8.1).

Microorganisms have the natural potency to clean up harmful matter, decompose pollutants, and maintain the ecological balance through significant contribution in the biological processes such as bioremediation (Bala et al., 2022). Microbes-based environmental bioremediation offers a crucial approach to clean up pollutants, which may promote better health. Recent advances in biotechnology, such as genetically modified microbial consortia, are enhancing the efficiency (Puranik et al., 2023). The combination of microbes-based processes with physical and chemical methods

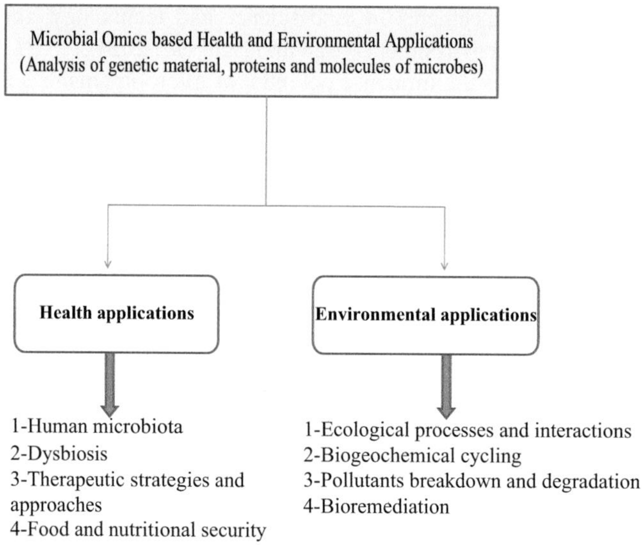

FIGURE 8.1 Diverse applications of microbial omics.

may further result in efficient outcomes (Sharma et al., 2023). The present chapter highlights the role, potential, and importance of microbes as invaluable tools for mitigating environmental pollution, developing waste management, and regulating the interaction with plants to maintain nutrient availability and energy flow (Choudhary et al., 2022; Singh et al., 2025a). Many biofertilizers, like symbiotic microbes, mycorrhizal fungi, are capable to improve and enhance soil quality and fertility, thereby supporting the plant growth (Singh et al., 2025b). Composting, a sustainable important method, converts organic waste into valuable compost. They can degrade organic matter into compost, turning it into a stable, soil- and plant-friendly material (Rai et al., 2025; Aravindharajan et al., 2025).

8.3 IMPACT OF MICROBES IN HUMAN AND VETERINARY WELLNESS

The microbiome of gastrointestinal tract is a vital component of human health, because these microbes conduct a wide range of significant key functions such as production of essential fatty acids, metabolites, and vitamins, which regulate numerous physiological, biological, and metabolic processes in the body (Beltran-Velasco and Clemente-Suarez, 2025). Human gut consists of a large number of microbes, which reside in the gastrointestinal tract of an individual (Hussain et al., 2025). These gut microbes influence various biological mechanisms related to gastrointestinal functions, immunity, metabolism, and neurological functions. They produce various enzymes that are capable to break down complex molecules such as carbohydrates, thereby aiding in digestion and absorption (Suchiita et al., 2025). These microbes have the capacity to produce metabolites and interact with immune cells which play an important role in provoking the immunity against infection. Gut microbiota have

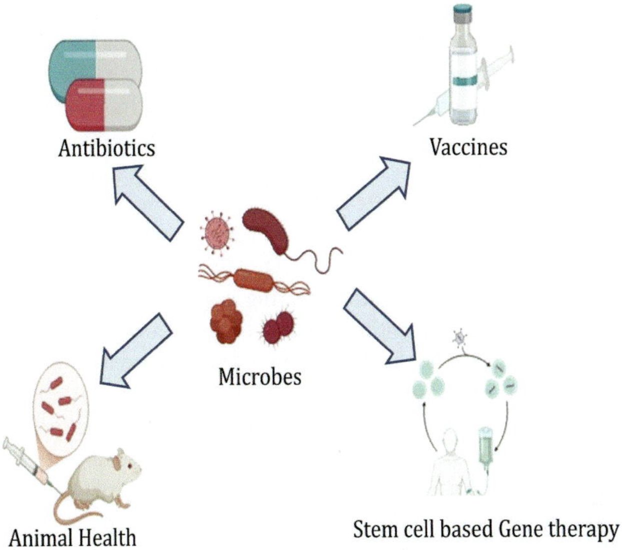

FIGURE 8.2 Microbes in health: human and animal applications.

immunomodulatory functions that may have a significant impact on the immune system. Thus, gut microbiome is reported to have a pivotal relationship with immune system and plays a key role in the maintenance of human health (Wu et al., 2025). The diseases like diabetes and inflammatory bowel syndrome in humans are the result of alteration in composition of gut microbiome (Murugesan et al., 2025). In addition, studies have reported that gut microbiome imbalances may lead to autoimmune disorders, cancers, and neuropsychiatric disorders such as depression and anxiety (Ajibola et al., 2025) (Figure 8.2).

The present therapeutic modalities such as probiotics, prebiotics, and faecal microbiota transplantation have depicted the promising potential of gut microbiome to treat various diseases and provide alternatives to restore microbial diversity in patients (Sartor, 2025). Various microbiome-based approaches such as post-biotic therapeutics and targeted antimicrobials aim to regulate microbiome modulation for health. The factors such as varied microbial composition, functional variability, and time instability are responsible for creating specific challenges for the applications of microbiome-based therapies (Lin et al., 2025). However, the ongoing global research on high-throughput technologies like next-generation sequencing, metagenomics, metatranscriptomics, artificial intelligence, and machine learning has assisted to understand host–microbe interactions and develop new methods for the treatment of various diseases (Dakal and Kumar, 2025). These research-based innovations and outcomes have helped in clearly understanding the concept of microbiome-based therapies and their diverse applications like developing therapeutics and products beneficial for human use.

Scientists having research expertise in the areas of microbiology, biotechnology, and medical science are crucially necessary for integrating microbiome-based research with clinical practices (Mir, 2025). Sufficient research fundings need to be allocated for advanced microbiome research in order to unlock its potential for developing novel therapeutic treatments (Shukla et al., 2024). Microbiome-based research has led to the introduction of novel drugs, medicine, and therapeutics for the treatment of different diseases, while antimicrobial resistance remains a serious global public health concern, which requires comprehensive attention and plan of action (Cheng et al., 2024; Ahmad et al., 2024). The molecular mechanisms such as enzyme production, alteration in bacterial genome, and gene transfer are responsible for antimicrobial resistance (Jhalora and Bist, 2024). Multiple drug-resistant bacteria are spreading at an alarming rate globally, creating a global health issue, especially in the developing countries, where antibiotics and other life-saving drugs are used irrationally (Gajic et al., 2025). The drug-resistant infections are fatal in humans leading to high mortality and morbidity rate. In order to mitigate the ill effects of multiple drug resistance, adaptation of rational use of antibiotics and strict precautionary and infection control measures are the need of the hour (Jafari Jozani et al., 2024).

The introduction of novel approaches to mitigate drug resistance involves phage therapy, special peptides, and combination treatments. The primary requirements are the appropriate research investment and a uniform approach throughout all the living organisms in the environment (Horvat and Kovačević, 2025).

8.4 USE OF MICROBES IN INDUSTRIES AND GREEN ENERGY TECHNOLOGY

Microorganisms are vital in industrial and green technologies, employed in the fabrication of various chemicals such as antibiotics and organic acids. They may act as biocatalysts in different important processes such as fermentation to produce food, bio-energy, and beverages. Microbes significantly contribute in environmental bioremediation through removal of contaminants like pesticides from soil, decomposing organic matter and enhancing sustainable agricultural development by the introduction of different biofertilizers and biopesticides. These approaches are sustainable, cost-effective, and may also promote green energy technology (Shah et al., 2025a).

Various microbial processes and their novel applications have contributed largely towards rise in the global bio-economy, revealing the importance of microbial community. The unique ability of microbes to provide a significant outcome using a minimal input makes them invaluable, versatile, and highly efficient organisms, which can be altered or engineered to produce different products of diverse sectors (Qumsani, 2024). Microorganisms have substantially contributed to economic growth and employment generation through their extensive applications in industries ranging from food processing to pharmaceuticals. Microbes have opened a new gateway of career opportunities in the present era (Akinsemolu et al., 2023). Currently, many beneficial bacteria and yeasts are employed for producing various food products, providing sustainable substitutes that may promote health (Nascimento and Barros, 2025).

Microbes have served as living biological factories in the pharmaceutical sector, as they are primarily used in developing various cost-effective efficacious vaccines, life-saving drugs and therapeutics (Shreya and Verma, 2024). In addition, microorganisms have served as biocatalysts in converting various agri-wastes into valuable products, replacing conventional processes in the chemical industries (Akhtar et al., 2025). Thus, microorganisms have transformed the employment sector by generating new livelihood opportunities and enhancing the socio-economic well-being of individuals and communities. Various new related areas like metabolic engineering and synthetic biology have contributed in promoting the economic importance of these unseen organisms.

8.5 FUTURE RESEARCH PERSPECTIVES

Microbial ecology is the study of complex relationship of microbes and their environment. It assists in understanding the relationship between microbes and various components of ecosystem like soil, water, and air. Microorganisms are beneficial and one of the most demanding players affecting humans and ecosystem, as they render many benefits through different processes such as biogeochemical cycling, environmental bioremediation, nutrient availability, and decomposition of organic matter (Idowu, 2024). Recent studies have reported the influence of microbes in several environmental strategies that offer significant contribution towards future sustainability. In recent times, significant development in the field of molecular biology and genetic engineering has revolutionized the field of microbial ecology. At present, researchers are capable to identify numerous microbes residing in a specific environment with techniques such as 16S rRNA gene sequencing, highlighting the diversity present in soil, water, and atmosphere (Maheshwari et al., 2024).

Metagenomics has contributed for interpretation of genetic material of microorganisms, which offers scientists to discover different functions and metabolic abilities of microbes (Nandal et al., 2025). This approach aids researchers to reveal the role of microbes in different applications such as environmental monitoring and renewable energy resources production. Metatranscriptomics is another advanced branch of omics that allows scientists to understand the response of microorganisms to different environmental alterations, interpret gene expression, and regulate host–microbe interactions (Shah et al., 2025b). The application of these high-throughput techniques produces a large set of data. Many different in silico tools and strategies such as annotation, sequencing, and visualization tools facilitate researchers to uncover the functions and interactions of microbes with ecosystem (Sikkander et al., 2025).

These scientific innovations and advanced techniques of biotechnology may lead to transformation of microbes-based ecology, which may enable scientists to uncover the importance and diverse functions of microbes for resolving the environmental challenges and providing a support to maintain ecological balance and life on earth. Furthermore, microbial research in different areas such as synthetic biology, microbiome-based development of personalized medicines and therapeutics for mitigation of severe diseases, and sustainable ecological development and management is the need of the hour.

8.6 DIVERSE INDUSTRIAL APPLICATIONS OF MICROBIOME

The microbiome plays a key role in sustainable agriculture development and maintaining the ecological balance using different microbes. Exploring microbiome offers many benefits in different domains. It makes a substantial contribution to improving human health and plays an important role in preserving environmental conservation (Andreou and Papaneophytou, 2025). Crop yield and productivity may be enhanced by employing beneficial microbes, which may improve the socio-economic status of farming community and also provide insect-resistant and disease-free products for consumption (Verma et al., 2021).

Microorganisms have a vital role in sustaining the quality of soil, lowering the level of greenhouse gases, and facilitating the removal of contaminants from environment (Hu et al., 2024). These organisms also participate in nitrogen fixation process and play a crucial role in maintaining the nutrient availability in the soil as well as in promoting soil regeneration (Chen et al., 2024). Moreover, the ability of microbes to eradicate contaminants from soil, air, and water bodies make them more important in order to maintain favourable conditions for life in ecosystem. Microorganisms possess the astounding ability to transform agricultural waste into a wide range of economically significant products, such as biochemicals and biofuels (Ufitikirezi et al., 2024).

Various eco-friendly products, bio-based chemicals, bioethanol, and bioplastics may be produced using different microbes-based processes (Rajendran et al., 2025). The integration of microorganisms into diverse agricultural practices has significantly improved crop yield and productivity, thereby benefiting farmers and enhancing agricultural sustainability. Such microbe-driven processes are beneficial and have an immense impact on ecosystem and sustainable development.

REFERENCES

Ahmad, Shahzad, Sham Lohiya, Amar Taksande, Revat J. Meshram, Ashish Varma, Keta Vagha, Shahzad Ahmad Jr, and Ashish Varma Sr. "A comprehensive review of innovative paradigms in microbial detection and antimicrobial resistance: beyond traditional cultural methods." *Cureus* 16, no. 6 (2024).

Ajibola, Fagbemi Oluwaseyi, Esther Chigbaziru Nwojiji, Mustapha Lawal, Suara Temilola Barakat, Iyiola Aanuoluwa Temitayo, Victor Ekoche Ali, and Taiwo Bakare-Abidola. "The role of the gut microbiome in immune modulation: implications for autoimmune diseases and cancer therapy." *World Journal of Biology Pharmacy and Health Sciences* 22, no. 1 (2025): 564–590.

Akhtar, Muhammad Saeed, Muhammad Tahir Naseem, Sajid Ali, and Wajid Zaman. "Metal-based catalysts in biomass transformation: from plant feedstocks to renewable fuels and chemicals." *Catalysts* 15, no. 1 (2025): 40.

Akinsemolu, Adenike A., Helen Onyeaka, and Foluke V. Arijeniwa. "Harnessing green microbiology for sustainable water management, agriculture, and energy generation: current advances and future prospects." *World Water Policy* 10, no. 3 (2024): 649–697.

Akinsemolu, Adenike, Helen Onyeaka, Omololu Fagunwa, and Adewale Henry Adenuga. "Toward a resilient future: the promise of microbial bioeconomy." *Sustainability* 15, no. 9 (2023): 7251.

Andreou, Eleni, and Christos Papaneophytou. "Boosting immunity through nutrition and gut health: a narrative review on managing allergies and multimorbidity." *Nutrients* 17, no. 10 (2025): 1685.

Aravindharajan, S. T. M., Livleen Shukla, D. Vijaysri, Sandeep Kumar Singh, Reetesh Kumar, Rajesh Kumar, Shalini Gaur Rudra, Gautam Chawla, and Ajay Kumar. "Comparative paddy straw degradation potential of Psychrotrophic Fungal Consortia and Pusa Decomposer at low temperatures." *Bioresource Technology Reports* 32 (2025): 102314.

Bala, Saroj, Diksha Garg, Banjagere Veerabhadrappa Thirumalesh, Minaxi Sharma, Kandi Sridhar, Baskaran Stephen Inbaraj, and Manikant Tripathi. "Recent strategies for bioremediation of emerging pollutants: a review for a green and sustainable environment." *Toxics* 10, no. 8 (2022): 484.

Beltran-Velasco, Ana Isabel, and Vicente Javier Clemente-Suarez. "Harnessing gut microbiota for biomimetic innovations in health and biotechnology." *Biomimetics* 10, no. 2 (2025): 73.

Chen, Qingxia, Yingjie Song, Yuxing An, Yinglin Lu, and Guohua Zhong. "Soil microorganisms: their role in enhancing crop nutrition and health." *Diversity* 16, no. 12 (2024): 734.

Cheng, Yan, Xueting Wang, Yingying Ma, Tao Chen, Luying Xiong, Pan Chen, Ping Lu, Erjia Zhang, and Yonghong Xiao. "A multi-sectoral approach to antimicrobial resistance containment: insights from One Health." *One Health Bulletin* 5, no. 3 (2024): 10–4103.

Choudhary, Saroj, Arjun Singh, and Mukta Rani. "Rhizosphere: hotspot of soil-microbes-plant interaction." *A Monthly Peer Reviewed Magazine for Agriculture and Allied Sciences* 80, no. 1 (2022): 1–511.

Dakal, Tikam Chand, Caiming Xu, and Abhishek Kumar. "Advanced computational tools, artificial intelligence and machine-learning approaches in gut microbiota and biomarker identification." *Frontiers in Medical Technology* 6 (2025): 1434799.

Devgan, Suyash, Sparsh Singh, Abhilansh Pandey, and Rashmi Mathur. "Microbial solution for mitigating plastic pollution: from environmental hazard to sustainable management." In Bala, K., Ghosh, T., Kumar, V., and Sangwan, P. (Eds.), *Harnessing Microbial Potential for Multifarious Applications*, pp. 101–135. Singapore: Springer Nature Singapore, 2024.

Eren, A. Murat, and Jillian F. Banfield. "Modern microbiology: embracing complexity through integration across scales." *Cell* 187, no. 19 (2024): 5151–5170.

Gajic, Ina, Nina Tomic, Bojana Lukovic, Milos Jovicevic, Dusan Kekic, Milos Petrovic, Marko Jankovic, et al. "A comprehensive overview of antibacterial agents for combating Multidrug-Resistant bacteria: the current landscape, development, future opportunities, and challenges." *Antibiotics* 14, no. 3 (2025): 221.

Godbole, Suchitra, and Dhara P. Sachdev. *Basic Concepts and Recent Advances in Microbial Diversity, Taxonomy, Speciation and Evolution*, pp. 79–106. Newcastle upon Tyne: Cambridge Scholar Publishing, 2024.

Guerrieri, Francesca, and Cédric Libert. "The invisible life." *Frontiers in Microbiology* 15 (2024): 1401487.

Gupta, Juhi, and Akarsh Verma. "MICROBIOLOGY: the science to the world of microbes." In Gupta, Juhi, and Verma, Akarsh (Eds.), *Microbiology-2.0 Update for a Sustainable Future*, pp. 1–16. Singapore: Springer Nature Singapore, 2024.

Horvat, Olga, and Zorana Kovačević. "Human and veterinary medicine collaboration: synergistic approach to address antimicrobial resistance through the lens of planetary health." *Antibiotics* 14, no. 1 (2025): 38.

Hu, Pengfei, Yongliang Ma, and Yinlong Yin. "The mechanisms and application prospects of soil microorganisms in soil health." *Geographical Research Bulletin* 3 (2024): 606–624.

Hussain, Md Sadique, Gurusha Bahl, Riya Mishra, Asif Ahmad Bhat, Riya Thapa, Raihan Siddiqui, Rahul Sharma, Rashi Kulshrestha, Neeraj Patel, and Gaurav Gupta. "Introduction to microbiome." In Gupta, G., Hussain, M. S., Dua, K., Gilhotra, R., and Dhanasekaran, M. (Eds.), *Gut Microbiome and Environmental Toxicants*, pp. 23–40. Boca Raton, FL: CRC Press, 2025.

Idowu, Emmanuel. "Microbial ecology: understanding the dynamics of microbial communities and their impact on ecosystems." (2024).

Jafari Jozani, Raziallah, Mauida F. Hasoon Al Khallawi, Darren Trott, Kiro Petrovski, Wai Yee Low, and Farhid Hemmatzadeh. "Unravelling antimicrobial resistance in Mycoplasma hyopneumoniae: genetic mechanisms and future directions." *Veterinary Sciences* 11, no. 11 (2024): 542.

Jhalora, Vandana, and Renu Bist. "A comprehensive review of molecular mechanisms leading to the emergence of multidrug resistance in bacteria." *Indian Journal of Microbiology* 65, no. 2 (2024): 1–22.

Lin, Anqi, Aimin Jiang, Lihaoyun Huang, Yu Li, Chunyanx Zhang, Lingxuan Zhu, Weiming Mou, et al. "From chaos to order: optimizing fecal microbiota transplantation for enhanced immune checkpoint inhibitors efficacy." *Gut Microbes* 17, no. 1 (2025): 2452277.

Maheshwari, Keerti, Pankaj Musyuni, Amitava Moulick, Harshita Mishra, Adam Ekielski, Pawan Kumar Mishra, and Geeta Aggarwal. "Unveiling the microbial symphony: Next-Gen sequencing and bioinformatics insights into the human gut microbiome." *Health Sciences Review* 11 (2024): 100173.

Mallik, Bhagaban, Tarun Mishra, Poornima Dubey, Minu Kesheri, and Swarna Kanchan. "Exploring the secrets of microbes: unveiling the hidden world through microbial omics in environment and health." In Kesheri, M., Kanchan, S., Salisbury, T. B., and Sinha, R. P. (Eds.), *Microbial Omics in Environment and Health*, pp. 269–294. Singapore: Springer Nature Singapore, 2024.

Mir, Shazia S. "Harnessing microbial ecosystems in health and disease: innovations in medical microbiology for advancing precision medicine and therapeutic strategies." *Egyptian Academic Journal of Biological Sciences. C, Physiology and Molecular Biology* 17, no. 1 (2025): 41–67.

Murugesan, Ria, Janardanan Kumar, Kakithakara Vajravelu Leela, Sachdev Meenakshi, Appandraj Srivijayan, Shubhashree Thiruselvam, Abhishek Satheesan, and Venkata Chaithanya. "The role of gut microbiota and bacterial translocation in the pathogenesis and management of type 2 diabetes mellitus: mechanisms, impacts, and dietary therapeutic strategies." *Physiology & Behavior* 293 (2025): 114838.

Nandal, Khushwant, Jitender Rathee, Priyanka Dahiya, and Manjeet Kaur. "Unraveling microbial ecosystems: advances in metagenomics for biotechnology applications." *Journal of Advances in Microbiology Research* 6, no. 1 (2025): 262–277.

Nascimento, Amanda Priscila Silva, and Ana Novo Barros. "Sustainable innovations in food microbiology: fermentation, biocontrol, and functional foods." *Foods* 14, no. 13 (2025): 2320.

Puranik, Shriniketan, Livleen Shukla, Aditi Kundu, Deeba Kamil, Sangeeta Paul, Govindasamy Venkadasamy, Rajna Salim, Sandeep Kumar Singh, Dharmendra Kumar, and Ajay Kumar. "Exploring potent fungal isolates from sanitary landfill soil for in vitro degradation of dibutyl phthalate." *Journal of Fungi* 9, no. 1 (2023): 125.

Qumsani, Alaa T. "The contribution of microorganisms to sustainable development: towards a green future through synthetic biology and systems biology." *Journal of Umm Al-Qura University for Applied Sciences* 11, no. 4 (2024): 1–17.

Rai, Ashish, Rahul Mishra, Abhik Patra, Arvind Kumar Singh, Sachin Sharma, Ayush Bahuguna, Sumit Rai, et al. "Microbes and compost: an emerging role in climate resilience agriculture." In Shalini Rai, Abhishek Kumar Bhardwaj, and Luciane Colla (Eds.), *Sustainable Management of Agro-Food Waste*, pp. 105–125. Cambridge: Academic Press, 2025.

Rajendran, Sundarakannan, Ali Al-Samydai, Geetha Palani, Herri Trilaksana, Thanikodi Sathish, Jayant Giri, Rathinasamy Saravanan, Joshua Isaac JoshuaRamesh Lalvani, and Faouzi Nasri. "Replacement of petroleum based products with plant-based materials, green and sustainable energy—A review." *Engineering Reports* 7, no. 4 (2025): e70108.

Ray, Bibek, and Arun Bhunia. *Fundamental Food Microbiology*. Boca Raton, FL: CRC Press, 2025.

Sadasivan, Sreeja, Senthilnathan Pradeep, Jishnu Chandran Ramachandran, Jayaprakash Narayan, and Michał Jan Gęca. "Advances in droplet microfluidics: a comprehensive review of innovations, morphology, dynamics, and applications." *Microfluidics and Nanofluidics* 29, no. 3 (2025): 17.

Sartor, R. Balfour. "Beyond random fecal microbial transplants: next generation personalized approaches to normalize dysbiotic microbiota for treating IBD." *Gastroenterology Clinics* 54, no. 2 (2025): 333–350.

Shah, Kamleshkumar R., Leena Patel, and Gopalkumar G. Raol. *Introduction to Microbiology: Understanding the Invisible World*. Palwal: Chyren Publication, 2025a.

Shah, Manan, Guido Sieber, Aman Deep, Daniela Beisser, and Jens Boenigk. "Unravelling the temporal dynamics of community functions in protists induced by treated wastewater exposure using metatranscriptomics." *Scientific Reports* 15, no. 1 (2025b): 23957.

Sharma, Monika, Sangita Agarwal, Richa Agarwal Malik, Gaurav Kumar, Dan Bahadur Pal, Mamun Mandal, Abhijit Sarkar, et al. "Recent advances in microbial engineering approaches for wastewater treatment: a review." *Bioengineered* 14, no. 1 (2023): 2184518.

Sharma, Shikha, Abhishek Chauhan, Anuj Ranjan, Darin Mansor Mathkor, Shafiul Haque, Seema Ramniwas, Hardeep Singh Tuli, Tanu Jindal, and Vikas Yadav. "Emerging challenges in antimicrobial resistance: implications for pathogenic microorganisms, novel antibiotics, and their impact on sustainability." *Frontiers in Microbiology* 15 (2024): 1403168.

Shreya, Juhi Gupta, and Akarsh Verma. "Unlocking the therapeutic potential of microorganisms in revolutionizing pharmaceuticals." In Juhi Gupta, and Akarsh Verma (Eds.), *Microbiology-2.0 Update for a Sustainable Future*, pp. 121–157. Singapore: Springer Nature Singapore, 2024.

Shukla, Vani, Shikha Singh, Shrikant Verma, Sushma Verma, Aliya Abbas Rizvi, and Mohammad Abbas. "Targeting the microbiome to improve human health with the approach of personalized medicine: latest aspects and current updates." *Clinical Nutrition ESPEN* 63 (2024): 813–820.

Sikkander, Abdul Razak Mohamed, Tripathi, Suman Lata and Theivanathan, Govindan. "Extensive sequence analysis: revealing genomic knowledge throughout various domains." In *Computational Intelligence for Genomics Data*, pp. 17–30. Cambridge: Academic Press, 2025.

Singh, Gajendra B., Shubhankhi Dey, and Gaurav Mudgal. "Microbial contributions to a circular economy." In Sukhminderjit Kaur, Sunny Dhiman, and Manikant Tripathi (Eds.), *Microbial Metabolomics: Recent Developments, Challenges and Future Opportunities*, pp. 419–439. Singapore: Springer Nature Singapore, 2025a.

Singh, Subhendra Kumar, Roop Kishor Pachauri, Heena Khatoon, Deepika Katiyar, and Gaurav Agnihotri. "The role of biofertilizers in enhancing soil and productivity-a review." *International Journal of Plant & Soil Science* 37, no. 3 (2025b): 141–161.

Suchiita, Anuupama, Navya Gupta, Kajal Nandi, Subhash Sonkar, and Lal Chandra. "Harmony within: unravelling the microbiome–immune system symbiosis for health." *Advanced Gut & Microbiome Research* 2025, no. 1 (2025): 9927379.

Thirumalaivasan, Natesan, Lalitha Gnanasekaran, Suresh Kumar, Rajesh Durvasulu, Thanigaivel Sundaram, Saravanan Rajendran, Senthilkumar Nangan, and Kuppusamy Kanagaraj. "Utilization of fungal and bacterial bioremediation techniques for the treatment of toxic waste and biowaste." *Frontiers in Materials* 11 (2024): 1416445.

Ufitikirezi, Jean de Dieu Marcel, Martin Filip, Mohammad Ghorbani, Tomáš Zoubek, Pavel Olšan, Roman Bumbálek, Miroslav Strob, et al. "Agricultural waste valorization: exploring environmentally friendly approaches to bioenergy conversion." *Sustainability* 16, no. 9 (2024): 3617.

Velusamy, Sathya, Murugaiyan Sinduja, Ramasamy Vinothini, and Govindaraj Kamalam Dinesh. "Role of microbes and microbiomes in solid waste management for ecosystem restoration." In Shiv Prasad, Govindaraj Kamalam Dinesh, Murugaiyan Sinduja, Velusamy Sathya, Ramesh Poornima, and Sangilidurai Karthika (Eds.), *The Role of Microbes and Microbiomes in Ecosystem Restoration*, pp. 156–175. Singapore: Bentham Science Publishers, 2024.

Verma, Hariom, Dharmendra Kumar, Vinod Kumar, Madhuree Kumari, Sandeep Kumar Singh, Vijay Kumar Sharma, Samir Droby, Gustavo Santoyo, James F. White, and Ajay Kumar. "The potential application of endophytes in management of stress from drought and salinity in crop plants." *Microorganisms* 9, no. 8 (2021): 1729.

Wu, Kaile, Yang Xiao, Tianjun Zhang, Yunlong Bai, and Meiling Yan. "The role of the gut microbiota and its metabolites: a new predictor in diabetes and its complications." *European Journal of Medical Research* 30, no. 1 (2025): 601.

Xing, Yingying, Yunxia Xie, and Xiukang Wang. "Enhancing soil health through balanced fertilization: a pathway to sustainable agriculture and food security." *Frontiers in Microbiology* 16 (2025): 1536524.

9 Nurturing the Soil
Microbial Strategies for Soil and Plant Health

Anisha Rupashree, Nitin Swamy, and Punam Kumari

9.1 INTRODUCTION

Soil is vital for the success of sustainable agricultural practices. It impacts the growth and output and overall performance of the crops. Tahat et al. (2020) emphasize its role in preserving the environment and supporting ecosystem resilience. This layer is often known as the living skin of the Earth. Soil serves as a complex and dynamic ecosystem. It nurtures crop development while enhancing biodiversity. The productivity of crops depends directly on soil quality and health. They impact crop's ability to withstand stress and manage water effectively. They are vital contributors to carbon storage in soils and the overall health of ecosystems. Soil health is fundamental to achieving sustainable agricultural practices. It plays a crucial role in meeting the food needs of the expanding global population.

The health of the soil ecosystem is demonstrated through the effects of biological indicators on nutrient cycling, disease suppression, and organic matter decomposition (Garland and Mills 1991). Preserving soil health is essential for sustainable agriculture and can be achieved through practices such as conservation tillage, vermicomposting, crop diversification, crop residue management, organic amendments, balanced fertilizer application, efficient irrigation management, the use of naturally occurring mineral amendments, the introduction of effective microorganisms, and the application of biopesticides. Modern approaches, such as precision agriculture, biochar application, and cover cropping, along with the use of the advanced seed blends, microbe-based soil inoculants, soil monitoring with digital sensors, and digital soil mapping, are in use and increasingly adopted for the conservation of soil, improvement in nutrient cycling, and sustainability of long-term productivity. The microbiome of soil is a diverse and very dynamic community of microorganisms that influences almost all aspects of plant and soil health. The composition of soil microbiomes is shaped by various components such as soil type, species of plant, climate, and land-use practices. The microbial diversity and richness contribute in various aspects such as nutrient cycling (e.g., nitrogen fixation, phosphorus solubilization), suppression of the disease, development of root and its architecture and stress tolerance, including biotic and abiotic both. Food

DOI: 10.1201/9781003664451-9

security around the globe is under tremendous pressure due to various challenges such as climate variability, degradation of soil, diminishing biodiversity, and the demands of the continuously increasing population.

Microbiome-based approaches entail adding beneficial microbes to the soil. The goal of these approaches is to lessen dependence on agrochemicals. Developments in farming strategies have led to substantial increases in food output. But they have resulted in overreliance on chemical inputs. Excessive application has severely degraded key soil resources and disrupted the soil microbial ecosystem which ultimately reduce soil health. Across the globe, soil fertility is steadily declining. Unstable rainfall patterns and rising incidences of plant diseases are becoming significant issues. Such issues are the main cause for declining agricultural performance and reduced resilience. Challenges like socio-economic inequality still persist, as well as the shortage of sustainable agricultural approaches. These issues further widen the existing gaps in ensuring food access globally. Microbes are essential for safeguarding soil physical properties and promoting soil fertility by assisting in nutrient cycling.

Few beneficial microbes, including nitrogen-fixing *Rhizobium*, phosphate-solubilizing *Bacillus*, and mycorrhizal fungi, engage closely with plant roots. These microbes engage in symbiotic association with plant root systems. These microbial root systems facilitate the maximization of nutrient absorption. Therefore, they encourage strong and flourishing plant growth. Such microbial communities help maintain the stability of soil structure. These microbes synthesize extracellular polysaccharides and establish biofilms. Their secretions contribute to the formation and stabilization of soil aggregates. These advantageous microorganisms contribute to plant health through several mechanisms. They are involved in the production of phytohormones like indole-3-acetic acid (IAA). They also produce siderophores, enzymes, and antibiotics that supress diseases caused by soil. In addition, they promote the activation of ISR (induced systemic resistance) in different plants. As a result, the soil microbiome is essential for preserving ecological equilibrium. It further contributes to a substantial increase in agricultural output. It also supports vital processes such as pesticide transformation, carbon storage, and heavy metals immobilization. Additionally, it promotes nitrogen cycling and augments drought tolerance. These beneficial microbes further contribute to strengthening the defence of plants against different types of pathogens. They facilitate this effect by inducing systemic resistance in plants. Core microbial functions including fixing nitrogen, absorbing micronutrients, solubilizing phosphorus, and tolerating osmotic pressure are vital for the growth of plants during their life cycle. Their relevance intensifies under stressful and adverse environmental conditions. They help plants maintain growth and yield despite various stress conditions. It is clear evidence that soil management practices and microbial activities are mutually influential. Factors which affect the interactions are fewer chemicals, organic supplements, and crop rotation. Such measures promote the persistence of healthy microbial communities. This demonstrates that microbiome-centred approaches should be integrated into resilient agricultural systems. Such integration can enhance overall productivity and stability. It is clearly evident that ensuring soil's ability to sustain human life is of critical importance. It will always remain crucial for sustaining food and feed production for a growing population. El Chami et al. (2020) report that healthy soil protects vital ecosystem functions, mitigates the impacts of climate change, and enhances the resilience of Earth's landscapes.

9.2 PLANT GROWTH–PROMOTING MICROORGANISMS (PGPMs)

PGPMs serve as key agents in managing soil health. They act as the natural driving force behind important aspects like soil nutrient richness. They also improve crop productivity and help in the maintenance of ecological balance. They make this possible through a broad spectrum operating either through direct or indirect mechanisms. PGPMs occupy areas like rhizosphere, phyllosphere, endosphere, and spermosphere. They create a dynamic association with plants. PGPMs' diversity and multifunctionality allow them an essential asset in promoting sustainable farming practices. They significantly contribute to increased nutrient acquisition and plant growth. They also safeguard plants from biotic and abiotic challenges. They play an important role in emphasizing their capacity to cut down on chemicals and support soil health. Embedding PGPMs within the farming system can provide an effective means to boost agricultural productivity. This approach may facilitate enduring food security and the promotion of environmental sustainability globally. Figure 9.1 emphasizes the important activities of soil microorganisms. It also emphasizes on how it helps promoting plant health and the preservation of soil fertility.

Within the rhizosphere—roots are the vibrant zone, beneficial microorganisms, including bacteria, fungi, and actinomycetes are found in abundance. They engage intimately with plant roots to help in keeping plants healthy and growing. They facilitate essential processes, including nitrogen fixation and phosphate solubilization. They secrete phytohormones that support plant growth. They facilitate greater nutrient uptake and help in root development. They help uphold the plant's defence against harmful microbes. These microorganisms also secrete siderophores, enzymes, and organic acids making nutrients more accessible. These secretions help control the spread of soil-borne diseases. Together, these interactions form stable and resilient microbial system. This network establishes the cornerstone of sustainable and

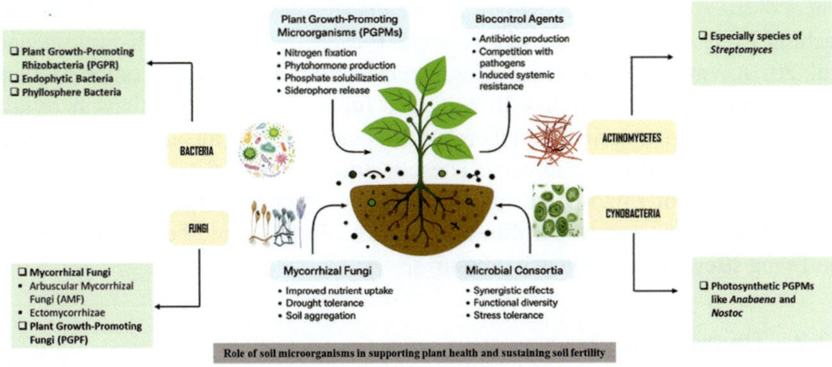

FIGURE 9.1 Role of soil microorganisms in supporting plant health and sustaining soil fertility.

high-yield agriculture. The key functional categories of soil microbes include bacteria, fungi, protozoa, and actinomycetes. All groups play individual or specific roles in maintaining plant and soil health. In bacteria, notable subgroups include plant growth–promoting rhizobacteria (PGPRs), endophytes, and phyllosphere bacteria (Pathak et al. 2022; Kashyap et al. 2023). These subgroups contribute significantly to plant growth. Phyla such as Actinomycetes, Firmicutes, and Proteobacteria fall within these groups. Fungal populations are dominated by mycorrhizal fungi. This fungi group comprises ectomycorrhizae, arbuscular mycorrhizae, ericoid mycorrhizae, and orchid mycorrhizae. Other growth-enhancing fungi are part of this group as well. Protozoa and actinomycetes are crucial to regulate microbial population. They assist in the process of maintenance of soil structure, nutrient transformation, and its bioavailability. As highlighted by Pandey and Saharan (2025), they help control plant pathogens, fostering both resilience and fertility in the soil. Microorganisms significantly contribute to the creation of stable soil aggregates. Fungi, arbuscular mycorrhizal fungi (AMF), and actinomycetes demonstrate this role most strongly. By releasing extracellular polysaccharides, they help stabilize and bind soil particles. Further, during the microbial breakdown of fresh organic matter, they form extracellular polysaccharides and lipids. These substances significantly contribute to soil structural integrity. They assist in forming soil aggregates and maintaining their stability (Forster 1990). AMF play a role in helping plants in producing a greater amount of organic acids, e.g., citric and malic acid. By chelating nutrients, they help regulate soil pH (Li et al. 2023). AMF help make organic acids like valeric acid more accessible in alkaline soils. This release helps break down polycyclic aromatic hydrocarbons. It also helps lower the pH of the soil (Figure 9.1).

9.2.1 The Rhizosphere

Rhizosphere is the narrow zone of the soil, which is influenced by root exudates. It is a dynamic hotspot of microbial activity where intense plant–microbe and microbe–microbe interactions take place. Roots release ~5% of their photosynthesis product that forms chemical gradients, which support beneficial microbes through various metabolic and signalling pathways. The rhizosphere soil is the most precious component of the soil microflora. Rhizosphere enhances the soil fertility and blocks soil leaching using nutrient cycles like N-cycle, P-cycle, Fe-cycle, and K-cycle (Canon et al. 2020). Various research based on metagenomics and next-generation sequencing has revealed that microbial diversity and functionality in the rhizosphere layer are directly linked to various factors related to plants. The factors include plant stress tolerance, yield, and soil resilience. These factors are interrelated with the activity and interactions of rhizosphere microbes. In the absence of these microbial partners, fertility in the soil would diminish quickly. Plants would be at greater risk to non-living stresses and pathogen pressures.

9.2.2 Diversity and Mechanism

Beneficial microorganisms, known as PGPMs, are crucial for fostering plant growth through several strategies, which cover both direct and indirect mechanisms. Direct mechanisms contribute to supporting plant nutrition and overall growth.

These mechanisms encompass the formation of key phytohormones such as auxins, gibberellins, and cytokinins. They influence overall plant developmental processes. One more vital process is biological nitrogen fixation in which microbes capable of fixing nitrogen transform atmospheric nitrogen (N_2) into ammonia (NH_3), thereby making it easier for plants to use them. Furthermore, many PGPMs synthesize siderophores which capture iron and carry it to the plant roots when iron is limited. PGPMs help make phosphate and potassium more available. Through this process of microbial nutrient solubilization through organic acid secretion and enzymatic weathering activities, these nutrients are converted to more accessible to plants which would normally very tough to be absorbed by the plants.

9.2.3 Direct Mechanisms

Direct mechanisms refer to the mechanism through which PGPMs directly enhance the growth of plant. They aid in plant growth by boosting through facilitating nutrient absorption and hormone production, and drive beneficial biochemical processes within the soil. These processes function without involving pathogen interactions. Primary mechanisms are key to ensuring proper plant nutrition and growth.

9.2.3.1 Phytohormone Production

A range of plant-friendly microbes known as PGPMs secrete plant hormones. They are necessary for several key developmental processes. Plant hormones play a key role in monitoring and controlling these developmental processes. Auxins drive the growth and elongation of lateral roots, helping the plant take up nutrients and water more efficiently. Gibberellins encourage seeds to germinate and support stem growth by driving cell division and triggering enzyme activity. Cytokinins regulate the division of cells and delay leaf development and senescence. They also help in maintaining the chlorophyll content and extending photosynthetic activity. Additionally, through 1-aminocyclopropane-1-carboxylate deaminase activity (ACC) deaminase activity, PGPMs modulate ethylene levels by degrading its precursor ACC, which effectively lowers stress-induced ethylene accumulation and allows healthier plant growth even under adverse environmental conditions.

9.2.3.2 Biological Nitrogen Fixation

Nitrogen-fixing bacteria, such as *Rhizobium*, *Azospirillum*, and *Azotobacter*, convert atmospheric nitrogen (N_2) into ammonia (NH_3) in a form which can be used by plants. This process increases the nitrogen content in the soil, which in turn reduces the utilization of synthetic fertilizers.

9.2.3.3 Phosphate Solubilization

Phosphorus is often present in insoluble forms in the soil. Nearly 90% of soil lack available phosphorus because it readily reacts with elements like aluminium (Al), calcium (Ca), and iron (Fe), forming insoluble compounds that limit its bioavailability and reduce phosphorus-use efficiency in plants.

Phosphate-solubilizing bacteria (PSB) play a very important role in increasing phosphorus availability for plants by solubilizing and mineralizing insoluble forms of phosphorus. The mechanisms for this involve secretion of organic acids, proton extrusion, and production of phosphatases, followed by other associated pathways. In this manner, the insoluble phosphatases are converted into free phosphate ions that plants can absorb easily. The organic acids produced in this mechanism are gluconic acid, 2-ketogluconic acid, oxalic acid, citric acid, acetic acid, malic acid, lactic acid, formic acid, succinic acid, propionic acid, and tartaric acid. These acids then chelate metal ions and lower the pH, due to which the phosphate solubilization takes place (Marra et al. 2019). Microbial respiration generates bicarbonate (HCO_3^-) and ammonium (NH_4^+), which release protons (H^+), further acidifying the environment and promoting phosphorus release. Apart from this active solubilization, microorganisms can also store bioavailable phosphorus during their life cycle. Upon cell lysis, the content is released back into the soil and enriches the phosphorus pool. A few PSB produce chelating compounds which bind with the metal ions and free the phosphate ions from their insoluble complexes. Some PSB like *Pseudomonas* and *Bacillus* secrete organic acids (e.g., gluconic and citric acid) and enzymes (e.g., phosphatases) which solubilize inorganic phosphate, making it readily available for plant uptake (Mohanty et al. 2021).

9.2.3.4 Potassium and Zinc Solubilization

Some PGPMs convert insoluble potassium- and zinc-containing minerals into soluble forms using acidification and chelation, thus improving the availability of these essential nutrients for plant metabolism and enzyme activation. Potassium-solubilizing bacteria facilitate the release of potassium from feldspar and aluminosilicate minerals through various mechanisms, including acidolysis, chelation, ion exchange, and complexation (Etesami, Emami, and Alikhani 2017). These bacteria also contribute to potassium availability by decomposing organic matter and crop residues. *Bacillus circulans*, *Bacillus mucilaginosus*, and *Bacillus edaphicus* are a few key potassium-solubilizing species which secrete organic and inorganic acids as a part of their metabolic processes. Hence, produced acids help disintegrate silicate mineral structures and mobilize potassium into the soil (Chen et al. 2020).

9.2.3.5 Siderophore Production

PGPMs secrete siderophores under iron-limited conditions. Siderophores are low-molecular weight compounds that have a high affinity towards iron. They chelate the iron available in the soil to generate a complex. These iron-siderophore conjugates are conveyed to the plant roots. In the root zone, these complexes assist in acquiring iron and at the same time restrict iron availability to disease-causing microbes.

9.2.3.6 Organic Acid Production

The production of organic acids like acetic, lactic, and oxalic acids aids modification in soil acidity or alkalinity, thereby boosting the solubilization and accessibility of key nutrients. Alternatively, indirect mechanisms facilitate plants to stay protected from microbial attacks and other environmental challenges. Microbial interactions initiate ISR, priming the plant's immune system to combat future threats with greater speed and efficiency. They generate lytic enzymes such as glucanases and chitinases, which hydrolyze the cell walls of pathogenic organisms.

Some plant growth–promoting microbes create antibiotics that limit or neutralize harmful pathogens. Some others release volatile organic compounds (VOCs) that both curb pathogen growth and initiate plant defence pathways. Furthermore, the production of protective enzymes, notably superoxide dismutase (SOD) and catalase (CAT), allows plants to mitigate oxidative stress and withstand adverse environmental conditions.

The indirect modes of action involve strategies by which PGPMs enhance plant growth by easing the impact of biotic and abiotic stresses. They also control pathogen activating plant defence responses. These actions do not deliver nutrients directly but they directly support plant health and increase resilience during adverse conditions.

9.2.4 Indirect Mechanism

9.2.4.1 Induced Systemic Resistance (ISR)

ISR is a plant's natural way of defending itself induced by certain plant-friendly microorganisms. These microorganisms are particularly strains of *Pseudomonas*, *Bacillus*, and *Trichoderma*. Unlike the immediate effects of antimicrobial agents, ISR activates and enhances the plant's defence mechanism. It also helps in improving its readiness to combat a broad array of pathogens commonly without detectable symptoms. This defence mechanism is managed by interconnected signalling networks. It gives special emphasis on pathways involving jasmonic acid (JA) and ethylene (ET), that synchronizes protective responses at local sites and across the whole plant. Harmless microbes can stimulate ISR, external environmental factors or microbe-derived molecular patterns (like flagellin, cyclic lipopeptides, β-glucans, and chitin). These elicitors turn on genes that mediate defence responses and related metabolic and signalling pathways, thereby boosting systemic resistance in plants. Importantly, ISR provides defence beyond the local site of microbial activity, as these molecules circulate through the plant, boosting immunity in all parts. Through hormone regulation, this defence strategy offers plants a low-cost and robust defence strategy allowing plants to cope with pathogen threats in both agriculture field and natural environments.

9.2.4.2 Antibiotic Production

Strains of bacteria, including *Pseudomonas*, *Streptomyces*, and *Bacillus*, generate antimicrobial agents including compounds like phenazines, pyoluteorin, and iturins that control or eliminate disease-causing fungi in plants and root-associated bacteria. It helps in decreasing the spread and seriousness of disease.

9.2.4.3 Lytic Enzyme Secretion

PGPMs secrete multiple hydrolytic enzymes:

- *Chitinases*: By chitin hydrolysis, these enzymes limit fungal expansion, and also, they protect plants from pathogen invasion.
- *β-1,3-Glucanases*: Their activity interferes with pathogen resilience contributing to stronger plant immunity.
- *Proteases*: These enzymes degrade the cell walls of fungal pathogens, thus acting as natural biocontrol agents.

9.2.4.4 Competition for Nutrients and Space

Beneficial microbes inhabit actively within the rhizosphere. It plays a vital role in filling ecological niches and exhausting key nutrients. If not done, it would have supported the growth of various pathogens. Through competitive exclusion, it becomes possible to suppress the ability of harmful microbes to take hold and proliferate.

9.2.4.5 Volatile Organic Compound Production

Some PGPMs release VOCs such as acetoin, metabolite 2,3-butanediol ($C_4H_{10}O_2$), and hydrogen cyanide (HCN). These volatile compounds help to prevent the growth of adjacent disease-causing microbes and may contribute to plant growth and tolerance to adverse conditions.

9.2.4.6 Production of Protective Enzymes

Through PGPM activity, plants may increase the production of protective enzymes, which include the following:

- *SOD*: It safeguards plants by turning harmful reactive superoxide molecules into oxygen and hydrogen peroxide.
- *Peroxidases*: They shield plants by converting H_2O_2 (hydrogen peroxide) into non-toxic molecules. These enzymes enhance cell wall integrity by promoting lignin buildup.
- *CAT*: These enzymes assist plants in defending against oxidative stress triggered by pathogens or harsh conditions like water scarcity and high soil salinity.

9.2.4.7 Biofilm Formation

Many beneficial microbes grow as biofilms on the root epidermis offering defence via physical and chemical mechanisms against disease-causing microbe invasion. Biofilms facilitate microbial colonization and adaptability to harsh soil conditions.

9.3 MYCORRHIZAL ASSOCIATIONS AND FUNGAL PARTNERS

9.3.1 Arbuscular Mycorrhizal Fungi Systems

AMF residing in the rhizosphere that serve as common partners in mutually beneficial interactions with plants of terrestrial ecosystems also contribute to enhanced nutrient uptake. Within the rhizosphere, AMF are common symbionts having mutual relationships with terrestrial plants. AMF are found near the root cortical cells forming branched structures called as arbuscules which act as a mediator for nutrient exchange between plant and fungus. This relationship is vital in conditions where nutrient uptake is critical. Arbuscules are known to reach the deep areas of soil where roots cannot and thus help in macro- and micronutrient uptake, like zinc and copper. Also, AMF release into the soil a glycoprotein known as glomalin that helps in developing soil aggregates, thus improving soil structure, porosity, and water-holding capacity.

During environmental stresses like drought, salinity, and heavy metal toxicity, AMF are observed to increase plant resilience. It also supports other microbial community in rhizosphere. This enables in plant ISR against pathogens.

AMF typically form *linear* or *coiling* hyphal structures within the root cortex, whereas ECM fungi form *epidermal* or *cortical* sheath associations. As mentioned earlier about AMF colonization and development of arbuscules and their contribution towards plant growth. Also, AMF are seen to interact with PGPRs and *Rhizobium* spp., thus enhancing plant growth.

9.3.2 *Trichoderma* and Other Beneficial Fungi

Trichoderma along with other fungi in soil microbiota help in promoting plant health, enhancing soil fertility, and suppressing soil-borne pathogens. *Trichoderma* acts as both a biocontrol agent and a biofertilizer, colonizing in the root rhizosphere where it synthesizes a variety of bioactive compounds, including antibiotics, hydrolytic enzymes such as chitinases and glucanases, and a range of secondary metabolites. It suppresses pathogenic fungi via mycoparasitism, antibiosis, and resource competition. It also elicits plant innate immune system. In addition to pathogen suppression, *Trichoderma* promotes plant growth by producing plant growth–regulating hormones such as auxins, stimulating root elongation, and enhancing nutrient solubilization—notably phosphorus and iron by secreting siderophores. Apart from it, *Piriformospora indica* (reclassified as *Serendipita indica*) is an endophytic fungus, which also promotes plant growth, known to colonize in plant roots which are non-mycorrhizal hosts and enhance nutrient uptake and stress tolerance.

9.4 MICROBIAL CONSORTIA AND COMMUNITY ENGINEERING

Microbial consortia for soil enhancement are a mixture of microorganisms that work together in synchrony to promote nutrient content and plant growth, and also develop tolerance against environmental stresses (Puranik et al. 2023). As compared to monoculture system, consortia get advantage of intertwined interactions for carrying out metabolism and signalling that could offer multidimensional benefits to agricultural sector. Complementary organisms—such as nitrogen-fixing bacteria, phosphate-solubilizing fungi, biocontrol actinomycetes, and hormone-producing rhizobacteria—engineered consortia can coordinate nutrient acquisition, stimulate root architecture, suppress pathogens, and enhance plant stress tolerance in a single, tailored formulation. Recent development in omics has helped in developing consortia through rationalization of taxa that helps in visualizing cross-kingdom communication, a guide for programming artificial communities with desired functions (Woo and Pepe 2018).

To achieve such outcomes, designs should ensure functional complementarity, ecological compatibility, and resilience to environmental fluctuations; for instance, co-culture assays and network analysis can identify synergistic pairings while avoiding antagonisms (Shayanthan, Ordoñez, and Oresnik 2022). Transition of such consortial prototypes from lab to commercial production faces several restrictions, like whole-genome sequencing (WGS) to prevent misidentification (Brown et al. 2019), viability under dehydrating and UV-exposed conditions (Fadiji et al. 2024), and maintenance of genomic pool (Seenivasagan and Babalola 2021). Addressing these operational challenges is essential for realizing the potential of microbial consortia to reduce chemical inputs, rehabilitate degraded soils, and foster sustainable crop production.

9.4.1 Natural and Synthetic Microbial Communities

In nature, soil consists of a diversified set of microbes that networks through different types of interactions viz., mutualistic, commensal, and competitive interactions. These interactions are key to nutrient cycling, disease suppression, and soil structure stabilization. We know that nitrogen-fixing bacteria (*Rhizobium leguminosarum*) fix atmospheric nitrogen to plant-absorbable NH_4^+, while *Bacillus megaterium* produces phosphatases to release bound phosphorus from mineral complexes (Jacoby et al. 2017). Within this system only, AMF help in increasing the reach of plant roots to deeper soil (Fall et al. 2022). Actinomycetes such as *Streptomyces griseus* produce a spectrum of antibiotics and hydrolytic enzymes that suppress fungal and bacterial pathogens, while also mineralizing organic matter to liberate nutrients (Boubekri et al. 2022).

Synthetic microbial communities (SynComs) are rationally assembled consortia of well-characterized strains chosen for complementary functional traits. By stacking capabilities—phytohormone synthesis, nutrient solubilization, biocontrol, and stress alleviation—SynComs can outperform single-strain inoculants under challenging field conditions. In one greenhouse study, a five-strain SynCom composed of *Azospirillum brasilense* (auxin production), *Pseudomonas fluorescens* (siderophore-mediated biocontrol), *Bacillus subtilis* (iturin lipopeptides), *Trichoderma harzianum* (cellulase secretion), and *Rhizophagus irregularis* increased maize biomass by 28% and improved water-use efficiency under drought stress (Armanhi et al. 2021).

9.4.2 Design Principles for Effective Consortia

Engineering robust microbial consortia requires adherence to key principles that maximize synergistic interactions while minimizing potential conflicts among community members.

9.4.2.1 Functional Complementarity

Consortia is a mix of strains that complement each other over several functions like nutrient solubilization, phytohormone synthesis, and pathogen suppression. It is seen that a consortium of auxin-producing *A. brasilense*, phosphate-solubilizing *B. megaterium,* and chitinase-secreting *T. harzianum* improves both plant (root elongation and protection against fungal pathogens) and soil quality (availability of phosphorus) without any metabolic overlap (Woo and Pepe 2018).

9.4.2.2 Ecological Compatibility

A combination of co-culture assays and interaction analyses employed to investigate and understand the inter-compatibility also helps prevent functional limitations within the consortium. Temkin et al. (2019) showed that *B. subtilis* lipopeptides can lower the nitrogenase activity of *Azotobacter* species. Studying the metabolites and volatile chemicals involved helps make these designs more effective.

9.4.2.3 Stability and Resilience

Functional stability under environmental abiotic stresses (temperature, drought, salinity) is generally achieved by strong consortia. Using a metagenomic approach, *Paenibacillus polymyxa* was identified from a drought-affected area that emerged as an important species capable of maintaining strong metabolic performance even under water stress conditions (Muzami et al. 2025).

9.4.2.4 Scalability

Commercialization of consortia demands for strains that could adapt to formulation process like freeze-drying and encapsulation so they can proliferate at a controlled pace. They should also possess viability above 70% under storage conditions (Fadiji et al. 2024).

9.4.3 Quality Control and Production Challenges

The process of commercialization of microbial consortia as bioinoculants requires strict quality control and requires tackling several production challenges to ensure effectiveness regulatory compliance and safety.

9.4.3.1 Strain Authentication

Product quality maintenance depends largely upon consortium strains. In this regard, WGS gives an insight about horizontal gene transfer events and taxonomic position that helps in identifying authentic strains for the consortia. An advantage of WGS is its ability to identify non-productive or functionally redundant strains within microbial consortia. For example, in biofertilizer formulations based on *Bacillus* consortia, approximately 30% non-productive strains were reported to reduce overall efficiency and pose potential environmental risks (Yared González-León et al. 2024), which could have been otherwise avoided by WGS.

9.4.3.2 Formulation Stability

Processing, storage, and field application are critical to microbial consortia viability. Drying forces like heat, UV radiation, and desiccation pose serious problems to cell survival. Few research studies have shown that encapsulating microbes using alginate bead gel entrapment method can enhance viability of cells up to 40% at 30°C. Several methods can be employed to act as protective layer, polymer coating (e.g., gelatin, skim milk), carrier matrices (e.g., biochar), and moisture content optimization for increased shelf life (Korlid Thinkohkaew et al. 2024).

9.4.3.3 Regulatory Compliance

Clear biosafety rules and guidelines are crucial to ensure the safe use of products made from microbial symbiotic communities. These strains need to be carefully checked to make sure they are not pathogenic—or devoid of pathogenicity-related factors—and environment-friendly. For example, *Streptomyces* species may unintentionally affect the soil microbiome by promoting the spread of antibiotic-resistant strains. Vurukonda, Giovanardi, and Stefani (2018) highlighted that harmonizing regulatory guidelines across global markets remains an ongoing challenge.

9.4.3.4 Batch Consistency

For optimal outcome, it is essential to scrutinize the applicable ratio and cell viability of consortia. qPCR targeting markers are a standard practice for post-production verification. This assay helps in identifying composition deviations and is sensitive up to 10^3 cells/gm. Additionally, flow cytometry combined with fluorescent in situ hybridization is employed to monitor cell viability and relative abundance in real time, facilitating rapid quality assessments during scale-up (Borchardt et al. 2021).

9.5 ADVANCED TECHNOLOGIES AND METHODOLOGICAL INNOVATIONS

With the convergence of high-throughput sequencing, synthetic biology, and microfabrication, a new era for précised microbiome research and ecosystem studies has started. Shotgun metagenomics and long-read sequencing technologies (e.g., PacBio, Oxford Nanopore) have allowed us to recover high-quality metagenome-assembled genomes from complex soil communities. These techniques have enabled us to uncover microbial taxa and specialized functional gene clusters which were not known previously (Diamond et al. 2019). These include novel nitrogenase variants and biosynthetic gene clusters for secondary metabolites.

In rhizosphere, when meta-transcriptomics and untargeted metabolomics are used together, they help in revealing active microbial functions and assist in metabolic exchange. Multiple approaches show that various practices related to farming significantly influence microbial activities. Strategies to manage agriculture such as like crop rotation or adding organic matter can change how nutrients move and cycle in the soil (Jansson and Hofmockel 2018).

Eco-FABs are engineered ecosystem devices that integrate microfluidic technology and transparent chamber, which allows scientists to see colonization between various microbes and root along with nutrient gradients that can be finely adjusted and closely studied in greater detail (Zengler et al. 2019).

In synthetic biology, scientists now use easy-to-apply genetic module. This module or kit helps to modify and design microbes that work with plants or crops. Strains like *Pseudomonas putida* KT2440 has been modified genetically for different plant growth–supporting functions. Costa-Gutierrez and co-workers (2020) found that the modified strain, which can dissolve phosphate and produce auxin, showed better crop growth.

9.5.1 Metagenomics and Microbiome Analysis

It is easy to analyze and characterize soil microbiome environment using metagenomics by sequencing entire genetic material present in the sample. Research studies have shown that shotgun metagenomics relate easily microbial gene and their specific function. The microbial genetic material identified through sequencing

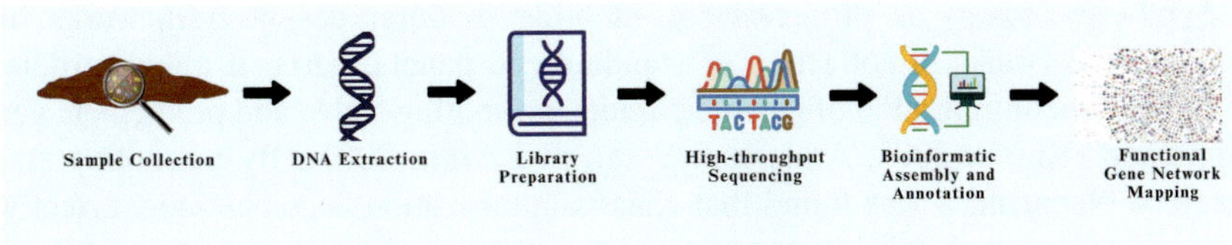

FIGURE 9.2 Workflow of a typical soil metagenomics pipeline.

method might be crucial for maintaining healthy soils. It also helps in the processes like nitrogen cycling and carbon storage. If we see some examples like how microbial genetic constituent controls N_2O nitrous oxide emissions and affect its nitrogen metabolism. It has been seen that emission throughout the process of natural organic waste composting is detected and monitored throughout the process. Also, it has helped to identify some bacterial pathways that naturally help to lower N_2O. These pathways ultimately become main target to developing soil management strategies aimed at reducing greenhouse gas emissions during composting (Xu et al. 2024) (Figure 9.2).

Advances in long-read sequencing technologies, including Oxford Nanopore, now enable the recovery of near-complete microbial genomes directly from soil metagenomes. This allows the discovery of previously uncharacterized taxa and biosynthetic gene clusters important for novel bioactive compound production and ecosystem services (Sánchez-Navarro et al. 2022). Integration of meta-transcriptomics and metabolomics provides real-time monitoring of microbial gene expression and metabolite production, offering insights into microbial activity and signalling dynamics in the rhizosphere under a variety of agronomic treatments. For instance, such multi-omics approaches have clarified how changes in agricultural practices—such as organic amendments or crop rotation—alter gene expression profiles and metabolite fluxes linked to soil nutrient cycling and disease suppression (Aguiar-Pulido et al. 2016).

Recently shotgun metagenomic analyses have revealed the impact of nitrogen amendments on microbial community structure and nitrogen cycling gene abundance. This analysis has revealed the functional link between microbial diversity and soil nutrient dynamics (Li et al. 2024).

9.5.2 Synthetic Biology Applications

Bioengineering (synthetic biology) offers advanced tools for modifying and designing genomes of microbes. Microbes which are modified are specifically designed to perform different types of agricultural roles like monitoring soil health condition using biosensors and nutrient solubilization enhancement. Studies have emphasized the potential of *P. putida* strains, whether occurring naturally or via genetic modification. These strains help in the production of phytohormones (IAA) and also solubilization of phosphates. Under greenhouse ambience, these traits together promote plant growth. Studies indicate that some *P. putida* strains significantly increase various growth characters in tomato plants (shoot, root, and fruit yield). Hernández-Montiel and colleagues (2017) linked these improvements to efficient phosphate solubilization combined with substantial IAA production.

Synthetic biology is progressively adopting modular design frameworks, for example Bio-Bricks, a collection of standardized genetic parts. It allows efficient design and reconfiguration of genetic circuits, promoting stable and predictable gene expression (Knight 2003; Anderson, Voigt, and Arkin 2007). By promoting standardized blueprint, it was found that chassis optimization becomes more effective, enhancing microbial stability and biosafety.

From the biosafety perspective, it has been seen that frameworks are emerging globally to analyze genetically modified microbial products. Essential biosafety strategies involve gene containment like modified auxotroph—which restrict microbial growth by making survival dependent on specific nutrients. The designs are made to lower the chance of genes spreading between microbes in horizontal gene transfer. Hayashi and colleagues (2024) reported that recent progress in CRISPR-based technologies offers reliable and safe ways to control engineered microbes in complex environments.

9.5.3 Fabricated Ecosystem Devices (Eco-FABs)

Eco-FABs are custom-designed microfluidic platforms produced with 3D printing methods and soft lithography techniques designed to simulate plant–soil–microbe interface processes in a controlled, sterile, standardized lab environment condition (Gao et al. 2018). They generally feature transparent chambers that enable direct visualization of root development. Here chambers are filled with solid materials like soil particles or hydrogels, or sometimes with liquid media. Under controlled condition, plant roots grow using these substrates provided inside these chambers. Real-time or periodic sampling during experimentation is easily carried out using these setups. Microelectrodes and optical sensors are embedded within the device. The device incorporates embedded microelectrodes and optical sensors to check upon various parameters like oxygen, pH, nutrient condition.

They also make it easier to carry out minute detail analysis and the colonization behaviour of microbes. Massalha et al. (2017) explain that this system makes detailed space- and time-based analysis possible. In addition, it examines interactions among roots, microbes, and environmental factors.

Confocal microscopy and fluorescence light-sheet imaging offer high-resolution live-imaging options facilitated by Eco-FABs. Handakumbura, Rivas-Ubach, and Battu (2021) noted that these techniques enable non-invasive monitoring of root-associated microbial communities and biofilm development. They showcase precise regulation of rhizo-deposits and microbial application. This microenvironment helps researchers better understand plant–microbe dynamics, It also helps scientists understand the cooperative interactions between plants–microbes and the surrounding rhizosphere community. Scientists conducted several experiments using an Eco-FAB setup inoculated with *T. harzianum*, *S. griseus*, and *A. brasilense*. A study showed auxin synthesis in the area surrounding the roots. Some scientists reported that microbe-treated plants produced nearly 45% more lateral root branches within 72 hours than the untreated controls (Moreira et al. 2022). This indicates that Eco-FABs are capable of measuring how different microbes affect the function of plants. This facilitates the investigation of controlled root architectural traits.

Eco-FAB designs are advancing the development of lab-grown ecosystems that closely resemble natural systems, including soil aggregates and water–air transition zones. As noted by Gao et al. (2018), these conditions enable a more accurate recreation of natural diversity and better reflect the structural richness of the rhizosphere. These features make it easier to study and compare how different groups of microbes perform.

9.6 TARGETED AND UNTARGETED INTERVENTION STRATEGIES

Next-generation sustainable farming will employ soil microflora to elevate crop productivity and strengthen the stability of environment. These intervention strategies change the microbial environment in the soil. There are two main types of these strategies: targeted microbial approaches, which focus on using specific bioinoculants or engineered consortia to change things, and untargeted ecosystem approaches, which focus on using agroecological methods to change the native microbiome and soil health as a whole. Both strategies have benefits, which work well together and are necessary for moving away from farming that uses a lot of chemicals.

9.6.1 Targeted Microbial Approaches

These approaches deploy naturally beneficial microbes, genetically engineered strains, or SynComs strategically assembled for complementary functions to improve crop performance (Figure 9.3).

9.6.1.1 Plant Growth–Promoting Rhizobacteria

A. brasilense is a well-studied free-living nitrogen-fixing bacterium. *A. brasilense* synthesizes phytohormones such as IAA, cytokinins, and gibberellins. Production of these hormones leads to enhanced root development, improved nutrient and water uptake, and increased biomass in cereals and legumes. Recent studies have shown an increase in the yield of crops such as maize. In various studies, tolerance to drought and salinity has been increased (Fukami, Cerezini, and Hungria 2018; Steenhoudt and Vanderleyden 2000).

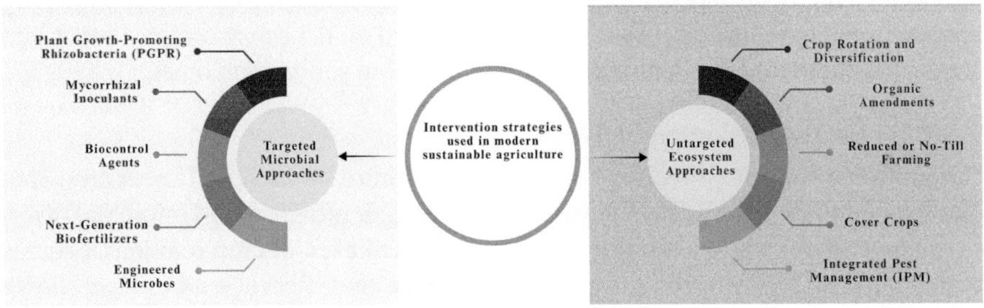

FIGURE 9.3 Targeted and untargeted intervention strategies used in modern sustainable agriculture to manage soil microbiomes.

9.6.1.2 Mycorrhizal Inoculants

AMF, especially *R. irregularis*, establish symbiotic relationships with plant roots. This association has significantly improved phosphorus uptake and water retention under drought stress. This improves plant biomass, photosynthesis efficiency, and antioxidant enzyme activity, thereby reducing drought effects in crops such as maize, soybean, wheat, and tomato (Tang et al. 2022; Hu, Xie, and Chen 2020).

9.6.1.3 Biocontrol Agents

Trichoderma and *B. subtilis* help to protect plants by producing antibiotics that prevent harmful microbes from spreading. They show fungal antagonism activity along with siderophores and lytic enzymes that suppress harmful pathogens derived from soil. Yao et al. (2023) and Feliciano et al. (2016) demonstrated that these practices reduce the dependence on synthetic fungicides while enhancing plant growth and resilience.

9.6.1.4 Next-Generation Biofertilizers

Engineered microbial consortia have been developed for multifunctional purposes—combining pathogen suppression, nitrogen fixation, and phosphate solubilization. They have demonstrated promising results in field trials outcomes with yield increase of 25% (Parray et al. 2023).

9.6.1.5 Engineered Microbes

Synthetic biology has enabled us to engineer strains for novel traits, such as *P. putida* to express phosphate-solubilizing enzymes. *P. putida* also expresses the auxin biosynthesis pathway, which leads to the enhancement of nutrient availability and improvement in plant growth. The fruit yield has also increased under greenhouse conditions. Though reported yield improvements vary, significant biomass increases and growth promotion are consistent findings (Nicolás et al. 2016; Singh et al. 2017).

9.6.2 Untargeted Ecosystem Approaches

Untargeted approaches utilize agronomic and ecological practices. These practices enhance the native soil microbial diversity, promote soil organic matter accumulation, and support multiple ecosystem functions. They don't require the direct introduction of specific microbial inoculants. For the purpose of maintaining crop productivity and environmental health, these approaches depend on the ecosystem resilience and services that are inherent within properly functioning soil microbiomes.

9.6.2.1 Crop Rotation and Diversification

Crop rotation is a very effective practice in agriculture. Rotating different crop species interrupts the life cycles of pests and pathogens. It promotes soil microbial diversity and improves the availability of nutrients. When used in crop rotation, legumes enrich nitrogen levels in soil through biological fixation. Recent studies have shown that diversified crop rotations, including cover crops, lead to significant increases in microbial richness and improved soil function compared to monoculture or simple rotations (Venter, Jacobs, and Hawkins 2016; Williams et al. 2022). For instance, planting different crops in a sequence across different seasons elevated microbial (bacterial and fungal) richness by 30%. leading to more stable soil aggregates formation.

9.6.2.2 Organic Amendments

Through the use of biochar: organic nutrients and natural soil covering enhance its organic richness. Furthermore, it elevates microbial biomass, reinforces soil architecture, and increases moisture-holding capacity. According to the work of Ouyang, Reeve, and Norton (2022), nutrient-rich organic amendments impact the microbial makeup and nitrogen-cycling genetic traits with a greater impact than compost, resulting ultimately in improved nutrient dynamics and increased disease control potential. These amendments encourage high levels of coordinated microbial interactions. This not only aids in preserving soil health but also enhances functionality in long run.

9.6.2.3 Reduced or No-Till Farming

A decrease in soil disruption contributes to the protection of soil aggregates and microbe-rich niches. It leads to better organization of microbial species clusters resilience and carbon capture. No-till practices favour a microbial consortium adapted to slower growth and stress-tolerant conditions, with higher fungal to bacterial ratios (F:B), promoting soil structure stability and preservation of nutrients. Meta-analyses confirm that no-till systems increase overall microbial diversity and biomass compared to conventional tillage (Schmidt et al. 2018).

9.6.2.4 Cover Crops

Maintaining continuous living plant cover through cover crops supports rhizosphere microbial diversity year-round, protects soil from erosion, and enriches populations of disease-suppressive microbes. Cover cropping significantly increases bacterial and fungal abundance, with microbial biomass rising alongside improvements in soil organic carbon and pH regulation (Gao et al. 2022). These microbial shifts are linked with enhanced nutrient cycling and soil health.

9.6.2.5 Integrated Pest Management (IPM)

The IPM strategies reduce the reliance on chemical pesticides by utilizing and promoting natural biocontrol agents, including *T. harzianum* and *P. fluorescens*. The IPM strategies help the beneficial microbial population to flourish. A number of field experiments have revealed that IPM practice helps to preserve beneficial ecological balance among plots, and sustain greater populations of beneficial microbes while suppressing pathogen abundance. Khokhar and his team (2024) reported that, compared to conventional practices, IPM promotes long-term disease suppression and enhances overall crop health.

9.6.2.6 Case Study: Midwestern U.S. Long-Term Field Trial

A decade-long experiment was conducted across the Midwestern region of the United States focusing on crop rotations that incorporated organic inputs and cover crops with the objective of comparing diversified cropping systems against conventional monocultures and tillage practices. The diversified systems exhibited a 35% enhancement in bacterial and fungal richness, enhancing soil stability and resulting in an 18% improvement in crop yield under stress (drought) conditions. These findings highlight the need for comprehensive management of soil microbial ecosystems. Lupwayi et al. (2024) emphasize that untargeted strategies can simultaneously enhance agricultural productivity and environmental sustainability.

9.7 ECONOMIC AND ENVIRONMENTAL IMPLICATIONS

9.7.1 Market Trends and Economic Viability

The rising demand for sustainable and greener crop production practices have led to significant growth in the global biofertilizer market in recent years. The biofertilizer industry is made up of multiple microbial strains and group. It includes PGPMs, biopesticides, and microbial consortia and communities.

9.7.2 Market Size and Growth

During 2021–2024, the fertilizer sector was estimated to have a market value of approximately USD 1.88–2.7 billion. Between 2022 and 2030, its market is projected to grow at an estimated CAGR of 8.5%–12.8%. By 2030, the market is expected to grow from approximately USD 4.5 billion to 6.3 billion, driven by stringent regulations on chemical pesticides and fertilizers, heightened consumer interest in organic produce, and expanding awareness about environmental sustainability and soil health; Mordor Intelligence 2025; Fortune Business Insights 2025; Grand View Research 2024). Biofertilizer market growth can be motivated by government initiatives promoting its production, technological advancements improving shelf life, product efficacy, and a global shift towards climate-resilient and regenerative approaches to agriculture. Private sector investment in innovative delivery formats like microbial seed coatings and subscription-based inoculants is also expanding the commercialization pipeline (Strategic Market Research 2024; S&P Global Commodity Insights 2022).

9.7.3 Economic Benefits for Farmers

Both small and commercial farmers have a lot of benefits with the use of microbial inoculants in their agricultural practices. The benefits include enhancement in the nutrient-use efficiency, improvement in the yield, and increase in resilience in response to stress. Along with these benefits, a reduction in chemicals use, which ultimately can reduce the associated cost, is also reported.

9.7.3.1 Yield and Input Savings

Field trials integrating *Rhizobium* and *Pseudomonas* microbial consortia on wheat crops have been conducted in the context of Indian farming system, leading to yield enhancement of 20%–30%, and contributing to reduced chemical fertilizer use by approximately 25%–30%. Findings reveal an annual net profit of about USD 200 per hectare via lowering fertilizer inputs and enhancing crop yields (Vidya and Sharma 2024; Santos, Nogueira, and Hungria 2019).

9.7.3.2 Reduction in Pesticide Use

Employing precise microbial biocontrol agents (*B. subtilis* strains and *Trichoderma* species) has played a significant role in lowering dependence on chemical pesticide use. Investigations across horticultural production systems in Europe and North America indicated a 40%–60% decrease in pesticide usage. Zion Market Research (2024) highlights that this shift leads to substantial production cost savings and also lowers the need for manual labour.

9.8 CONCLUSION

Microorganisms play a vital role in sustainable agriculture practice, fortifying the soil ecosystem, improving plant health, and reducing dependency on chemical fertilizers and pesticides. They increase soil fertility and plant resilience. PGPMs help plants to absorb nutrients more efficiently and also produce phytohormones that support healthy growth. These phytohormones activate systemic resistance, increasing the ability of different crops to withstand wide-level environmental stresses (Noman et al. 2021). Well-balanced and diverse rhizosphere helps in healthy growth of the plant. It also increases resiliency against various diseases and lessens the need for synthetic agrochemicals. These strategies support long-term crop productivity, help to reduce the negative impacts of climate change, and align with global sustainability initiatives worldwide (El Chami et al. 2020). To meet future needs, it is essential to integrate different modern technological approaches such as synthetic biology, metagenomics, and metabolomics by gaining a deeper understanding of microbial diversity and its metabolic capacity. In addition to innovative strategies, agricultural technologies and artificial intelligence–powered predictive modelling will help create tailored microbial inoculants that focus on specific crops, climate and soil and climatic condition necessary for efficient function. Ex- In the context of changing climatic condition, creating microbial solutions that improve drought tolerance, carbon sequestration, and salinity resilience capacity will greatly be increased. The success of these technologies will depend on a number of factors, such as effective cooperation between researchers, agronomists, farmers, and policymakers to make these technologies scalable, affordable, and environmentally safe. Enhancing sustainable agriculture system practices through microbial ecology offers promising avenues for achieving regenerative and sustainable farming systems. This approach can contribute to the development of sustainable farming systems resilient to changing climatic conditions, to meet the rising global food demand. Future developments in this area will aid in restoring biodiversity enrichment, soil health, and the mitigation of greenhouse gas emissions. Investing in soil microbiome research can safeguard farming yields and preserve ecological surroundings and benefits for future generations.

REFERENCES

Aguiar-Pulido, Vanessa, Wenrui Huang, Victoria Suarez-Ulloa, Trevor Cickovski, Kalai Mathee, and Giri Narasimhan. 2016. "Metagenomics, Metatranscriptomics, and Metabolomics Approaches for Microbiome Analysis." *Evolutionary Bioinformatics* 12s1 (January): EBO-S36436. https://doi.org/10.4137/ebo.s36436

Anderson, J. Christopher, Christopher A. Voigt, and Adam P. Arkin. 2007. "Environmental Signal Integration by a Modular and Gate." *Molecular Systems Biology* 3 (1): 133. https://doi.org/10.1038/msb4100173

Armanhi, Jaderson Silveira Leite, Rafael Soares Correa de Souza, Bárbara Bort Biazotti, Juliana Erika de Carvalho Teixeira Yassitepe, and Paulo Arruda. 2021. "Modulating Drought Stress Response of Maize by a Synthetic Bacterial Community." *Frontiers in Microbiology* 12: 747541. https://doi.org/10.3389/fmicb.2021.747541

Borchardt, Mark A., Alexandria B. Boehm, Marc Salit, Susan K. Spencer, Krista R. Wigginton, and Rachel T. Noble. 2021. "The Environmental Microbiology Minimum Information (EMMI) Guidelines: QPCR and DPCR Quality and Reporting for Environmental Microbiology." *Environmental Science & Technology* 55 (15): 10210–23. https://doi.org/10.1021/acs.est.1c01767

Boubekri, Kenza, Abdoulaye Soumare, Ilham Mardad, Karim Lyamlouli, YedirOuhdouch, Mohamed Hafidi, and Lamfeddal Kouisni. 2022. "Multifunctional Role of Actinobacteria in Agricultural Production Sustainability: A Review." *Microbiological Research* 261 (August): 127059. https://doi.org/10.1016/j.micres.2022.127059

Brown, Eric, Uday Dessai, Sherri McGarry, and Peter Gerner-Smidt. 2019. "Use of Whole-Genome Sequencing for Food Safety and Public Health in the United States." *Foodborne Pathogens and Disease* 16 (7): 441–50. https://doi.org/10.1089/fpd.2019.2662

Canon, Fanny, Thibault Nidelet, Eric Guédon, Anne Thierry, and Valérie Gagnaire. 2020. "Understanding the Mechanisms of Positive Microbial Interactions That Benefit Lactic Acid Bacteria Co-Cultures." *Frontiers in Microbiology* 11 (September). https://doi.org/10.3389/fmicb.2020.02088

Chen, Yan-hui, Xiao-zhu Yang, Zhuang Li, Xiu-hong An, Ren-peng Ma, Yan-qing Li, and Cun-gang Cheng. 2020. "Efficiency of Potassium-Solubilizing *Paenibacillus mucilaginosus* for the Growth of Apple Seedling." *Journal of Integrative Agriculture/Journal of Integrative Agriculture* 19 (10). Elsevier BV: 2458–69. https://doi.org/10.1016/s2095-3119(20)63303-2

Costa-Gutierrez, Stefanie B., María Jesús Lami, Carolina Caram-Di, Ana M. Zenoff, Paula A. Vincent, María Antonia Molina-Henares, Manuel Espinosa-Urgel, and Ricardo. 2020. "Plant Growth Promotion by *Pseudomonas putida* KT2440 under Saline Stress: Role of EptA." *Applied Microbiology and Biotechnology* 104 (10). Springer Science+Business Media: 4577–92. https://doi.org/10.1007/s00253-020-10516-z

Diamond, Spencer, Peter F. Andeer, Zhou Li, Alexander Crits-Christoph, David Burstein, Karthik Anantharaman, Katherine R. Lane, et al. 2019. "Mediterranean Grassland Soil C-N Compound Turnover Is Dependent on Rainfall and Depth, and Is Mediated by Genomically Divergent Microorganisms." *Nature Microbiology* 4 (8): 1356–67. https://doi.org/10.1038/s41564-019-0449-y

El Chami, Daniel, André Daccache, and Maroun El Moujabber. 2020. "How Can Sustainable Agriculture Increase Climate Resilience? A Systematic Review." *Sustainability* 12 (8): 3119. https://doi.org/10.3390/su12083119

Etesami, Hassan, Somayeh Emami, and Hossein Ali Alikhani. 2017. "Potassium Solubilizing Bacteria (KSB): Mechanisms, Promotion of Plant Growth, and Future Prospects – A Review." *Journal of Soil Science and Plant Nutrition* 17 (4): 897–911. https://doi.org/10.4067/S0718-95162017000400005

Fadiji, Ayomide Emmanuel, Chao Xiong, Eleonora Egidi, and Brajesh K. Singh. 2024. "Formulation Challenges Associated with Microbial Biofertilizers in Sustainable Agriculture and Paths Forward." *Journal of Sustainable Agriculture and Environment* 3 (3). https://doi.org/10.1002/sae2.70006

Fall, Abdoulaye Fofana, Grace Nakabonge, Joseph Ssekandi, HassnaFounoune-Mboup, Samuel Obeng Apori, Abibatou Ndiaye, Arfang Badji, and Khady Ngom. 2022. "Roles of Arbuscular Mycorrhizal Fungi on Soil Fertility: Contribution in the Improvement of Physical, Chemical, and Biological Properties of the Soil." *Frontiers in Fungal Biology* 3 (March). https://doi.org/10.3389/ffunb.2022.723892

Feliciano, Ricardo, Leise Inês Heckler, Marília Lazarotto, Lucas, Cecilia Rego, and Elena Blume. 2016. "*Trichoderma* spp. and *Bacillus subtilis* for Control of *Dactylonectria macrodidyma* in Grapevine." *DOAJ (DOAJ: Directory of Open Access Journals)*, July. https://doi.org/10.14601/phytopathol_mediterr-18048

Forster, Susan M. 1990. "The Role of Microorganisms in Aggregate Formation and Soil Stabilization: Types of Aggregation." *Arid Soil Research and Rehabilitation* 4 (2): 85–98. https://doi.org/10.1080/15324989009381236

Fortune Business Insights. 2025. "Biofertilizers Market Size, Share & Analysis Report [2020–2027]." www.fortunebusinessinsights.com. https://www.fortunebusinessinsights.com/industry-reports/biofertilizers-market-100413

Fukami, Josiane, Paula Cerezini, and Mariangela Hungria. 2018. "*Azospirillum*: Benefits that Go Far beyond Biological Nitrogen Fixation." *AMB Express* 8 (May). https://doi.org/10.1186/s13568-018-0608-1

Gao, Huan, Gangming Tian, Muhammad Khashi u Rahman, and Fengzhi Wu. 2022. "Cover Crop Species Composition Alters the Soil Bacterial Community in a Continuous Pepper Cropping System." *Frontiers in Microbiology* 12 (January). https://doi.org/10.3389/fmicb.2021.789034

Gao, Jian, Joelle Sasse, Kyle M. Lewald, Kateryna Zhalnina, Lloyd T. Cornmesser, Todd A. Duncombe, Yasuo Yoshikuni, John P. Vogel, Mary K. Firestone, and Trent R. Northen. 2018. "Ecosystem Fabrication (EcoFAB) Protocols for the Construction of Laboratory Ecosystems Designed to Study Plant-Microbe Interactions." *Journal of Visualized Experiments*, 134 (April). https://doi.org/10.3791/57170

Garland, Jay L., and Aaron L. Mills. 1991. "Classification and Characterization of Heterotrophic Microbial Communities on the Basis of Patterns of Community-Level Sole-Carbon-Source Utilization." *Applied and Environmental Microbiology* 57 (8): 2351–59. https://doi.org/10.1128/aem.57.8.2351-2359.1991

Grand View Research. 2024. "Biofertilizers Market Size, Share & Growth Report, 2020–2027." www.grandviewresearch.com. https://www.grandviewresearch.com/industry-analysis/biofertilizers-industry

Handakumbura, Pubudu P., Albert Rivas Ubach, and Anil K. Battu. 2021. "Visualizing the Hidden Half: Plant-Microbe Interactions in the Rhizosphere." *MSystems* 6 (5). https://doi.org/10.1128/msystems.00765-21

Hayashi, Naoki, Yong Lai, Jay Fuerte-Stone, Mark Mimee, and Timothy K. Lu. 2024. "Cas9-Assisted Biological Containment of a Genetically Engineered Human Commensal Bacterium and Genetic Elements." *Nature Communications* 15 (1): 2096. https://doi.org/10.1038/s41467-024-45893-w

Hernández-Montiel, Luis G., César J. Chiquito Contreras, Bernardo Murillo Amador, Librado Vidal Hernández, Evanjelina E. Quiñones Aguilar, and Roberto G. Chiquito Contreras. 2017. "Efficiency of Two Inoculation Methods of *Pseudomonas putida* on Growth and Yield of Tomato Plants." *Journal of Soil Science and Plant Nutrition* 17 (4): 1003–12. https://doi.org/10.4067/S0718-95162017000400012

Hu, Yanbo, Wei Xie, and Baodong Chen. 2020. "Arbuscular Mycorrhiza Improved Drought Tolerance of Maize Seedlings by Altering Photosystem II Efficiency and the Levels of Key Metabolites." *Chemical and Biological Technologies in Agriculture* 7 (1). https://doi.org/10.1186/s40538-020-00186-4

Jacoby, Richard, Manuela Peukert, Antonella Succurro, Anna Koprivova, and Stanislav Kopriva. 2017. "The Role of Soil Microorganisms in Plant Mineral Nutrition—Current Knowledge and Future Directions." *Frontiers in Plant Science* 8 (1617). https://doi.org/10.3389/fpls.2017.01617

Jansson, Janet K., and Kirsten S. Hofmockel. 2018. "The Soil Microbiome—From Metagenomics to Metaphenomics." *Current Opinion in Microbiology* 43 (June): 162–68. https://doi.org/10.1016/j.mib.2018.01.013

Kashyap, Nikhil, Sandeep Kumar Singh, Nisha Yadav, Vipin Kumar Singh, Madhuree Kumari, Dharmendra Kumar, Livleen Shukla, Kaushalendra, Nikunj Bhardwaj, and Ajay Kumar. 2023. "Biocontrol Screening of Endophytes: Applications and Limitations." *Plants* 12 (13): 2480. https://doi.org/10.3390/plants12132480

Khokhar, Mukesh Kumar, Rakesh Kumar, Anoop Kumar, Mukesh Sehgal, Satya Prakash Singh, Prabhu Narayan Meena, Niranjan Singh, et al. 2024. "Impact of IPM Practices on Microbial Population and Disease Development in Transplanted and Direct-Seeded Rice." *Frontiers in Microbiology* 15 (July). Frontiers Media SA. https://doi.org/10.3389/fmicb.2024.1388754

Knight, Thomas. 2003. "Idempotent Vector Design for Standard Assembly of Biobricks." Dspace.mit.edu. https://dspace.mit.edu/handle/1721.1/21168

Korlid Thinkohkaew, Veasarach Jonjaroen, NuttaweeNiamsiri, Atikorn Panya, InthawootSuppavorasatit, and Pranut Potiyaraj. 2024. "Microencapsulation of Probiotics in Chitosan-Coated Alginate/Gellan Gum: Optimization for Viability and Stability Enhancement." *Food Hydrocolloids*, February. Elsevier BV, 109788–88. https://doi.org/10.1016/j.foodhyd.2024.109788

Li, Kexin, Hongmei Lin, Mei Han, and Limin Yang. 2024. "Soil Metagenomics Reveals the Effect of Nitrogen on Soil Microbial Communities and Nitrogen-Cycle Functional Genes in the Rhizosphere of Panax Ginseng." *Frontiers in Plant Science* 15 (August). Frontiers Media. https://doi.org/10.3389/fpls.2024.1411073

Li, Xia, Xiaofei Kang, Junzhu Zou, Jiahui Yin, Yuancheng Wang, Ao Li, and Ma Xiao-dong. 2023. "Allochthonous Arbuscular Mycorrhizal Fungi Promote *Salix viminalis* L.–Mediated Phytoremediation of Polycyclic Aromatic Hydrocarbons Characterized by Increasing the Release of Organic Acids and Enzymes in Soils." *Ecotoxicology and Environmental Safety* 249 (January). Elsevier BV: 114461–61. https://doi.org/10.1016/j.ecoenv.2022.114461

Lupwayi, Newton Z., T. Kelly Turkington, Breanne D. Tidemann, Hiroshi Kubota, and Rodrigo O. Polo. 2024. "Crop Rotation and Fungicides Impact Microbial Biomass, Diversity and Enzyme Activities in the Wheat Rhizosphere." *Frontiers in Agronomy* 6 (November). Frontiers Media. https://doi.org/10.3389/fagro.2024.1455448

Marra, Leandro Marciano, Silvia Maria de Oliveira-Longatti, Cláudio Roberto Fonsêca Sousa Soares, Fábio Lopes Olivares, and Fatima Maria de Souza Moreira. 2019. "The Amount of Phosphate Solubilization Depends on the Strain, C-Source, Organic Acids and Type of Phosphate." *Geomicrobiology Journal* 36 (3). Informa UK Limited: 232–42. https://doi.org/10.1080/01490451.2018.1542469

Massalha, Hassan, Elisa Korenblum, Sergey Malitsky, Orr H. Shapiro, and Asaph Aharoni. 2017. "Live Imaging of Root–Bacteria Interactions in a Microfluidics Setup." *Proceedings of the National Academy of Sciences* 114 (17): 4549–54. https://doi.org/10.1073/pnas.1618584114

Mohanty, Pratikhya, Puneet Kumar Singh, Debosmita Chakraborty, Snehasish Mishra, and Ritesh Pattnaik. 2021. "Insight into the Role of PGPR in Sustainable Agriculture and Environment." *Frontiers in Sustainable Food Systems* 5 (June). https://doi.org/10.3389/fsufs.2021.667150

Mordor Intelligence. 2025. Biofertilizers Market Size & Share Analysis – Growth Trends and Forecast (2026 – 2031). https://www.mordorintelligence.com/industry-reports/global-biofertilizers-market-industry

Moreira, Vitoria deAlmeid, Carlos Eduardo, Arshad Jalal, Isabela Martins, Sena Oliveira, Guilherme, Victoria Moraes Giolo, Letícia Schenaide Vitória, Karen Vicentini Tamburi, and Marcelo Carvalho. 2022. "Inoculation with *Trichoderma harzianum* and *Azospirillum brasilense* Increases Nutrition and Yield of Hydroponic Lettuce." *Archives of Microbiology* 204 (7). Springer Science+Business Media. https://doi.org/10.1007/s00203-022-03047-w

Muzami, Eugene Mwanza, George Kitundu, Oscar Mwaura Muriithi, Agnes Mumo Kavoo, Virginia Gathoni Gichuru, and Wilton Mwema Mbinda. 2025. "Metagenomic Insights to Bacterial Communities, Functional Traits, and Soil Health in Banana Smallholder Agroecosystems of Kenya." *Frontiers in Microbiology* 16 (May). Frontiers Media. https://doi.org/10.3389/fmicb.2025.1582271

Nicolás, Alejandro, Pastor, Susana Rosas, Maria Luna, and Marisa Rovera. 2016. "Inoculation with *Pseudomonas putida* PCI2, a Phosphate Solubilizing Rhizobacterium, Stimulates the Growth of Tomato Plants." *Symbiosis* 62 (3): 157–167.

Noman, Muhammad, Temoor Ahmed, Usman Ijaz, Muhammad Shahid, Azizullah, Dayong Li, Irfan Manzoor, and Fengming Song. 2021. "Plant–Microbiome Crosstalk: Dawning from Composition and Assembly of Microbial Community to Improvement of Disease Resilience in Plants." *International Journal of Molecular Sciences* 22 (13): 6852. https://doi.org/10.3390/ijms22136852

Ouyang, Yang, Jennifer R. Reeve, and Jeanette M. Norton. 2022. "The Quality of Organic Amendments Affects Soil Microbiome and Nitrogen-Cycling Bacteria in an Organic Farming System." *Frontiers in Soil Science* 2 (August). https://doi.org/10.3389/fsoil.2022.869136

Pandey, Komal, and Baljeet Singh Saharan. 2025. "Soil Microbiomes: A Promising Strategy for Boosting Crop Yield and Advancing Sustainable Agriculture." *Discover Agriculture* 3 (1). Springer Science and Business Media LLC. https://doi.org/10.1007/s44279-025-00208-5

Parray, Javid A., Nowsheen Shameem, Dilfuza Egamberdieva, and Riyaz Z. Sayyed. 2023. *Rhizobiome*. Elsevier.

Pathak, Parul, Vineet Kumar Rai, Hasan Can, Sandeep Kumar Singh, Dharmendra Kumar, Nikunj Bhardwaj, Rajib Roychowdhury, et al. 2022. "Plant-Endophyte Interaction during Biotic Stress Management." *Plants* 11 (17): 2203. https://doi.org/10.3390/plants11172203

Puranik, Shriniketan, Livleen Shukla, Aditi Kundu, Deeba Kamil, Sangeeta Paul, Govindasamy Venkadasamy, Rajna Salim, Sandeep Kumar Singh, Dharmendra Kumar, and Ajay Kumar. 2023. "Exploring Potent Fungal Isolates from Sanitary Landfill Soil for In Vitro Degradation of Dibutyl Phthalate." *Journal of Fungi* 9 (1): 125. https://doi.org/10.3390/jof9010125

S&P Global Commodity Insights. 2022. "Sharing Insights Elevates Their Impact." https://www.spglobal.com/commodity-insights/en/research-analytics/eu-member-states-approve-new-rules-for-biocontrol-agents

Sánchez-Navarro, Roberto, Matin Nuhamunada, Omkar S Mohite, Kenneth Wasmund, Mads Albertsen, Lone Gram, Per Halkjær Nielsen, Tilmann Weber, and Caitlin M Singleton. 2022. "Long-Read Metagenome-Assembled Genomes Improve Identification of Novel Complete Biosynthetic Gene Clusters in a Complex Microbial Activated Sludge Ecosystem." *MSystems* 7 (6). American Society for Microbiology. https://doi.org/10.1128/msystems.00632-22

Santos, Mariana Sanches, Marco Antonio Nogueira, and Mariangela Hungria. 2019. "Microbial Inoculants: Reviewing the Past, Discussing the Present and Previewing an Outstanding Future for the Use of Beneficial Bacteria in Agriculture." *AMB Express* 9 (1). https://doi.org/10.1186/s13568-019-0932-0

Schmidt, Radomir, Kelly Gravuer, Anne V. Bossange, Jeffrey Mitchell, and Kate Scow. 2018. "Long-Term Use of Cover Crops and No-till Shift Soil Microbial Community Life Strategies in Agricultural Soil." edited by Rita Grosch. *PLoS One* 13 (2): e0192953. https://doi.org/10.1371/journal.pone.0192953

Seenivasagan, Renganathan, and Olubukola Oluranti Babalola. 2021. "Utilization of Microbial Consortia as Biofertilizers and Biopesticides for the Production of Feasible Agricultural Product." *Biology* 10 (11): 1111. https://doi.org/10.3390/biology10111111

Shayanthan, Ambihai, Patricia Ann C. Ordoñez, and Ivan John Oresnik. 2022. "The Role of Synthetic Microbial Communities (SynCom) in Sustainable Agriculture." *Frontiers in Agronomy* 4 (June). https://doi.org/10.3389/fagro.2022.896307

Singh, Shalu, Sarita Singh, PK Sharma, and Deepak Sharma. 2017. "Isolation, Identification and Molecular Characterization of Phytase Producing Bacteria, Pseudomonas Sp. Aazad." *Journal of Pure and Applied Microbiology* 11 (4): 1845–50. https://doi.org/10.22207/jpam.11.4.24

Steenhoudt, Oda, and Jos Vanderleyden. 2000. "*Azospirillum*, a Free-Living Nitrogen-Fixing Bacterium Closely Associated with Grasses: Genetic, Biochemical and Ecological Aspects." *FEMS Microbiology Reviews* 24 (4): 487–506. https://doi.org/10.1111/j.1574-6976.2000.tb00552.x

Strategic Market Research. 2024. "Biofertilizers Market Size ($8.5 Billion) 2030." Strategicmarketresearch.com. https://www.strategicmarketresearch.com/market-report/biofertilizers-market

Tahat, Monther Mohumad, Kholoud Alananbeh, Yahia Othman, and Daniel I. Leskovar. 2020. "Soil Health and Sustainable Agriculture." *Sustainability* 12 (12). MDPI: 1–26. https://doi.org/10.3390/su121248590

Tang, Haiying, Muhammad Umair Hassan, Liang Feng, Muhammad Nawaz, Adnan Noor Shah, Sameer H. Qari, Ying Liu, and Jianqun Miao. 2022. "The Critical Role of Arbuscular Mycorrhizal Fungi to Improve Drought Tolerance and Nitrogen Use Efficiency in Crops." *Frontiers in Plant Science* 13 (July). https://doi.org/10.3389/fpls.2022.919166

Temkin, Mia I., Caitlin M. Carlson, Aaron L. Stubbendieck, Cameron R. Currie, and Reed M. Stubbendieck. 2019. "High Throughput Co-Culture Assays for the Investigation of Microbial Interactions." *Journal of Visualized Experiments* 152 (October). https://doi.org/10.3791/60275

Venter, Zander Samuel, Karin Jacobs, and Heidi-Jayne Hawkins. 2016. "The Impact of Crop Rotation on Soil Microbial Diversity: A Meta-Analysis." *Pedobiologia* 59 (4): 215–23. https://doi.org/10.1016/j.pedobi.2016.04.001

Vidya, Annabatula Sree, and Jaidev Sharma. 2024. "The Effect of Combination of *Azotobacter*, *Trichoderma* and Chemical Fertilizers on Wheat (*Triticum aestivum* L.) Yield in Gwalior Region, Madhya Pradesh, India." *International Journal of Plant & Soil Science* 36 (9). Science domain International: 293–98. https://doi.org/10.9734/ijpss/2024/v36i94979

Vurukonda, Sai Shiva Krishna Prasad, Davide Giovanardi, and Emilio Stefani. 2018. "Plant Growth Promoting and Biocontrol Activity of Streptomyces Spp. As Endophytes." *International Journal of Molecular Sciences* 19 (4): 952. https://doi.org/10.3390/ijms19040952

Warra, Aliyu Ahmad, and Prasad, Majeti Narasimha Vara. 2020. *Agrochemicals Detection, Treatment and Remediation*. Amsterdam: Butterworth-Heinemann.

Williams, Alwyn, Henry W. G. Birt, Anil Raghavendra, and Paul G. Dennis. 2022. "Cropping System Diversification Influences Soil Microbial Diversity in Subtropical Dryland Farming Systems." *Microbial Ecology*. https://doi.org/10.1007/s00248-022-02074-w

Woo, Sheridan L., and Olimpia Pepe. 2018. "Microbial Consortia: Promising Probiotics as Plant Biostimulants for Sustainable Agriculture." *Frontiers in Plant Science* 9 (December). https://doi.org/10.3389/fpls.2018.01801

Xu, Zhicheng, Xingzu Gao, Guoxue Li, Long D Nghiem, Wenhai Luo, and Fusuo Zhang. 2024. "Microbial Sources and Sinks of Nitrous Oxide during Organic Waste Composting." *Environmental Science & Technology* 58 (17). American Chemical Society: 7367–79. https://doi.org/10.1021/acs.est.3c10341

Yao, Xin, Hailin Guo, Kaixuan Zhang, Mengyu Zhao, Jingjun Ruan, and Jie Chen. 2023. "Trichoderma and Its Role in Biological Control of Plant Fungal and Nematode Disease." *Frontiers in Microbiology* 14 (May). https://doi.org/10.3389/fmicb.2023.1160551

Yared González-León, Esaú De, Rocío Ramírez-Vargas, Miguel Angel Anducho-Reyes, and Yuridia Mercado-Flores. 2024. "Whole Genome Analysis of *Bacillus velezensis* 160, Biological Control Agent of Corn Head Smut." *Microbiology Spectrum* 12 (4). American Society for Microbiology. https://doi.org/10.1128/spectrum.03264-23

Zengler, Karsten, Kirsten Hofmockel, Nitin S. Baliga, Scott W. Behie, Hans C. Bernstein, James B. Brown, José R. Dinneny, et al. 2019. "EcoFABs: Advancing Microbiome Science through Standardized Fabricated Ecosystems." *Nature Methods* 16 (7). Nature Portfolio: 567–71. https://doi.org/10.1038/s41592-019-0465-0

Zion Market Research. 2024. "Biocontrol Agents Market in North America and Europe Worth $11.84 Bn by 2032." 2024.. https://www.zionmarketresearch.com/news/biocontrol-agents-market

10 Mycorrhizal Associations
Enhancing Nutrient Acquisition

Rajendra Prasad, Kunal Madhav, Shivani Mehra, Shweta Pandey, Neha Pundir, Palak Chauhan, and Shikha Thakur

10.1 INTRODUCTION

Mycorrhiza represents a mutualistic relationship between the roots of many land plants and specific soil fungi. The term originates from the Greek words mykes (fungus) and rhiza (root) and refers to a relationship that has been present for over 400 million years (Smith & Read, 2010). In the mycorrhizal-plant association, the plant supplies the fungus with carbohydrates obtained through photosynthesis, while the fungus enhances the plant's absorption of water and essential mineral nutrients, particularly nitrogen and phosphorus (Van der Heijden et al., 2015). Mycorrhizal relationships are generally categorized into arbuscular mycorrhiza (AM), ectomycorrhiza (EcM), and ericoid mycorrhiza (ErM). AM, created by fungi from the phylum Glomeromycota, is found in most plant species and is recognized by the development of arbuscules within root cortical cells. ECM, prevalent in various trees like pines, oaks, and birches, features fungal sheaths encasing roots and a Hartig net among root cells. ERM primarily occurs in plants belonging to the Ericaceae family, facilitating their survival in acidic, nutrient-deficient soils (Brundrett, 2009). Ecologically, mycorrhizae improve nutrient uptake beyond the root's depletion area by extending fungal hyphae deep into the ground. They can enhance drought resistance, safeguard roots from pathogens, and aid in soil structure through substances such as glomalin (Rillig, 2004b). On a larger scale, mycorrhizal networks link various plants, promoting nutrient exchange and affecting plant community interactions (Simard et al., 1997). In agriculture, mycorrhizal fungi are being more frequently investigated as biofertilizers to decrease chemical usage, enhance nutrient efficiency, and rehabilitate degraded soils. Their efficacy relies on the compatibility among plant and fungal species, alongside soil and environmental factors (Gianinazzi et al., 2002).

In conclusion, mycorrhiza is an essential, ancient, and extensive alliance that promotes plant well-being, soil health, and ecosystem resilience, becoming increasingly important for sustainable land stewardship.

DOI: 10.1201/9781003664451-10

10.2 TYPES OF MYCORRHIZAL FUNGI

Mycorrhizal fungi establish a mutualistic relationship with plant roots to increase the uptake of nutrients and water and are supplied with carbohydrates. These are categorized into arbuscular mycorrhizae (AM), ectomycorrhizae (EcM), ericoid mycorrhizae (ErM), and orchid mycorrhizae (OM), with each having a different structure, host specificity, and ecological role. The most common are the arbuscular mycorrhizae (Glomeromycota) that develop intracellular structures known as arbuscules, which help a plant acquire phosphorus and small quantities of micronutrients, especially in agricultural production (Smith & Read, 2010). Ectomycorrhizae (Basidiomycota, Ascomycota) are external sheathing and form membranes around the roots and Hartig nets between cortical cells, mostly in association with forest trees, *Pinus* and *Eucalyptus* (Tedersoo et al., 2020). The so-called ericoid mycorrhizae are found primarily on plants belonging to the family Ericaceae and are adapted to acid tolerance and can grow in the impoverished soils. These fungi enter the root hair cells and increase the uptake of nitrogen and organic nutrients (Cairney & Meharg, 2003). The most distinctive aspect of orchids is that their seeds are not endospermic, but totally dependent on fungal partners (usually of *Rhizoctonia* type) to germinate and upon which they develop in early life (Dearnaley et al., 2012). The types of such symbioses vary in their ecological distribution, nutrient exchange, and adaptation, and they play a crucial role in the existence of plants in other ecosystems. They are important in ecosystem management, sustainable agriculture, and conservation biology, and it is important to understand their diversity and services. The main functions of mycorrhizal associations with plants in nutrient uptake and mobilization, stress tolerance, and ecosystem function are summarized in Figure 10.1.

10.2.1 ARBUSCULAR MYCORRHIZAL FUNGI

Arbuscular mycorrhizal fungi (AMF) are a type of fungi that belong to the phylum Glomeromycota. These fungi engage in partnership with plant roots in a helpful way, found in over 80% of terrestrial plants. These fungi possess fine thread-like structures called hyphae that explore large volumes of soil, thereby enhancing nutrient acquisition. Within plant roots, they form specialized structures known as arbuscules, which facilitate the exchange of nutrients and other metabolites between the fungus and the host plant. Arbuscules are hyphal structures that are highly branched within root cortical cells. AMF absorb nutrients from the soil that they need, and in return, plants supply the chemical energy to AMF, usually in the form of sugar, through photosynthesis, by which the plants make energy. The nutrient transfer, which can occur in both directions between the plants roots and the AMF, is shown in Figure 10.2. The species of AMF are extensive and they play an important role in enhancing plant performance and ecosystem services. Moreover, AMF are also known for providing additional benefits to crops by enhancing the yield (Schaefer et al., 2021).

Across the ecosystem, AMF are being widely distributed, from arid deserts to tropical forests. These fungi have been associated with plants for a long time, over 400 million years, as they co-evolved with the first plants ever to appear on land (Keddie & Orchall, 2000). Arbuscular mycorrhizal fungi (AMF) are essential for maintaining healthy and well-structured soils and play a vital role in supporting the

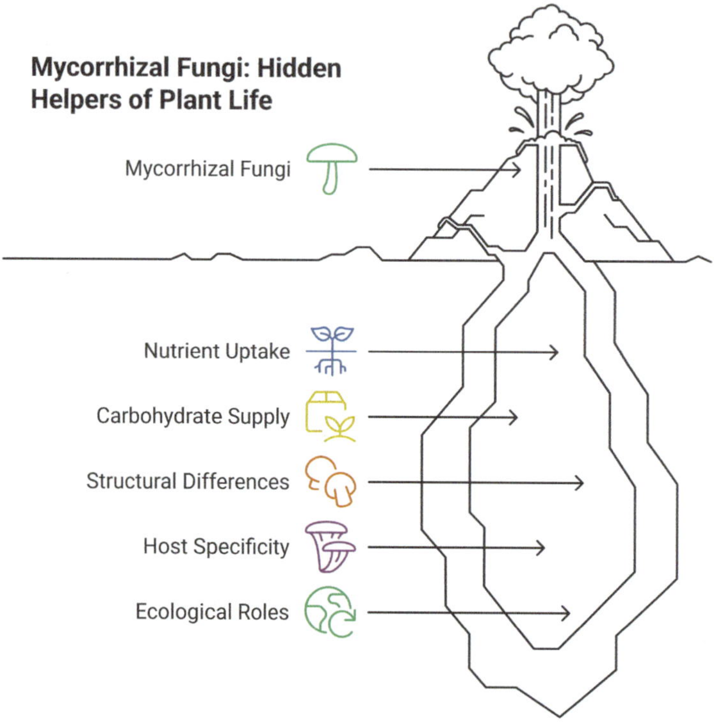

FIGURE 10.1 Principal functional roles played by mycorrhizae in the growth and survival of plants.

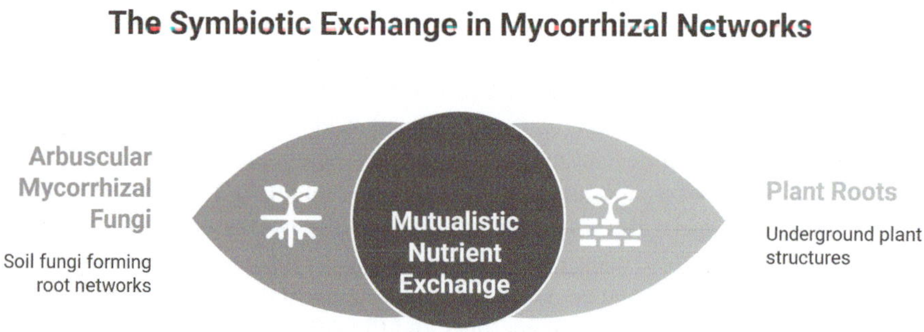

FIGURE 10.2 Mutual nutrient exchange between plant roots and arbuscular mycorrhizal fungi.

survival and coexistence of diverse plant species within an ecosystem. AMF produce a sticky substance called glomalin, which helps the particles of soil to aggregate together. Glomalin is a glycoprotein that makes the soil structure healthier and more stable (Rillig, 2004b). In relevance to agriculture, AMF are used as biofertilizers, which improve the productivity of crops, make them resistant to tough conditions, and reduce the dependency on chemical fertilizers. AMF abundance and diversity are affected by the agricultural practices; specifically, the negative impact of frequently and deeply plowing the soil and adding phosphate increasingly can harm these helpful fungi (Lekberg & Koide, 2005). Plant signaling pathways regulate the interaction between the AMF and plants, which involves important chemicals called strigolactones and common symbiosis signaling genes known as Sym. The chemical interaction between AMF and plants shows how plant-derived signals initiate the symbiotic development and arbuscule formation. It is important to understand this communication as it is crucial for designing crops that are compatible with AMF (Gutjahr & Parniske, 2013). AMF helps the plants to become resilience and more able to survive under abiotic stresses like salinity, drought, and heavy metals. The use of water more effectively under dry conditions when AMF are connected with plants; these fungi facilitate phytoremediation by enhancing the tolerance to metals and make them useful for restoring contaminated soils (Augé, 2001) (Figure 10.2).

AMF are helpful and important fungi; their diversity is in danger because of intense farming practices, growing only one type of crop (monocultures), and land-use change. For sustainable framing practices and conservation strategies, AMF should be included, which means that development and application of AMF bioinoculants can be a viable solution for crop varieties (Basiru & Hijri, 2022). The fundamental components of the plant–soil systems are AMF which influence the plant nutrition, the functioning of ecosystem, and productivity in agriculture. AMF help in the achievement of sustainable development by integrating different fields such as studying nature, understanding tiny biological processes, and agronomy.

10.2.2 Ectomycorrhizal Fungi

Ectomycorrhizal fungi (EcM) are a specialized group of fungus that have symbiotic relation with the roots of woody plants, especially the trees in the temperate and boreal forests including oaks, pines, birches, and eucalyptus. For the forest's ecology, this symbiotic association is important because it help in development of plants, enhancing the structure of soil, and against various environmental stresses it enhances the host plant resistance. Ectomycorrhizal fungi are different from other root fungi because of how they interact with plant's cells. Where AMF goes insides the plant's root cells, and EcM fungi stays between the root cells. A complex web structure of fungal thread, called the "Hartig net", is created by EcM in the spaces. This net is where the fungus and plants facilitate nutrient and carbon exchange without breaching the host cell walls (Courty et al., 2010). This partnership, called EcM symbiosis, is highly efficient in nutrient uptake from plants, especially those nutrients that poorly mobilize in the soil. Such as phosphorus and nitrogen. Extracellular enzymes are produced by EcM fungi like proteases, phosphatases, and laccases. This enzymatic activity helps fungi to release nutrients that plants can't normally get on their own such as in forest litter complex organic matter. In exchange, the plant gives the carbohydrates derived through photosynthesis; it's a regulated exchange system. The efficiency of this mutualism depends on which specific fungus and which specific plants are involved. Some EcM fungi exhibit high host specificity, while others can team up with a broad range of tree species (Marschner & Dell, 1994).

EcM fungi are taxonomically diverse group of fungi. Includes, *Amanita, Boletus, and Russula,* are fungi belong to the group called *Basidiomycota, Tuber, and Cenococcum* are from *Ascomycota* group, and some come from the *Zygomycota* group. Their fruiting bodies, what we commonly called mushroom are often seen on the forest floor, are major component of fungal biodiversity. The structure and function of EcM fungal communities can change depending on soil type, climate, vegetation, and anthropogenic factors (like pollution and deforestation). The ecological differences imply that various EcM fungi accumulate various nutrient exploration approaches, for example, *Piloderma* or *Cortinarius* form extensive mycelial network that spread far and wide through the soil, for nutrients over long distances, while others remain closer to the plant root, forming small network. These different exploration styles have significant ecological implications, which affect nutrient cycle, soil aggregation, and forest resilient (Sharma, 2017). In addition to nutrient acquisition, EcM fungi also help plants to deal with abiotic stresses by enhancing the uptake of water and increasing tolerance to dry spells, which they achieve by expanding the root anchorage surface area and modifying the structure of the roots. In polluted and metal-contaminated soils, EcM fungi can immobilize metals in their fungal tissues, by reducing the toxicity for host plants, making these fungi valuable for phytoremediation. By affecting the decomposition rates of organic matter, EcM fungi also contribute to carbon cycling, through "Gadgil effect". Some EcM species possess saprotrophic microbial activity, thereby promoting soil carbon storage and slowing decomposition. This ability has gained interest in the context of change in global climate, as EcM fungi might store more carbon than non-mycorrhizal or arbuscular mycorrhizal systems (Fernandez & Kennedy, 2016). Ectomycorrhizal fungi are important symbionts in most forest ecosystems, and they are key players in driving nutrient cycling, promoting plant health, and influencing the carbon dynamics. Their ecological importance extends beyond individual plant hosts; they actually shape the whole forest communities and contribute to the resilience of terrestrial ecosystems.

10.2.3 Orchid and Ericoid Mycorrhizae

A special type of mycorrhizal association that is present between orchids and ericoid mycorrhizae is essential because in places where soil is poor or a challenging environment, they help certain plants for nutrient acquisition. Both form a mutualistic relation between specific plant families and groups of fungi. Enabling their host to survive in ecosystems such as dense tropical forests, acidic soils, or nutrient-depleted habitats, OM work with orchidaceae and ericoid mycorrhizae with Ericaceae. Even though these two types of mycorrhizae have differences in morphology and evolved separately, both forms of mycorrhizae have a crucial role in the development of plants, nutrient cycling, and ecosystem resilience (Chauhan & Attri, 2024). OM are unique because the seeds of orchids are incredibly tiny and lack sufficient nutrient reserves to independently support germination. This means that orchids are completely reliant on certain fungi from the genera *Tulasnella, Ceratobasidium*, and *Sebacina* during the early developmental stages for carbon and nutrients. This early dependency is called mycoheterotrophy, where the baby orchid called as protocorm obtains carbon from the fungus instead of photosynthesis. As the orchid matures, the relationship might change to a mutualistic or partially autotrophic association, and in return for continued nutrient uptake, the plant photosynthesizes and supplies carbohydrates to the fungus (Sathiyadash et al., 2020).

Inside the orchid's root cells, unique coiled structures are formed by fungus, called pelotons, which are digested periodically by the host to release nutrients. In this process, fungus is growing inside and digestion process is unique to OM and essential for nutrient turnover. OM are also important for conservation, as in the ecosystem without the fungal symbionts many endangered orchid species cannot be grown or restored. Besides this, ericoid mycorrhizae are typically found in plants that belong to the Ericaceae family, which includes well-known genera such as *Vaccinium, Rhododendron*, and *Calluna*. These plants are grown in challenging environments like heathlands, peat bogs, and mountain forests, where acidic soil is present, rich in organic material, and nutrient-poor soil. These fungi associated with ericoid mycorrhizae are primarily ascomycetes, which are commonly *Rhizoscyphus ericae* and *Oidiodendron maius,* which grow into the roots of their host plant and within the epidermal cells forming intracellular coils (Grelet et al., 2016). Unlike "pelotons" which is more organized in OM or the "Hartig net" in ectomycorrhizal, the coils in ericoid roots are less structured. But they are still highly effective in facilitating nutrient exchange as ericoid fungi produce a range of powerful enzymes such as phosphatases, cellulases, and proteases. These enzymes allow them to degrade complex organic matter and release valuable nitrogen and phosphorus from humic substances. This ability provides a competitive advantage to ericoid plants in an ecosystem where the availability of mineral nutrients is extremely low. To address specific ecological challenges, both orchid and ericoid mycorrhizae have evolved. OM help orchid seedlings to grow in shady environments and in nutrient-deficient environments, while ericoid mycorrhizae allow plants to get nutrients from the organic matter present in soil that is unavailable for non-mycorrhizal species (Roy & Choudhury, 2022).

It is important to know that both types of mycorrhizal fungi can act as endophytes, can live inside plant tissues without harming them, and are capable of surviving as saprotrophs in soil even when their host plant is absent. This ability to live in two ways lifestyle increases their ecological versatility and contributions, helping their partnership with plants under changing environmental conditions.

10.2.4 Comparison of Their Functional Roles

A crucial component of terrestrial ecosystems is mycorrhizal associations that support a symbiotic relationship between fungi and plants. This diverse partnership varies in their structural and functional characteristics, and significantly influencing the dynamics of the entire ecosystems as well as nutrient uptake and stress tolerance. They have distinct roles in functioning and are highly adapted to their host plant ecological niches and the current environmental conditions, while a bidirectional exchange of resources is included in all mycorrhizal types (Bortolot et al., 2024). The most ancient and widespread type of mycorrhizal partnership is of arbuscular mycorrhizal fungi that approximately associate with 80% of terrestrial plants, which includes many agricultural farm crops, grasses, and trees in warm climates. AMF grows inside the plant's root cells and forms arbuscules that are the primary site of nutrient exchange and vesicles for storage. AMF are good at helping plants to uptake nutrients which are immobile in soil, especially phosphorus (P). They also help in nitrogen, zinc, and copper by extending their hyphal network far out beyond the root depletion zone (Smith & Read, 2010). These extraradical hyphae are grown by these fungi to enhance the soil structure through the production of glycoproteins called glomalins, which act as glue to hold the soil particles together. This helps in enhancing water infiltration, prevents soil erosion, and let the air move through the soil. Furthermore, these fungi also help plants to tolerate various abiotic stresses such as drought, salinity, toxic metals, and abiotic stress like pathogens by changing host plant physiological and molecular responses (Ahanger et al., 2019). Since they possess less host specificity and almost found everywhere make them vital in recycling and keeping ecosystems healthy.

Ectomycorrhizal fungi primarily associate with trees in cooler forests, like oak, pines, and birches. Unlike AMF, these fungi do not enter the plant cells. Instead, they form mantle, which is a thick covering around the root tips and grows between the root cells to form the Hartig net. The main strength of the – fungi is that they have the ability to uptake nutrients like nitrogen and phosphorus from the complex organic material in the soil. They achieve this by secretion of extracellular enzymes, such as cellulases, proteases, phosphatases, and phenol oxidase, that break down the organic polymers, which helps the trees to uptake the nutrients from soil with low nutrients but high organic matter (Tedersoo et al., 2010). It also protects plant roots from pathogens and enhances the plant tolerance to heavy metals, by trapping and neutralizing the metals (Phillips et al., 2013). Overall, their influence on how carbon and nutrient cycling move through forest ecosystems makes them important for ecosystem health and productivity (Chot & Reddy, 2022). OM are a highly specialized type of partnership found in only orchidaceous family. Orchid seeds are extremely minute and lack endosperm, so they depend on certain fungi for early germination and seedling development (Rasmussen, 1995). These fungi are usually from the *Tulasnellaceae, Ceratobasidiaceae, or Sebacinaceae* families, which provide germinating protocorms with carbon compounds and nutrients they need for growth. The fungi essentially function as saprotrophs by breaking down dead organic matter in the soil. As the orchid grows mature, the relationship changes or varies. Some orchids continue to rely on the fungus for carbon, and this condition is known as mycoheterotrophy, while others begin to photosynthesize but still receive nutrients from the fungus (Li et al., 2021). This interaction between orchids and fungi is very specific, because each type of orchid may require a particular type of fungus to survive. That's why the specificity of orchid-fungal association is generally higher than that of AMF or EcM. The major ecological and physiological functions of AMF, ECM, orchid, and ericoid are shown in Figure 10.3.

In Ericaceae family, ericoid mycorrhizae forms association with the plants which usually grow in acidic, low-nutrient soils like those found in heathy lands, boreal forests, and montane ecosystems. Ericoid mycorrhizal fungi make a loose network around the root cells and also grow intracellularly, forming characteristic "hair roots" (Wei et al., 2022). Functionally, ericoid mycorrhizae exhibit saprotrophic capability to produce a broad spectrum of extracellular enzymes that efficiently degrade the complex organic compounds, like nitrogen and phosphorus. Ericaceous plants

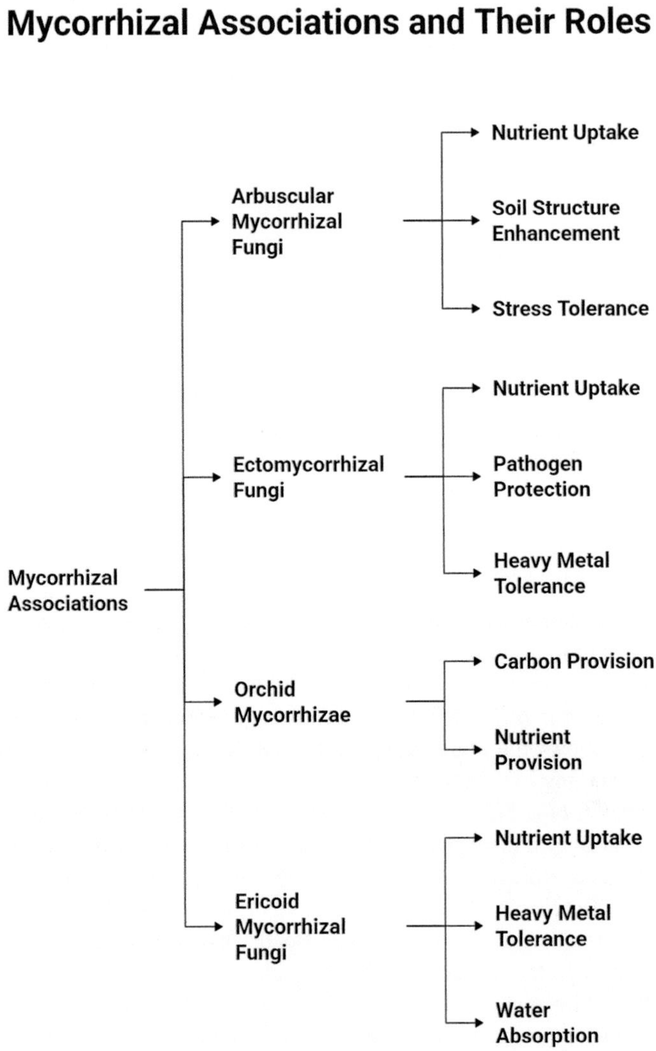

FIGURE 10.3 Plant functional roles of diverse mycorrhizal associations.

possess the ability to survive in harsh conditions where other plants cannot survive. Ericoid mycorrhizal fungi also enhance the tolerance to heavy metals and improve water absorption and make them even more suitable for poor soil (Forgy, 2012). They play a role as both decomposers and partners to benefit plants in nutrient-poor conditions.

10.3 NUTRIENT ACQUISITION MECHANISM BY MYCORRHIZAE

The mycorrhizal fungi improve nutrient uptake in plants by creating a specialized symbiotic interfaces which have resulted in providing larger surface area for the exchange of nutrients. The arbuscular mycorrhizae (AM) invade root cortical cells and undergoes arbuscular on the root and vice versa, where there is exchange of nutrients. AM fungi enhance phosphorus (P) acquisition through high-affinity phosphate transporters, especially in soil deficient in P (Smith & Smith, 2011). They also improve the utilization of micronutrients, including zinc (Zn) and copper (Cu) by means of extraradical hyphae growth into nutrient-deficient areas outside of the rhizosphere (Lehmann et al., 2014). Ectomycorrhizal fungi (ECM) wrap the root cells with a Hartig net, through which they can obtain nitrogen (N) efficiently. They produce extracellular enzymes (proteases, phosphatases, etc.) to mineralize organic matter and liberate amino acids as well as inorganic P, they acquire (Courty et al., 2010). Ericoid mycorrhizae (ErM) grow under an acidic and low-nutrient environment, and they have robust enzymatic functions to unlock nitrogen and phosphorus complex organic molecules (Read & Perez-Moreno, 2003). Additionally, mycorrhizae enhance the acquisition of nutrients indirectly by modifying the rhizosphere pH, enhancing root branching and promoting beneficial microbial communities (Van der Heijden et al., 2015; Singh et al., 2022). This symbiotic effectiveness goes a long way in plant productivity, ecosystem stability, and chemical fertilizers dependence.

10.3.1 Nutrient Mobilization and Uptake Pathways

Mycorrhizal association plays an important role for plants to get nutrients, especially for the nutrients that are immobile, like phosphorus, nitrogen, and some micronutrients like zinc and copper. AMF live in the outer layers of roots and send their hyphae out into the soil around them. This creates a very effective network for taking up nutrients. This allows the plants to get nutrients that are in the areas around their roots that are running low on them.

Phosphorus Mobilization and Transportation: The mycorrhizal pathway allows the plants to take up phosphorus, which is one of the most important macronutrients. Phosphorus is typically found in soils in forms that don't dissolve well or move around well. AMF release acid phosphatase and organic acids that help release phosphate bound to the soil particles (Smith & Read, 2010). High-affinity fungal phosphate transporters in the extraradical hyphae take up phosphate once it is dissolved (Bucher, 2006). Phosphate moves through the fungal hyphae toward the root and is given at the arbuscule interface, specifically in the peri-arbuscular region, where plant phosphate transporters take up the nutrient.

Nitrogen Mobilization and Uptake: AMF help plants to get nitrogen as well as phosphorus, especially from soils that don't have much nitrogen in them. AMF take nitrogen as ammonium, nitrate, and sometimes organic nitrogen such as amino acids. This uptake is made easier by fungal nitrate and ammonium transporters (Govindarajulu et al., 2005). AMF typically turn nitrogen into arginine, which is a stable form that may be stored and moved. Then, the fungal hyphae move arginine toward the arbuscule, where it is broken down and ammonium is released so that the plant may take it through its ammonium transporters. The major steps in nutrient mobilization, transport, uptake, and utilization through mycorrhizal association are presented in Figure 10.4.

Micronutrient Uptake: AMF structures assimilate nutrients for plants to assimilate nano materials such as zinc, copper, and manganese that are usually distorted in farm dust. Fungal hyphae release siderophores and chelating chemicals to move

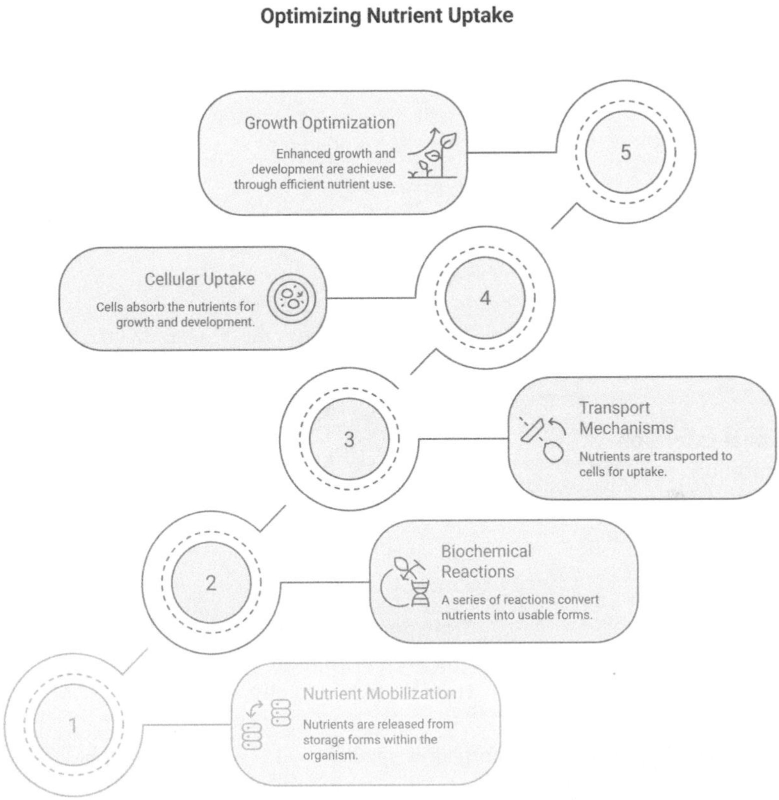

FIGURE 10.4 Major processes of the uptake and utilization of nutrients in plants.

these metals around. Certain metal transporters in fungal membranes help the fungi take up metals and move them to host plants (Mohan et al., 2014).

Symbiotic Interface and Resource Exchange: The arbuscule–plant interface in AM symbiosis is the most important element of the symbiotic interaction for exchanging nutrients. In this case, carbon molecules, mostly hexoses, move from the plant to the fungus, whereas minerals move in the other direction. Transporter expression, arbuscule longevity, and environmental parameters, including soil richness and moisture, all affect how well nutrients are exchanged.

10.3.2 Enhancement of Phosphorus Acquisition

Phosphorus is an important macronutrient for the growth of plants. It is important for transferring energy, sending signals, and making large molecules. But it doesn't dissolve well and sticks strongly to the soil particles; phosphorus is commonly known as a limiting nutrient in agricultural soil. AMF helps plants to get more phosphorus by changing the biochemical environment of the rhizosphere and increasing the surface area of the roots that can absorb it. The major biological and physical barriers limiting phosphorus uptake by plants are shown in Figure 10.5.

FIGURE 10.5 Plant barriers to phosphorus uptake.

Mycorrhiza Hyphal Extension and Soil Exploration: The main way that AMF help plants to take up more phosphorus is by generating a large network of extraradical hyphae that goes beyond the areas where phosphorus is low near the root surface. These hyphae search a larger area of soil and find phosphate ions that can't reach plant roots (Smith & Read, 2010). The high surface area to volume ratio and small diameter ensure it is far easier for the hyphae to absorb phosphorus present within the micropores of the soil.

Solubilization of Inorganic Phosphate: AMF change the chemistry of the rhizosphere, which helps move phosphorus around. They release the organic acids like oxalic acid, citric acid, and malic acids, which bind to the cations like calcium, iron, and aluminum that hold the phosphorus in a form that don't dissolve. This bonding makes inorganic phosphate easier to dissolve, which makes it easier for fungal hyphae to uptake (Richardson et al., 2009). Also, AMF make acid phosphatases, which break down organic phosphorus molecules and release inorganic phosphate into the rhizosphere (Tarafdar & Marschner, 1994).

Transport of Assimilated Phosphate and Symbiotic Interface: The taken-up phosphate should be transported to the arbuscule by way of the hypha, and this is a symbiotic organ located in the root cortical cells. Phosphate is discharged from the arbuscule into the peri-arbuscular space, where plant-specific phosphate transporters on the peri-arbuscular membrane help the root cell cytoplasm to take up phosphate (Javot et al., 2007). During AM association, certain PT genes, like MtPT4 in *Medicago truncatula*, are turned on more than usual. They are also necessary for the transfer of phosphate between the plants.

Reduced Fertilizer Dependency and Improved Efficiency: AMF improve the phosphorus acquisition and make plants grow better, especially when there isn't much phosphorus available. Because of this, crops that are associated with AMF often need less phosphate fertilizer. For example, adding *Rhizophagus irregularis* to maize has been proven to enhance the amount of phosphorus it takes up and the amount of biomass it produces, even in soils that are low in phosphorus. This means that fertilizer use can be cut by 30%–50% without hurting yields (Ramírez-Flores et al., 2019). The symbiotic relationship between plants and AMF is an important adaptation for phosphorus acquisition in nutrient-poor soils. This synergy of the physical exploration of the soil and biochemical solubilization of phosphorus, together with the considerable symbiotic transfer of nutrients provides a sustainable solution to the absence of phosphorus limitation in agriculture.

10.3.3 Importance and Mechanism of Uptake of Nitrogen, Zinc, Copper, and Other Micronutrients

Mycorrhizal fungi help the plant to get phosphorus and other micronutrients such as nitrogen, zinc, copper, manganese, iron, and boron. Plants need these nutrients to enhance their structure, regulate their metabolism, and support strong enzymatic activity. The way these AMF and their host plants associate together allows them to get nutrients from the soil that they are not able to reach on their own. This makes these plants healthier and more productive, especially in soils that are poor in nutrients. Nitrogen is an essential component of amino acids, nucleotides, and chlorophyll, and it is not linked to AMF as compared to phosphorus. Some studies indicate that AMF play an important role in nitrogen absorption, particularly in low nitrogen environments. AMF can uptake nitrate and ammonium through fungal nitrate and ammonium transporters and transfer them to the plant through the arbuscular interface. The transfer of nitrogen takes place mostly within the fungal hyphae as arginine, which is later catabolized at the arbuscule and releases ammonium to be absorbed by the plants (Govindarajulu et al., 2005).

Zinc is important for the stability of protein structures and the functions of enzymes, while copper is crucial for redox reaction, photosynthesis, and lignification. Both are absent in calcareous or alkaline soils because of limited solubility. AMF enhance their acquisition by exuding low-molecular-weight organic acids and chelators, which augment metal ion solubility in the rhizosphere (Dhalaria et al., 2020). Upon mobilization, fungal metal transporters enable the absorption and transfer of Zn and Cu to the plant. Inoculation of Glomus intraradical enhances the zinc uptake and yield in maize under low zinc conditions. Comparatively, copper in citrus and tomato plants outcomes is noted, where AMF enhanced absorption efficiency and reduced deficient symptoms (Marschner & Dell, 1994). Iron is essential for the production of chlorophyll and the transportation of electrons. AMF enhance iron uptake by improving the root access to the iron-rich micro zones and maybe through the interactions with siderophore-producing rhizobacteria that further solubilize the iron (Jansa et al., 2013). Manganese and boron play important roles in enzyme activation and cell wall production, and they are absorbed more efficiently in mycorrhizal connections. The increase in nitrogen and micronutrient absorption through mycorrhizal associations is required for obtaining the most effective plant nutrition. Such mechanisms increase the metabolic activities of the plants besides reducing the use of chemical fertilizers. Therefore, the mycorrhizal pathway functions as a combined strategy for presenting both agronomic and ecological benefits, especially in soil with low nutrients.

10.3.4 Alteration in Root Architecture and Rhizosphere Chemistry

Mycorrhizal association particularly involves AMF, which induce substantial modifications in the root architecture and rhizosphere chemistry. These changes enhance nutrient and water absorption. To enhance the functional zone of the plant roots and optimize nutrient absorption from the soil, these morphological and biochemical alterations are important. The root architecture is highly adaptable and capable of altering the response to environmental factors, particularly nutrition availability. For example, when phosphorus is limited, plants frequently modify their root architecture to generate elongated root hairs and enhance the density of lateral roots, especially in the upper soil layers where phosphorus typically concentrates. This "topsoil foraging" method optimizes the surface of the root area for nutrient absorption. Nitrogen availability can similarly affect the primary root elongation and the lateral root branching, which affects the difference based on the form of nitrogen and concentration. These architectural modifications are not only physical reactions. They are frequently regulated by plant hormones such as auxins, cytokinin, and ethylene, whose production, transit, or responsiveness can be affected by nutrition availability (López-Bucio et al., 2003). Additionally, these roots significantly modify their rhizosphere composition by releasing a range of chemicals, and these chemicals comprise a complex combination of organic acids, sugars, amino acids, phenolics, enzymes, and other secondary metabolites. They also provide multiple functions such as signaling beneficial microbes, binding nutrients to improve the availability, and preventing the growth of competitors or pathogens (Basiru & Hijri, 2022). Figure 10.6 shows the effect of mycorrhizal colonization on the root system and rhizosphere chemistry.

An example of the relationship between the root architecture and rhizosphere chemistry is the association with AMF. AMF significantly improve the nutrition absorption, particularly for immobile nutrients such as phosphorus, while also affecting the root architecture. AMF colonization may induce alterations in the root branching patterns, specific to the root length, and the overall root to shoot biomass ratio, but the exact impacts may differ (Cheng et al., 2023). The structural modifications, together with the fungal hyphae, reach beyond the root's reach, creating more extensive and efficient absorptive network. In addition to that, the existence of AMF may alter the consistency of the root exudates and similarly influence the broader microbial community within the rhizosphere, hence institute climate positive feedback on nutrient cycling and plant health (Schaefer et al., 2021).

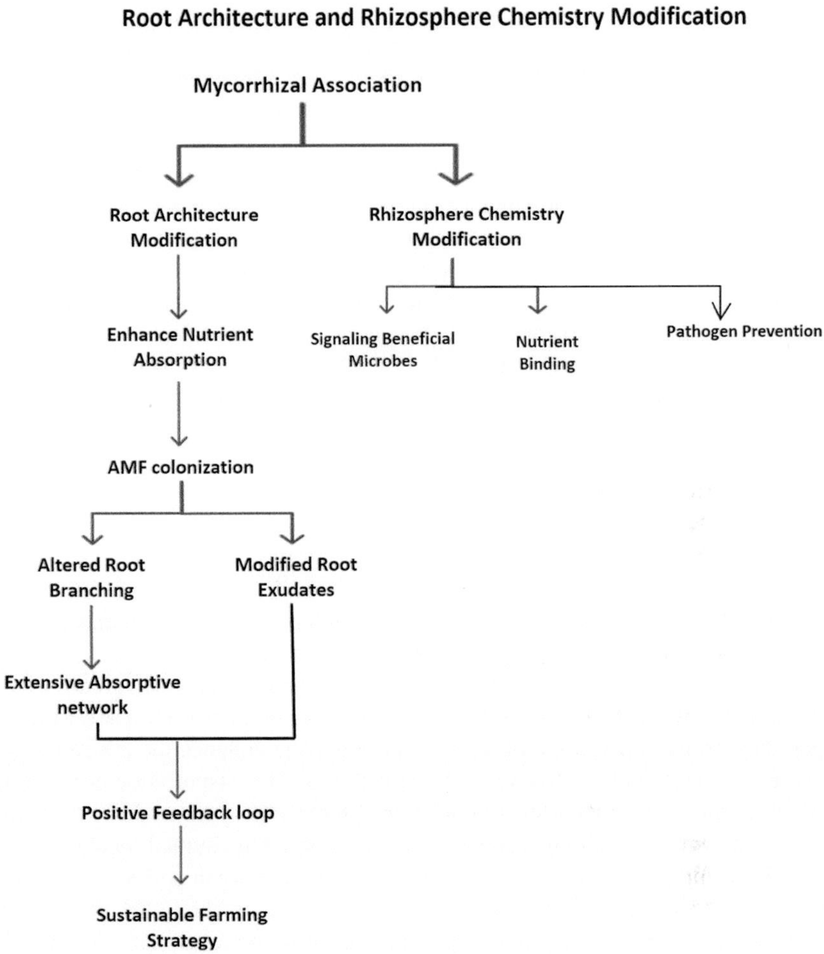

FIGURE 10.6 Influence of mycorrhizal associations on root architecture and the rhizosphere chemistry.

The architecture and chemical modification of these rhizospheres will allow them to adapt to the diverse soil conditions and also enhance the nutrient uptake. It is also important to understand that this complex partnership is important for formulating a sustainable farming strategy that will be utilized by the natural plant–soil processes to enhance the productivity and resilience of the crop.

10.4 MYCORRHIZAL CONTRIBUTION TO WATER STRESS RESISTANCE

Water stress is a serious abiotic stress factor that dramatically inhibits physiological activities of plants, damages metabolism, and slows down growth and productivity, a primary challenge to global agriculture. Mycorrhizal fungi, especially AMF, establish a mutualistic symbiotic relationship with the roots of the majority of terrestrial plants and are among the most effective natural partners of plants in the struggle against drought stress. The extraradical hyphae of these fungi penetrate the soil well beyond the zones of root depletion, greatly expanding the absorptive surface area and helping the fungi access water that lies outside the root depletion zones. Besides improving water uptake, AMF also change root architecture, upregulate aquaporins (water channel proteins), and enhance the production of glomalin-related soil proteins (GRSPs), which ameliorate soil aggregation and water-holding capacity. At the biochemical level, they control the concentration of plant stress hormones, including abscisic acid (ABA), which is involved in stomatal control, limiting the excessive transpiration during water shortage. Additionally, all the modifications of root gene expression and enzyme activity stimulated by AMF also contribute to drought tolerance through the preservation of cellular homeostasis and osmotic balance. All these structural and molecular adjustments largely contribute to the acquisition of plant resilience, enabling them to survive, grow, and even reproduce during extended durations of reduced water supply.

10.4.1 Improved Water Absorption and Retention

AMF have a key role in enhancing the water relations of plants (particularly in drought conditions) due to their physiological and structural peculiarities. The ability of AMF to produce extensive extraradical hyphal networks, which greatly extend into the soil beyond the depletion zones of roots, is one of the main ways through which they promote hydration of plants. These hyphae enhance the absorptive surface area by many folds and serve as direct pathways of transportation of water and nutrients to the plant root cortex, hence keeping the plants hydrated despite less than ideal soil moisture conditions (Smith & Read, 2008). This hyphal-mediated water transport is important, especially in the drought period when soil water is scarce and unevenly distributed. Being involved in direct water uptake, AMF also produce glomalin-related soil proteins (GRSPs) which are glycoproteins that AMF binding soil particles together, forming stable soil aggregates (Rillig, 2004b). These stable aggregates improve the soil structure by creating porosity and better water infiltration and retention, therefore providing a micro environment that promotes moisture retention and root penetration. The overall result of these biological and biochemical activities not only enhances the physical quality of the soil but also gives the plants a continuous supply of water, enabling them to better withstand water stress.

10.4.2 Role in Drought Tolerance and Hydration Efficiency

Mycorrhizal fungi, especially AMF, increase plant drought tolerance through increasing root system hydraulic properties. One of the important processes is the upregulation of aquaporins, that is, specialized water channel proteins bound to membranes found in root cortical and endodermal cells, which enable the regulated transfer of water across cell membranes. The AMF colonization leads to the induction of multiple aquaporin genes, including PIP1 and PIP2, and, as a result, to the enhancement of root hydraulic conductivity and more efficient internal water transport despite the poor external water supply (Porcel et al., 2006; Ruiz-Lozano et al., 2012). The given physiological adaptation plays a crucial role in reducing the effects of drought because it enables plants to continue the water transport process between roots and shoots even when the soil dries out. As a result, plants inoculated with AMF have an increased efficiency of hydration that is characterized by higher relative water content (RWC), enhanced leaf turgor pressure, and the delayed symptoms of wilting in comparison with non-mycorrhizal controls when both groups are subjected to comparable water-limiting conditions (Wu & Xia 2009; Aroca et al., 2007). This improved water condition allows sustenance of photosynthetic capacity, cell expansion, and stomatal control, extending plant survival and productivity under extended drought conditions. The additivity of the structural (increased hyphal absorption) and physiological (regulation of aquaporins) changes highlights the central position of AMF in the plant water homeostasis in variable and demanding conditions.

10.4.3 Modulation of Stomatal Conductance and Hormonal Balance

The other mycorrhizal-dependent key mechanism involved in increasing drought tolerance is abscisic acid (ABA) signaling in regulating stomatal behavior. ABA is a major phytohormone in plant reactions to water deficit, especially in the induction of stomatal closure to reduce water loss through transpiration and save water status inside the plant. AMF have been reported to trigger an increase in endogenous ABA levels in colonized plants than non-mycorrhizal control plants during water deficit, mainly in root and leaf tissues (Ruiz-Lozano et al., 2016; Dodd et al., 2010). This increase is, however, not sudden or extreme, but rather it is exquisitely controlled, such that plants can minimize transpiration, but still ensure a gas exchange necessary to carry out photosynthesis. This optimized stomatal control maintains a prolonged carbon assimilation and water use efficiency, even in less than ideal hydration status. In addition to ABA, AMF also regulate the dynamics of other major plant hormones that adjust drought responses. As an example, cytokinins, normally decreased in drought to minimize growth and preserve energy, are more sustained in AMF-colonized plants, in favor of shoot biomass and senescence postponement (Zhang et al., 2018). Auxins, critical to root extension and lateral develop, are also induced leading to a greater root exploration and soil coverage, which further enhances water absorption (Ludwig-Müller, 2010). Also, AMF regulate jasmonates, related to stress signaling and defense, to increase plant plasticity during drought and to decrease the price of stress-related damages. The result of the crosstalk between these hormones coordinates by the symbiotic interaction and is a holistic physiological response, which enhances the adaptive capacity of the plant in quickly varying moisture conditions, not merely its survival but its functional resilience.

10.5 ECOLOGICAL SIGNIFICANCE AND ECOSYSTEM SERVICES

Mycorrhizal fungi are among the most important ecological engineers in terrestrial ecosystems, whose role goes well beyond plant nutrition and drought tolerance. These symbiotic fungi create complicated networks, also called the wood wide web, which connect several plant species underground via widespread hyphal networks.

Mycorrhizal fungi increase nutrient cycling, increase soil structure, and mediate plant-microorganism interactions through these networks. They also help in the movement of several crucial nutrients, such as phosphorus and nitrogen, that not only contribute to the well-being of individual plants but also boost higher productivity in the ecosystems across diverse habitats (Smith & Read, 2008). With respect to conservation of biodiversity, mycorrhizal-plant associations reduce the competitive exclusion and hence equal distribution of resources occurs among the near neighbor plant species. This may create and encourage the occurrence of rare or slow-growing species that enhance above- and below-ground species richness (Van der Heijden et al., 1998). Also, the fungi act as ecological buffers to climatic stresses such as drought, temperature variations, and nutritionally impoverished soils. Mycorrhizae improve plant survival and physiological performance through enhancing nutrient acquisition of plants and water use efficiency, thus vegetation dynamics are stabilized and ecosystems become less vulnerable to degradation (Rillig et al., 2019b). Mycorrhizal fungi play a number of roles in the long-term resilience and sustainability of ecosystems by their roles in soil aggregation, carbon sequestration, nutrient recycling, and even inter-plant communication. Their ecological functions consequently cannot be spared in the ensemble of the global environmental transformations, including intensification of land-use, fragmentation of habitats, and climate changes.

10.5.1 Soil Health and Structure Improvement

Mycorrhizal fungi in association with plant roots are very promising in enhancing physical and biological properties of soils, thereby making them stronger, fertile, and capable of sustaining agricultural productivity even under adverse environmental conditions. AMF are among the most significant contributors to this process, because they secrete a unique type of glycoproteins, so-called glomalin-related soil proteins (GRSPs). These proteins are a natural soil glue, which holds soil particles and organic matter together to make stable aggregates. This aggregation plays a very vital role in the soil since it enhances or improves the stability of the soil aggregates resistant to water erosion as well as compaction, which in the long run contributes to soil productivity and health (Wright & Upadhyaya, 1998). Besides the biochemical stabilization, AMF presence in the soil in the form of hyphal networks also plays a role in enhancing soil texture. These extraradical hyphae make a large web-like matrix in the soil, and effectively, this makes a sponge-like structure that improves soil porosity, infiltration rates, and aeration. Consequently, soils colonized by AMF will have an increased capacity to retain water as well as enhance gaseous exchange, which is essential in root respiration as well as microbial activity (Rillig et al., 2001). These fine hyphal filaments additionally act as micro-channels which enhance the transport of water to root areas, thus boosting hydration efficiency and minimizing the exposure to surface run-offs and erosion. AMF also enhance the accumulation of soil organic matter by depositing fungal biomass into the soil system biologically. The dying of fungal hyphae and the secretion of organic matter add to the soil organic carbon (SOC) pool, which increases the microbial diversity and enzymatic activity – both of which are marks of a healthy soil microbial ecosystem (Treseder & Allen 2000). The process is also important in the sequestration of carbon where greenhouse gas emissions are mitigated by having the carbon locked in to more stable soil compartments over time (Wilson et al., 2009). And by these various processes, mycorrhizae not only maintain plant growth but also engender the properties of ecosystem engineers, reclaiming degraded lands and bringing about sustainable land management regimes.

10.5.2 Impact on Plant Biodiversity and Community Dynamics

Mycorrhizal associations, especially those established by AMF, are known to have a pivotal role in structuring and dynamics of plant communities. Such symbiosis relations adjust plant-plant competition, enhance nutrient use efficiency, and lead to the coexistence of species since they can redirect belowground resources in ways that cannot be attained by roots acting alone. One of the key ecological processes that adds to this influence is the common mycorrhizal network (CMN) – a subsoil web of fungal hyphae which physically connects the root systems of multiple plants, often even across different species and functional groups (Van der Heijden & Horton, 2009). In the CMN, the mycorrhizal fungi facilitate the exchange of nutrients, such as phosphorus and nitrogen and even carbon compounds between plants relative to plant needs and nutrition. Via this network, more equal access to soil resources may be achieved, particularly benefiting those species which are less competitive, or early successional, which would otherwise be excluded in rapidly growing, or aggressive neighbor communities. In this way, mycorrhizal symbioses bring ecological leveling, which imparts plant species richness and stability of communities that grow in resource-limiting or disturbed conditions (Simard et al., 2012). In addition, CMNs can also transport chemical and hormonal messages that can warn adjacent plants of environmental stress, including herbivory or drought, and orchestrate defense reactions at the community scale. Such plant-to-plant signaling via hyphae of fungal species increases the functional connectivity of ecosystems and adds to the collective resilience to biotic and abiotic stresses (Barto et al., 2012). Mycorrhizal fungi are keystone elements of terrestrial diversity and stability by facilitating nutrient exchange, stress resilience, and multispecies incorporation.

10.5.3 Influence on Ecosystem Functions under Changing Water Availability

With the accumulation of challenges related to climate change, specifically the acceleration of variability in water supply, mycorrhizal associations can be described as crucial biological mediators, stabilizing ecosystems. With the increase of drought and anomalous precipitation caused by climate change, mycorrhizal fungi, particularly AMF, ameliorate the negative effects of precipitation on the plants by increasing water absorption, enhancing root hydraulic conductivity, and modulating plant responses to stress. The overall impact of these effects is seen in decreasing the water stress of plants, and this couples to affect ecosystem-level processes including evapotranspiration, carbon cycling, and nutrient turnover (Augé, 2001). In stabilizing plant physiological performance under stress, AMF contribute to stabilizing the hydrological balance of ecosystems, with more reliable water fluxes during dry spells. In addition, mycorrhizal symbiosis increases the efficiency of the use of resources, especially water and nutrients, which is of great concern in arid and semi-arid ecosystems. Such areas are quite vulnerable to climatic extremities like permanent drought and land deterioration. Plants colonized by AMF in such habitats tend to be more resilient and take longer to die; net primary productivity is thus stabilized across years despite climate anomalies (Barzana et al., 2012). This buffering effect minimizes inter-annual variance in vegetation cover and biomass production, thereby exerting a dominant role in alleviating desertification and ecosystem sustainability. Also, due to their ability to preserve the soil structure and organic matter via extensive hyphal networks and their secretion of glomalin-related soil proteins (GRSPs), AMF indirectly help fight climate change by sequestering carbon in soil and preventing soil erosion (Leifheit et al., 2014). The mycorrhizal fungi, therefore, act as ecological stabilizers enabling the plant community and the soil system to adjust and remain functional under the sudden changes in the environment.

10.6 APPLICATION IN SUSTAINABLE AGRICULTURE

Mycorrhizal fungi are a key element in sustainable agriculture, enhancing nutrient intake of the plant, soil organization, and reliance on chemical fertilizers (Kumar et al., 2021). Arbuscular mycorrhizae (AM) expand the supply of phosphorus (P), nitrogen (N), and micronutrients, e.g., zinc (Zn), through their wide hyphal networks, which minimizes the use of chemical inputs (Smith & Read, 2010). This helps reduce pollution of the environment due to excessive use of fertilizers. The mycorrhizae also contributed in such a way to plant capacity to bear abiotic stresses such as drought, salt and metal toxicity, increasing the water absorption and ion homeostasis. Climate-smart agricultural practices are facilitated by their carbon sequestration ability, which is facilitated by root mass growth and aggregation of soils (Rillig et al., 2019a). Typical of forests, ectomycorrhizal (ECM) plant-fungus associations facilitate the growth of trees in poor soil environments in order to support reforestation and land recovery (Martin et al., 2016). Also, mycorrhizae increase the biodiversity of the soil by enhancing beneficial microbial community, inhibiting pathogens of plants, and making ecosystems resilient (Cameron et al., 2013). Such environmental advantages predestinate mycorrhizal fungi as biofertilizers to become a valuable alternative in organic as well as regenerative farming systems. The idea of incorporating mycorrhizal inoculants in cropping systems is a possible approach of addressing the global food needs, considering that the environmental footprints are kept at minimal levels, which is within the objectives of sustainable agriculture.

10.6.1 Integration into Cropping System

Integration of mycorrhizae into farm fields is a significant stride toward growing crops in a more environmentally sustainable way. These mycorrhizal fungi, particularly AMF, that associate with the roots of crops improve the health and structure of soil and help plants for better absorption of nutrients and water, which will strengthen their ability in tough conditions like drought. These fungi help farmers to become less reliant on chemical fertilizers, which will improve soil health and increase productivity. This is especially beneficial for farms that use fewer resources or practice organic farming. In conventional farming methods, where excessive tillage, grow only one type of crop repeatedly, and high use of chemical fertilizers, the natural networks of these helpful fungi in the soil often get damaged. However, if farmers make a conscious effort to inoculate mycorrhizal to their fields or use conservation practices, these can rejuvenate networks. The effectiveness of this mycorrhizal integration depends on the type of crop being grown, soil characteristics, native mycorrhizal fungi are already present, and the overall farming methods used (Bender et al., 2016).

One successful example of using these helpful fungi is observed in maize-legume intercropping systems. Intercropping systems between maize and legumes provide a successful illustration of mycorrhiza integration. AMF improve the phosphorus acquisition of maize (*Zea mays*), a crop with a high P demand. Along with complementary nutrient use, intercropping mycorrhiza-friendly legumes such as cowpea (*Vigna unguiculata*) or soybean (*Glycine max*) increases mycorrhizal colonization because of legume root exudates and reduces soil disturbance. This results in higher system productivity and better nutrient cycling (Jansa et al., 2013).

Moreover, integration of mycorrhizal fungi helps the soil to store carbon and also aggregates the soil together for long-term fertility. These fungi produce a substance called glomalin, which acts like a binder and enhances the soil particles to clump together (Rillig et al., 2001). The advantages would be particularly useful on marginal or degraded lands where conventional practices are inefficient. To encourage these beneficial mycorrhizal fungi in farm fields, several methods can be used, such as minimal soil disturbance by less tilling, rotating the crops by growing different plants with mycorrhiza-compatible plants, planting cover crops to protect and improve the soil, and adding commercial mycorrhizal products directly to the soil. Nevertheless, the compatible functioning in the ecosystem and site-specific feedback should be considered to promote its broad use.

10.6.2 Mycorrhiza Bioinoculants and Formulation

Mycorrhizal bioinoculants contain live mycorrhizal fungi that are used by farmers to enhance plant growth, absorb more nutrients, and improve soil health. These bioinoculants, especially those based on AMF, are more beneficial as a sustainable and eco-friendly alternative to chemical fertilizers in modern agriculture. They are a key part of the ecological farming system by encouraging beneficial plant-microbe interaction and boosting crops against abiotic and biotic stresses.

The formulation of a successful mycorrhizal bioinoculants involves few key steps: the selection of fungal strains that are effective at helping plants grow huge amount of these fungi in a controlled way, and the formulation of an inoculant so it stays alive and effective on the shelf for a long time. The most popular mycorrhizal fungi used in these commercial inoculants include *Rhizophagus irregularis, Funneliforis mosseae*, and *Claroideoglomus*. These are chosen to form beneficial relationships with their broad host range and high colonization efficiency (Ijdo et al., 2011).

The formulation of mycorrhizal bioinoculants depends on how they'll be used; some are solid-based, and materials such as vermiculite, peat, or talc are widely used for coating seeds or applications to plants in nurseries. In liquid-based formation, they are gaining popularity due to ease of application into irrigation systems where fertilizers are delivered with water, a process called fertigation, and for better shelf stability. The bioinoculants are kept in an appropriate climate, and there is no direct exposure to UV radiation to retain the viable microbes.

For example, tomato plants are inoculated with AMF Glomus intraradical and *Glomus mosseae*. This showed a 35% increase in the absorption of phosphorus and 30% increase in fruit yield as compared to non-inoculated controls. Moreover, the inoculated plants had better root biomass and were more resistant to a common plant disease called Fusarium wilt. This shows that the mycorrhizal fungi have both nutritional and protective roles (Ortas, 2012). The role of mycorrhizal associations in enhancing plant productivity, nutrient-use efficiency, and sustainable agricultural practices is summarized in Figure 10.7.

Mycorrhizal bioinoculants are more effective when integrated with certain practices, such as disturbing the soil less and using less phosphorus fertilizer and crop rotation. AMF colonization may be suppressed due to nutrient redundancy because of high doses of chemical fertilizers, especially phosphorus (Smith & Smith, 2011). Therefore, for optimizing the plant–fungi partnership, managing nutrients input is important.

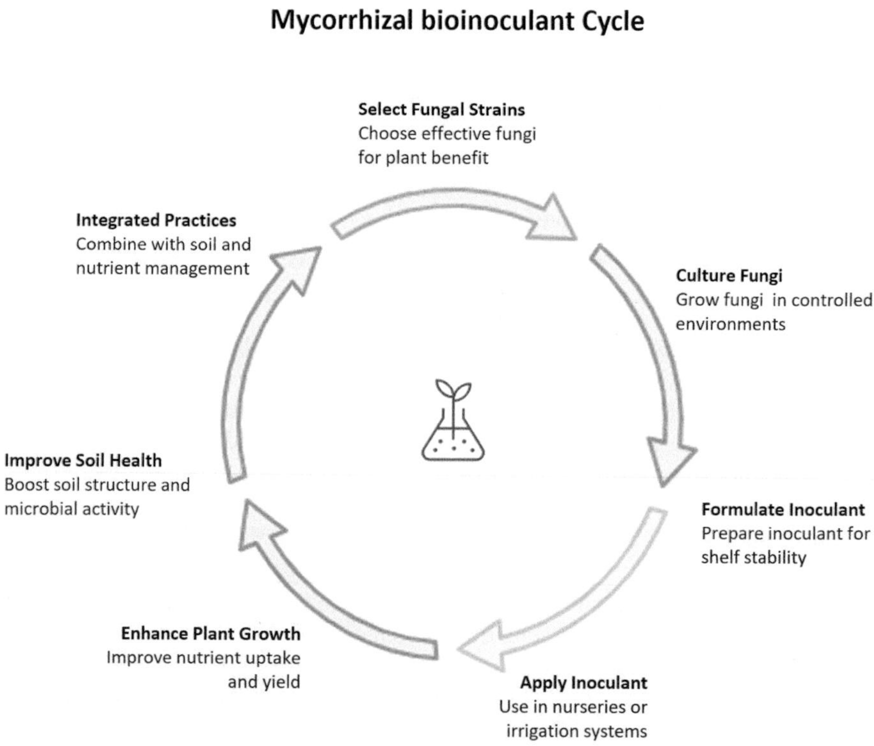

FIGURE 10.7 Mycorrhizal bioinoculant cycle.

The challenges in the development of bioinoculants include the obligate symbiotic nature of AMF, which also complicates their culturing in the laboratory without a host plant, and due to soil type, the existing microbes, and other environmental factors such as temperature and moisture, the effectiveness in fields varies. However, recent advancements in root organ culture (ROC) systems allow to grow AMF spores by using transformed carrot roots, and this advancement significantly improves the scalability and purity of inoculants (Declerck et al., 2005).

10.6.3 Challenges and Opportunities in Field Applications

Use of mycorrhizal inoculants in the farmlands provides good potential in regard to nutrient efficiency, in addition to improving the soil health, as well as the reduction in the application of chemical fertilizers. It suffers, however, significantly with laboratory efficacy to field passage. AMF in field applications are influenced by a wide range of factors such as biotic and abiotic variables, and site-specific and system-based approaches are truly effective.

10.6.3.1 Challenges in Field Applications

The challenges of using these mycorrhizal fungi inoculants have low consistency in field performance. Although the AMF inoculants exhibit good outcomes in a controlled laboratory setting, they can fail to show good outcomes in the field due to competition with the indigenous soil microorganisms, especially the natural strains of AMF since they might already be colonizing the root (Wagg et al., 2011). If the fungi are not more competitive or functionally high, they will fail to colonize the roots or contribute significantly to the plant growth. Additionally, AMF colonization and symbiosis can be suppressed by soil conditions such as pH, salinity, moisture, and nutrient levels, particularly the availability of high phosphorus (Smith & Smith, 2011). The inoculant production is complicated by the obligate biotrophic nature of AMF, for the spore generation requires live host plants or root organ cultures, that makes large-scale commercial production costly and technically demanding.

Also, some application techniques, such as seed coating and dipping of roots or soil, can alter the efficiency of colonization and viability of spores. Reducing field effectiveness and spore quality can be degraded by improper handling (Ijdo et al., 2011). In many nations, market growth and farmer adoption are further constrained by the lack of standardized regulatory frameworks for biofertilizers.

10.6.3.2 Opportunities and Advancements

Several opportunities are emerging in spite of these disadvantages. Using locally adapted AMF strains has shown improved performance by site-specific inoculant formulations and ensuring the compatibility with native soil conditions and crops (Berutti et al., 2016). The improvement in carrier materials such as hydrogel beads and alginate encapsulation has increased the shelf life, viability, and easy application of spores.

An important example is the use of AMF inoculants in wheat *Triticum aestivum* fields in India. In trials experiments conducted, wheat inoculated with a combination of *Rhizophagus fasciculatus* and *Glomus mosseae* increased grain yield by

20%–25% under rainfed conditions. AMF's role in improving resilience in low-input system is highlighted by the inoculated plant's improved phosphorus uptake and water-use efficiency (Shawei et al., 2022). Additional opportunities include farmers' education and policy support. Government subsidies and extension services can promote the use of bioinoculants in spite of the increased focus on sustainability and organic farming. There are also potential effects when AMF is integrated with other beneficial microbes such as phosphate-solubilizing bacteria in consortia-based biofertilizers.

10.6.4 Case Studies and Success Stories

Case Study 1: Mycorrhiza in Wheat Cultivation

In this study, AMF is a beneficial fungus, which helps plants to deal with drought conditions. Two types of wheat, WH 1025 and WH 1105, were grown in pots. Some seeds were treated with the fungus, and some were not. Plants are exposed to different water conditions. It was found that the wheat variety WH 1025 performed better in terms of growth and yield. They grew taller, produced more productive stems that bear grains, and have longer spikes with more grains. Under normal watering conditions, these benefits were seen, but they were particularly important when the plants were under drought conditions. This study concludes that the inoculation of mycorrhizal fungi in wheat helps the plants to grow better and produce more yield by making them more resilient to drought (Rani et al., 2018).

Case Study 2: Improvement of Nutritional Quality

In this study, commercially available tomato plants were grown without soil to check if mycorrhizae could improve the nutritional quality. Fungus *Funneliformis mosseae* is used and inoculated to some tomato plants, growing them in a mix of peat and vermiculite, with different amounts of phosphorus. Tomato fruits are analyzed to see how their genes are expressed and what metabolites are present. The finding showed that when the tomatoes are grown with the AM fungi and phosphorus at a low level, there is improvement in quality. Significant increase was observed in the important traits like free amino acids, lycopene (47%), and beta-carotene (29.6%), all without reducing the number of tomatoes produced. This study suggests that if these beneficial mycorrhizal fungi are added to tomato plants, then it is a great way to grow tastier and more nutritious tomatoes for people to eat (Allah et al., 2023).

Case Study 3: Mycorrhizae Enhancing the Economic Revenue of Water

If lack of water due to climate change poses a significant challenge for farming, it not only limits the irrigation but also prevents the ability to absorb nutrients. The application of mycorrhizal fungi will improve the capacity of crops to absorb nutrients and water, and it is a promising remedy. As a result, the production of crop becomes more sustainable and healthier. To find out how AMF affected the sweet corn, two field tests were conducted, which examined the yield, growth, uptake of nutrients, benefit-cost ratio, and irrigation water efficiency. The levels of phosphorus fertilizers (50%, 75%, and 100%) and water vary from what corn typically requires.

According to the findings, sweet corn grew significantly better and yielded more when given the AMF. Significantly, this improvement was observed even with a moderate amount of phosphorus fertilizer and less water. If plants are used with AMF and used water more effectively, farmers would create more profit from their crops and irrigation water. Example of inoculated plants showed that they were taller, cobs with increased number and weights of corn grains were larger and heavier, and the fresh and dry weights of shoots and roots soared high (Abdel-Fattah et al., 2023).

Case Study 4: Mycorrhizal Roots Improve the Growth of Tropical Tree Seedlings

In tropical mountain rainforests, most trees naturally live in partnership with AMF. A study in southern Ecuador found that out of 115 tree species, only three didn't have these fungal species. These tropical seedlings, when grown in a nursery, might need to associate with these AMF to have a better chance of surviving when they're planted out in the forest. There aren't clear methods for setting up plantations with native trees in Ecuador yet. In this study, the mycorrhizal roots are used from four specific tree species, *Inga acreana, Tabebuia chrysantha, Cedrela montana,* and *Heliocarpus americanus*, that had picked up these helpful fungi from forest soil in the nursery. Then, these "fungi-carrying" roots are used to inoculate seedlings of *Cedrela montana* and *Heliocarpus americanus* with their native AMF. Three methods are compared with inoculating a mix of mycorrhizal roots from all four species, inoculating only with roots from the same tree species, and using a moderate amount of fertilizer. After six months, the growth and fungal connections of the *Cedrela montana* and *Heliocarpus americanus* seedlings were checked. This study showed that inoculation of the seedlings with the mycorrhizal roots significantly improved their growth and helped them to develop more diverse fungal associations, even more so by just using moderate fertilizer. The moderate fertilizer didn't harm the mycorrhizal connection at all (Ugilles et al., 2009).

10.7 FUTURE RESEARCH DIRECTIONS

10.7.1 Genetic and Molecular Understandings of Mycorrhizal Symbiosis

The establishment of mycorrhizal fungi within the plant roots is genetically regulated, which involves a set of genes (Table 10.1) that coordinate recognition, signaling, and structural adaptation. These genes are often conserved across species but sometimes share similarities with other symbiotic relationships, such as rhizobium bacteria that form symbiotic relationships with plant roots. Before physical contact, plants communicate with AMF through molecular signaling, particularly by secreting strigolactones. The strigolactones are a group of plant hormones that regulate plant development and

TABLE 10.1
Important Genes Involved in Mycorrhizal Associations

Gene	Description	Origin	Function	Reference
RAM1	Required for Arbuscular Mycorrhization 1	Plant	GRAS-domain TF regulates arbuscule development, peri-arbuscular membrane formation, lipid transfer, and nutrient exchange	Ho-Plágaro et al. (2024)
RAM2	(Not expanded; GPAT family enzyme)	Plant	Required for hyphopodia formation in AM fungi, arbuscule formation, and cutin monomer synthesis	Gobbato et al. (2013)
CCaMK/ DMI3	Calcium/ Calmodulin-Dependent Protein Kinase	Plant	Decodes nuclear calcium spiking, activates CYCLOPS for downstream transcription	Ho-Plágaro and García-Garrido (2022)
CYCLOPS /IPD3	CYCLOPS (also known as IPD3)	Plant	Transcriptional activator, works with CCaMK and DELLA to induce RAM1 expression	Ho-Plágaro and García-Garrido (2022)
MiSSP7	Mycorrhiza-Induced Small Secreted Protein 7	Fungus	An effector protein, stabilizes Populus JAZ6, represses jasmonate signaling, remodels root transcriptome for symbiosis	Plett et al. (2014)
MiSSP8	Mycorrhiza-Induced Small Secreted Protein 8	Fungus	Small effector involved in hyphal aggregation and Hartig net or mantle formation in ectomycorrhiza	Pellegrin et al. (2019)

stimulate interactions with other organisms, like mycorrhizal fungi in soil. These hormones activate metabolic processes and trigger hyphal growth in specific mycorrhizal fungi. The AM fungi compatibly release Myc factors-lipo-chito-oligosaccharides (Myc-LCOs) and chito-oligosaccharides (Myc-COs) that trigger the plant's common symbiosis signaling pathway (CSSP). Subsequently, molecular receptors in plant cells like LysM receptor kinases and SYMRK (DMI2) perceive Myc signals, which are then transduced via nuclear calcium spiking, calcium/calmodulin-dependent protein kinase (CCaMK), CYCLOPS (a DNA-binding transcriptional activator), and GRAS-domain transcription factors (GRAS TFs), i.e., RAM1 and RAD1, that are essential for arbuscule formation (Figure 10.8).

The GRAS TFs, particularly RAM1 and RAD1, regulate other genes that are involved in lipid and carbohydrate metabolism and membrane function that increase the formation of arbuscules in the plant root tissues. RAM1 is involved in the transcriptional regulation of gene PT4 and lipid biosynthesis enzymes, and RAM2 accounts for fungal entry and arbuscule initiation (Gobbato et al., 2013). The DELLA proteins act as TFs (Table 10.2), modulate symbiosis by associating with CYCLOPS which initiate RAM1. However, in specific conditions, DELLA proteins interact with other regulators, e.g., NSP1 and MYB1, that control arbuscule degeneration. At the same time, phytohormones like ABA, gibberellins (GAs), and strigolactones contribute to

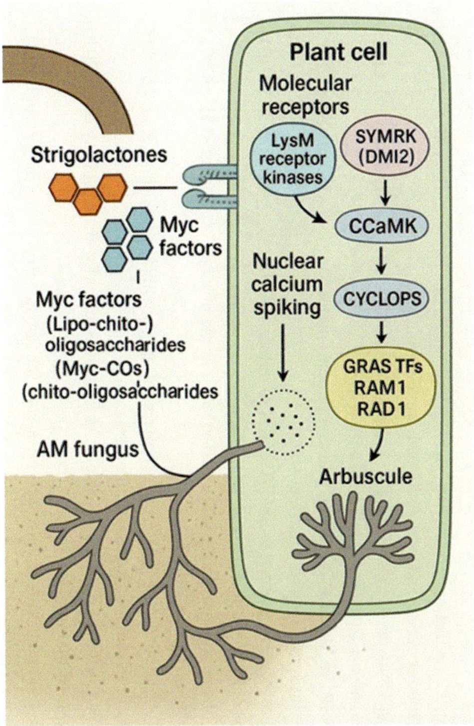

FIGURE 10.8 Strigolactones signaling cascade stimulates mycorrhizal symbiosis.

morphological changes for mycorrhizal symbiosis. The ABA at low concentrations promotes colonization of mycorrhizal fungi around and within the plant roots, but its high concentration inhibits this. Lower concentration of GAs promotes fungal symbiosis through stabilizing the DELLA proteins that interact with transcription factors CYCLOPS. However, high concentration of GA destabilizes DELLAs that suppress the symbiosis. Similarly, the strigolactones (SLs), which are carotenoid-derived phytohormones, play dual roles in plants. They act as rhizospheric signaling molecules, and their different concentrations regulate root and shoot branching. Under phosphorus (P) deficiency in soil, SLs synthesis increases in plant root cells, which secrete into the rhizosphere to attract mycorrhizal fungi.

Future research in the genetics and molecular biology of mycorrhizal symbiosis will depend on increasing resources of fungal genomics, delivering transcriptional and epigenetic regulation, and on the integration of molecular findings with changing ecological and environmental circumstances. By merging gene editing, chemical signaling studies, systems biology, and evolutionary genomics, the complexity of cellular and subcellular communications and dynamic symbiosis among the mycorrhizal fungi and plant can be well understood. Such knowledge is essential for translating molecular understandings into applied strategies that can enhance nutrient uptake, stress tolerance, and climate resilience in future agriculture. The molecular signalling network and important genes that are involved in establishing mycorrhizal symbiosis are summarized in Figure 10.8.

TABLE 10.2
Transcription Factors (TFs) and Regulatory Genes Involve in Mycorrhizal Associations

TF	Description	Origin	Function	Reference
RAD1	Required for Arbuscule Development 1	Plant	GRAS-domain TF regulating arbuscule branching, expression of nutrient transporter genes, and symbiotic membrane formation	Pimprikar et al. (2016)
NSP1	Nodulation Signaling Pathway 1	Plant	GRAS-domain TF involved in activating downstream symbiosis genes in response to Myc-LCO signals, interacts with RAM1	Delaux et al. (2013)
NSP2	Nodulation Signaling Pathway 2	Plant	Works with NSP1 in Myc-LCO-induced transcription, promotes strigolactone biosynthesis under nutrient stress	Delaux et al. (2013)
DELLA	Asp-Glu-Leu-Leu-Ala motif in protein	Plant	GA-signaling repressors, interact with CYCLOPS to induce RAM1, regulate arbuscule development and degeneration	Floss et al. (2013)
PT4	Phosphate Transporter 4	Plant	High-affinity phosphate transporter in the peri-arbuscular membrane, essential for P uptake from AM fungi	Javot et al. (2007)
STR / STR2	Stunted Arbuscule / Stunted Arbuscule 2	Plant	Half-size ATP-binding cassette (ABC) transporters forming a complex for lipid transfer to fungi during arbuscule formation	Jiang et al. (2017)
VAPYRIN	vpy mutant phenotype	Plant	Required for intracellular accommodation of AM fungi, associated with peri-arbuscular membrane formation	Murray et al. (2010)
RAM2	Required for Arbuscular Mycorrhization 2	Plant	GPAT enzyme producing cutin monomers, necessary for fungal entry and hyphopodia formation	Wang et al. (2012)
MTLb / MTLb2	Metal Tolerance Protein-like b	Fungus	Associated with nutrient transport and metal tolerance during symbiosis	Devers et al. (2011)
MiSSP7	Mycorrhiza-Induced Small Secreted Protein 7	Fungus	Suppresses host jasmonic acid responses and promote ectomycorrhizal colonization	Plett et al. (2014)

10.7.2 Biotechnological Innovations in Mycorrhizal Association

Biotechnological innovation can be used potentially for increasing the efficiency of mycorrhizal association and consequently for improving growth, nutrient uptake, and stress tolerance in plants (Kumar & Dubey, 2020). Genetic engineering can be used to develop plant varieties that may have improved genes for better recognition (e.g., *LysM receptor kinases*, *SYMRK*) and mycorrhizal arbuscule development (e.g., *RAM1*, *RAD1*). Similarly, genetic engineering has the capability to improve colonization ability, nutrient transport efficiency, and stress resilience of mycorrhizal fungi (Garg et al., 2025). Synthetic biology may be used to manipulate cellular signaling molecules such as strigolactone production in host plants that attract beneficial fungi more efficiently, irrespective of nutrient-rich conditions. Advanced genetic engineering may be applied to manipulate genes in mycorrhizal fungi related to Myc factors specialized for better host compatibility. Techniques of tissue culture and *in-vitro* root tip cultures can also be used for making artificial seeds inoculated with useful mycorrhizal fungi or mass production of biofertilizer. In addition, CRISPR/Cas-based genome editing may be applied for targeted gene modification of plant or mycorrhizal fungi that improve symbiotic efficiency of each without affecting their physiological processes. This technique may also be used for modulation in the synthesis and secretion of phytohormones in soil, which critically enhances the mycorrhizal efficiency (Hakim et al., 2021). Despite these developments, challenges remain in translating the laboratory findings to field applications, particularly because environmental variation not only exists across the different seasons but also varies from place to place and is further influenced by climate change and global warming. At the same time, different physiologies of plants may not respond equally to engineered fungi.

There is an immense potential for biotechnological innovations in mycorrhizal association, which can transform traditional agriculture into a climate-resilient, low-input, and sustainable system (Singh et al., 2018). The subsequent phase of research is likely to integrate the advanced techniques of molecular breeding, synthetic biology, bioinformatics, and microbial ecology to develop inclusive solutions for plant stress management through mycorrhizal fungi.

10.7.3 Impact of Climate Change on Mycorrhizal Associations

Although mycorrhizae often contribute to soil nutrient improvement, plant resilience to extreme environment, and carbon sequestration, the stability of these beneficial interactions depends mainly on biodiversity, geology and geographical properties, and specific climatic conditions of a location. With the accelerated rate of climate change and increasing environmental stressors in the past 50 years (Shivanna, 2022), mycorrhizal communities are shifting in their structure, function, and ecological roles significantly, which give substantial impacts on plant growth by modifying the efficiency of nutrient uptake and ability to resist drought (Bennett & Classen, 2020). Climate change is reshaping the plant–mycorrhizal relationships by altering temperature regimes, precipitation patterns, atmospheric CO_2 concentrations, and nutrient deposition across the world. As evidenced from the literature, increased temperature often changes the atmospheric concentration of carbon and alters the precipitation patterns (Yang et al., 2022). These deviations are now changing the carbon allocation to plants during photosynthesis and increasing the supply of carbonic acid in soil after frequent incidences of acid rain, and they are predominantly affecting mycorrhizal association with plants (Zhou et al., 2022). Recently, many climatic zones have been reported with warmer temperatures, which directly reduced the amount of AMF colonized within plant roots and affected nutrient absorption from the soil. Moreover, the impacts of stronger or weaker heat changes in soil moisture, nutrient levels, and the types of plant communities growing at a location also affected the mycorrhizal associations (Wilson et al., 2016). When AM fungi are less active, plants may also struggle to get adequate phosphorus and other nutrients from the soil in warmer conditions. Changes in nitrogen deposition in soil due to human activity associated with climatic factors affect mycorrhizal function. Elevated nitrogen deposition often reduces plant dependence on mycorrhizae and hence decreases AM colonization and other fungal diversity in rhizosphere, which causes declination of the essential nutrients and poor drought-protecting services. The geographical location, experiencing both warming and increased N inputs, is therefore at high risk of losing efficient plant–fungal partnerships (Maček et al., 2019). No doubt, mycorrhizal fungi also contribute to soil aggregation, hyphal-derived enrichment of organic carbon, and nutrient retention, and therefore, any declination in fungal biomass or diversity due to climate change may also reduce soil carbon persistence and increase vulnerability to nutrient losses during extreme weather events (Zhou et al., 2022).

To sustain the natural association between plant and mycorrhizae and benefit from this association under climate change, the strategies include conservation of native fungal diversity, promotion of using fungal inocula adapted for drought-prone lands, integration of mycorrhiza-friendly management (i.e., reduction of soil disturbance, and optimization of fertilizer uses), and breeding/selection of those crop varieties that have better responsiveness with mycorrhizal fungi. The long-term ecosystem-scale experiments are needed to develop and refine the predictive models that can link the climate trajectories to mycorrhizal community shifts and ecosystem services.

10.8 CONCLUSION

Mycorrhizal association is essential for optimum absorption of nutrients by plants, soil health, and ecological resilience and contributes to the pressing challenges in modern agriculture naturally. The mycorrhizae facilitate nutrient mobilization, particularly phosphorus, nitrogen, and essential micronutrients, and modify root architecture and rhizosphere chemistry. Besides these functional attributes, they enhance the capacity of plants for water absorption, improve drought tolerance, and modulate physiological processes like stomatal conductance. These excellent natural functions of mycorrhizae position them as critical allies for sustainable agri-productivity.

The understandings of mycorrhizal fungi and their ability to establish natural associations with plants clearly emphasize that they are strategic partners for agricultural crops as well as essential for environmental sustainability. The mycorrhizal fungi are the most important organisms contributing to enhance plant growth by improving nutrient use efficiency, reducing dependency on chemical fertilizers, and maintaining soil structure and biodiversity. In the context of hunger alleviation, the role of mycorrhizal fungi becomes vital in nature, predominantly for increasing crop yields without compromising environmental integrity.

With increasing demand for food every year, decreasing natural resources, and the consequences of climate change, the integration of mycorrhizal technology into mainstream agricultural practices becomes important for food security. There is no doubt that innovative approaches in genetics, molecular biology, and biotechnological innovations at the field-level, mycorrhizal associations can potentially bridge the gap between agricultural productivity and environmental sustainability. Moreover, scientific exploration of symbiotic relationships between mycorrhizal fungi and plants is a substantial step toward resilient crop growth, increasing crop yield, improving soil health, and contributing significantly to the global mission to eradicate hunger.

REFERENCES

Abdel-Fattah, D. A., M. Maza, B. A. Ali, and N. M. Awed. "Role of Mycorrhizae in Enhancing the Economic Revenue of Water and Phosphorus Use Efficiency in Sweet Corn (*Zea mays* L. var. *saccharata*) Plants." *Journal of the Saudi Society of Agricultural Sciences* 22, no. 3 (2023): 174–186. https://doi.org/10.1016/j.jssas.2022.10.001

Ahanger, M. A., P. Ahmad, M. Morad-Ali, A. Kumari, M. Ashraf, and M. Yasin. "Role of Arbuscular Mycorrhizal Fungi in Plant Growth Regulation: Implications in Abiotic Stress Tolerance." *Frontiers in Plant Science* 10 (2019): 1068. https://doi.org/10.3389/fpls.2019.01068

Allah, F. U., H. Ullah, M. Ishfaq, L. Gul, S. Kumar, and Z. Li. "Improvement of Nutritional Quality of Tomato Fruit with *Funneliformis mosseae* Inoculation under Greenhouse Conditions." *Horticulturae* 9, no. 4 (2023): 448. https://doi.org/10.3390/horticulturae9040448

Aroca, R., R. Porcel, and J. M. Ruiz-Lozano. "How Does Arbuscular Mycorrhizal Symbiosis Regulate Root Hydraulic Properties and Plant Hormone Status under Drought?" *New Phytologist* 173, no. 4 (2007): 670–681. https://doi.org/10.1111/j.1469-8137.2006.01959.x

Augé, R. M. "Water Relations, Drought and Vesicular-Arbuscular Mycorrhizal Symbiosis." *Mycorrhiza* 11, no. 1 (2001): 3–42. https://doi.org/10.1007/s005720100097

Barto, E. K., M. Hilker, J. M. Müller, J. Mohney, S. Wezel, and M. C. Rillig. "The Fungal Fast Lane: Common Mycorrhizal Networks Extend Bioactive Zones of Allelochemicals in Soils." *PLoS One* 7, no. 11 (2012): e49395. https://doi.org/10.1371/journal.pone.0049395

Barzana, G., A. Aroca, A. Paz, J. M. Chaumont, M. J. Martinez-Ballesta, M. C. Carvajal, and J. M. Ruiz-Lozano. "Arbuscular Mycorrhizal Symbiosis Increases Relative Apoplastic Water Flow in Roots of the Host Plant under Drought Stress." *Plant, Cell & Environment* 35, no. 2 (2012): 393–403. https://doi.org/10.1111/j.1365-3040.2011.02367.x

Basiru, S., and M. Hijri. "The Potential Applications of Commercial Arbuscular Mycorrhizal Fungal Inoculants and Their Ecological Consequences." *Microorganisms* 10, no. 10 (2022): 1897. https://doi.org/10.3390/microorganisms10101897

Bender, S. F., C. Wagg, and M. G. A. van der Heijden. "An Underground Revolution: Biodiversity and Soil Ecological Engineering for Agricultural Sustainability." *Trends in Ecology & Evolution* 31, no. 6 (2016): 440–452. https://doi.org/10.1016/j.tree.2016.03.008

Bennett, A. E., and A. T. Classen. "Climate Change Influences Mycorrhizal Fungal–Plant Interactions, But Conclusions Are Limited by Geographical Study Bias." *Ecology* 101, no. 4 (2020): e02978. https://doi.org/10.1002/ecy.2978

Berutti, J., E. Lumini, R. Balestrini, and V. Bianciotto. "Arbuscular Mycorrhizal Fungi as Natural Biofertilizers: Let's Benefit from Past Successes." *Frontiers in Microbiology* 6 (2016): 1559. https://doi.org/10.3389/fmicb.2015.01559

Bortolot, M., B. Buffoni, S. Mazzarino, G. Hoff, E. Martino, V. Fiorilli, and A. S. Di Fossalunga. "The Importance of Mycorrhizal Fungi and Their Associated Bacteria in Promoting Crops' Performance: An Applicative Perspective." *Horticulturae* 10, no. 12 (2024): 1326. https://doi.org/10.3390/horticulturae10121326

Brundrett, M. C. "Mycorrhizal Associations and Other Means of Nutrition of Vascular Plants." *Mycorrhiza* 19, no. 6 (2009): 357–81.

Bucher, M. "Functional Biology of Plant Phosphate Uptake at Root and Mycorrhiza Interfaces." *New Phytologist* 173, no. 1 (2006): 11–26. https://doi.org/10.1111/j.1469-8137.2006.01935.x

Cairney, J. W. G., and A. A. Meharg. "Ericoid Mycorrhiza: A Partnership that Exploits Harsh Edaphic Conditions." *European Journal of Soil Science* 54, no. 4 (2003): 735–740. https://doi.org/10.1046/j.1351-0754.2003.0555.x

Cameron, D. D., A. L. Neal, S. C. van Wees, and J. Ton. "Mycorrhiza-Induced Resistance: More than the Sum of Its Parts?" *Trends in Plant Science* 18, no. 10 (2013): 539–545. https://doi.org/10.1016/j.tplants.2013.06.004

Chauhan, P., and L. K. Attri. "Mycorrhizal Associations in Orchids: A Review." *Asian Journal of Biological and Life Sciences* 13, no. 2 (2024): 278–286. https://doi.org/10.5530/ajbls.2024.13.36

Cheng, H., G. Twagirayezu, F. Zhang, Y. Shi, C. Lu, Z. Wang, and D. Fan. "Arbuscular Mycorrhizal Fungi Adjusts Root Architecture to Promote Leaf Nitrogen Accumulation and Reduce Leaf Carbon-Nitrogen Ratio of Mulberry Seedlings." *Forests* 14, no. 12 (2023): 2448. https://doi.org/10.3390/f14122448

Chot, E., and M. S. Reddy. "Role of Ectomycorrhizal Symbiosis Behind the Host Plants Ameliorated Tolerance Against Heavy Metal Stress." *Frontiers in Microbiology* 13 (2022): 855473. https://doi.org/10.3389/fmicb.2022.855473

Courty, P.-E., M. Buée, A.-G. Diedhiou, P. Frey-Klett, F. Le Tacon, F. Rineau, M.-P. Turpault, et al. "The Role of Ectomycorrhizal Communities in Forest Ecosystem Processes: New Perspectives and Emerging Concepts." *Soil Biology and Biochemistry* 42, no. 5 (2010): 679–698. https://doi.org/10.1016/j.soilbio.2009.12.006

Dearnaley, J. D. W., F. Martos, and M.-A. Selosse. "12 Orchid Mycorrhizas: Molecular Ecology, Physiology, Evolution and Conservation Aspects." *Fungal Biology Reviews* 26, no. 2–3 (2012): 60–75. https://doi.org/ 10.1007/978-3-642-30826-0_12

Declerck, S., D. G. Strullu, and A. Fortin. *In vitro Culture of Mycorrhizas*. edited by A. Varma. Berlin: Springer Science & Business Media, 2005.

Delaux, P.-M., N. Sejalon-Delmas, G. Becard, and J.-M. Ané. "Evolution of the Plant–Microbe Symbiotic 'Toolkit'." *Trends in Plant Science* 18, no. 6 (2013): 298–304. https://doi.org/10.1016/j.tplants.2013.01.008

Devers, E. A., A. Branscheid, P. May, and F. Krajinski. "Stars and Symbiosis: MicroRNA- and MicroRNA*-Mediated Transcript Cleavage Involved in Arbuscular Mycorrhizal Symbiosis." *Plant Physiology* 156, no. 4 (2011): 1990–2010. https://doi.org/10.1104/pp.111.172627

Dhalaria, R., D. Kumar, H. Kumar, E. Nepomilova, K. Kuta, M. T. Islam, and R. Verma. "Arbuscular Mycorrhizal Fungi as Potential Agents in Ameliorating Heavy Metal Stress in Plants." *Agronomy* 10, no. 6 (2020): 815. https://doi.org/10.3390/agronomy10060815

Dodd, I. C., C. E. Ferreira, K. J. Dias de Cássia, and M. H. Peixoto. "ABA and Stomatal Conductance of Mycorrhizal Plants under Drought Stress: New Insights from Split-Root and Hormone Mutant Studies." *Plant, Cell & Environment* 33, no. 4 (2010): 617–625. https://doi.org/10.1111/j.1365-3040.2009.02091.x

Fernandez, C. W., and P. G. Kennedy. "Revisiting the 'Gadgil Effect': Do Interguild Interactions Control Carbon Cycling in Forest Soils?" *New Phytologist* 209, no. 4 (2016): 1382–1394. https://doi.org/10.1111/nph.13648

Floss, D. S., J. G. Levy, V. Lévesque-Tremblay, N. Pumplin, and M. J. Harrison. "DELLA Proteins Regulate Arbuscule Formation in Arbuscular Mycorrhizal Symbiosis." *Proceedings of the National Academy of Sciences* 110, no. 51 (2013): E5025–E5034. https://doi.org/10.1073/pnas.1308973110

Forgy, D. "Arbuscular Mycorrhizal Fungi Can Benefit Heavy Metal Tolerance and Phytoremediation." *Journal of Natural Resources and Life Sciences Education* 41, no. 1 (2012): 23–26. https://doi.org/10.4195/jnrlse.2012.0001se

Garg, S., M. Kim, and D. Romero-Suarez. "Current Advancements in Fungal Engineering Technologies for Sustainable Development Goals." *Trends in Microbiology* (2025). https://doi.org/10.1016/j.tim.2024.11.001

Gianinazzi, S., H. Schüepp, J. M. Barea, and K. Haselwandter, eds. *Mycorrhizal Technology in Agriculture*. Basel: Birkhäuser, 2002.

Gobbato, E., E. Wang, G. Higgins, S. A. Bano, C. Henry, M. Schultze, and G. E. Oldroyd. "RAM1 and RAM2 Function and Expression during Arbuscular Mycorrhizal Symbiosis and *Aphanomyces euteiches* Colonization." *Plant Signaling & Behavior* 8, no. 10 (2013): e26049. https://doi.org/10.4161/psb.26049

Govindarajulu, M., P. E. Pfeffer, H. Jin, J. Abubaker, D. D. Douds, J. W. Allen, H. Bücking, P. J. Lammers, and Y. Shachar-Hill. "Nitrogen Transfer in the Arbuscular Mycorrhizal Symbiosis." *Nature* 435, no. 7043 (2005): 819–823. https://doi.org/10.1038/nature03610

Grelet, G. A., E. Martino, I. A. Dickie, R. Tajuddin, and R. Artz. "Ecology of Ericoid Mycorrhizal Fungi." In *Molecular Mycorrhizal Symbiosis*, edited by F. Martin. Hoboken, NJ: Wiley, 2016. https://doi.org/10.1002/9781118951446.ch22

Gutjahr, C., and M. Parniske. "Cell and Developmental Biology of Arbuscular Mycorrhiza Symbiosis." *Annual Review of Cell and Developmental Biology* 29 (2013): 593–617. https://doi.org/10.1146/annurev-cellbio-101512-122413

Hakim, S., T. Naqqash, M. S. Nawaz, Iffat Laraib, M. J. Siddique, R. Zia, M. S. Mirza, and A. Imran. "Rhizosphere Engineering with Plant Growth-Promoting Microorganisms for Agriculture and Ecological Sustainability." *Frontiers in Sustainable Food Systems* 5 (2021): 617157. https://doi.org/10.3389/fsufs.2021.617157

Ho-Plágaro, T., and J. M. García-Garrido. "Molecular Regulation of Arbuscular Mycorrhizal Symbiosis." *International Journal of Molecular Sciences* 23, no. 11 (2022): 5960. https://doi.org/10.3390/ijms23115960

Ho-Plágaro, T., M. I. Tamayo-Navarrete, S. Ćavar Zeljković, P. Tarkowski, and J. M. García-Garrido. "A Dual Regulatory Role for the Arbuscular Mycorrhizal Master Regulator RAM1 in Tomato." *Journal of Experimental Botany* 75, no. 16 (2024): 5021–5036. https://doi.org/10.1093/jxb/erae210

Ijdo, M., S. Cranenbrouck, and S. Declereck. "Methods for Large-Scale Production of AM Fungi: Past, Present, and Future." *Mycorrhiza* 21 (2011): 1–16. https://doi.org/ 10.1007/s00572-010-0337-z

Jansa, J., P. Bukovská, and M. Gryndler. "Mycorrhizal Hyphae as Ecological Niche for Highly Specialized Hypersymbionts – or Just Soil Free-Riders?" *Frontiers in Plant Science* 4 (2013): 134. https://doi.org/10.3389/fpls.2013.00134

Javot, H., R. V. Penmetsa, N. Terzaghi, D. R. Cook, and M. J. Harrison. "A *Medicago truncatula* Phosphate Transporter Indispensable for the Arbuscular Mycorrhizal Symbiosis." *Proceedings of the National Academy of Sciences* 104, no. 5 (2007): 1720–1725. https://doi.org/10.1073/pnas.0608136104

Jiang, Y., W. Wang, Q. Xie, N. Liu, L. Liu, D. Wang, X. Zhang, et al. "Plants Transfer Lipids to Sustain Colonization by Mutualistic Mycorrhizal and Parasitic Fungi." *Science* 356, no. 6343 (2017): 1172–1175. https://doi.org/10.1126/science.aam9970

Keddie, M., and A. L. Orchall. "Glomalean Fungi from the Ordovician." *Science* 289, no. 5486 (2000): 1920–1921. https://doi.org/10.1126/science.289.5486.1920

Kumar, A., S. K. Singh, C. Kant, H. Verma, D. Kumar, P. P. Singh, A. Modi, S. Droby, M. S. Kesawat, H. Alavilli, and S. K. Bhatia. "Microbial Biosurfactant: A New Frontier for Sustainable Agriculture and Pharmaceutical Industries." *Antioxidants* 10, no. 9 (2021): 1472. https://doi.org/10.3390/antiox10091472

Kumar, P., and K. K. Dubey. "Biotechnological Interventions for Arbuscular Mycorrhiza Fungi (AMF) based Biofertilizer: Technological Perspectives." In *Plant-Microbe Interactions in Agro-Ecological Perspectives*, edited by D. K. Chaudhary, V. S. Mishra, and A. V. Singh, 161–191. Singapore: Springer, 2020. https://doi.org/10.1007/978-981-15-6895-4_9

Lehmann, A., S. D. Veresoglou, E. F. Leifheit, and M. C. Rillig. "Arbuscular Mycorrhizal Influence on Zinc Nutrition in Crop Plants—A Meta-Analysis." *Soil Biology and Biochemistry* 69 (2014): 123–131. https://doi.org/10.1016/j.soilbio.2013.11.001

Leifheit, E. F., E. A. Veresoglou, C. Lehmann, M. Morris, and M. C. Rillig. "Multiple Factors Influence the Role of Arbuscular Mycorrhizal Fungi in Soil Aggregation—A Meta-Analysis." *Plant and Soil* 374, no. 1–2 (2014): 523–537. https://doi.org/10.1007/s11104-013-1899-2

Lekberg, Y., and R. J. T. Koide. "Is Plant Performance Limited by Abundance of Arbuscular Mycorrhizal Fungi? A Meta-Analysis of Studies Published between 1988 and 2003." *New Phytologist* 168, no. 1 (2005): 189–204. https://doi.org/10.1111/j.1469-8137.2005.01490.x

López-Bucio, J., A. Cruz-Ramírez, and L. Herrero-Estrella. "The Role of Nutrient Availability in Regulating Root Architecture." *Current Opinion in Plant Biology* 6, no. 3 (2003): 280–287. https://doi.org/10.1016/s1369-5266(03)00035-9

Ludwig-Müller, J. "Auxin Conjugates: Their Role for Plant Development and in the Evolution of Land Plants." *Journal of Experimental Botany* 61, no. 5 (2010): 2259–2270. https://doi.org/10.1093/jxb/erq122

Maček, I., D. R. Clark, N. Šibanc, G. Moser, D. Vodnik, C. Müller, and A. J. Dumbrell. "Impacts of Long-Term Elevated Atmospheric CO_2 Concentrations on Communities of Arbuscular Mycorrhizal Fungi." *Molecular Ecology* 28, no. 14 (2019): 3445–3458. https://doi.org/10.1111/mec.15160

Marschner, H., and B. Dell. "Nutrient Uptake in Mycorrhizal Symbiosis." *Plant and Soil* 159, no. 1 (1994): 89–102. https://doi.org/10.1007/BF00000098

Martin, F., A. Kohler, C. Murat, C. Veneault-Fourrey, and D. S. Hibbett. "Unearthing the Roots of Ectomycorrhizal Symbioses." *Nature Reviews Microbiology* 14, no. 12 (2016): 760–773. https://doi.org/10.1038/nrmicro.2016.149

Mohan, J. E., C. C. Cowden, P. Baas, A. Dawadi, P. T. Frankson, K. Helmick, E. Hughes, S. Khan, A. K. Lang, M. B. Machmüller, M. K. Taylor, and C. A. Witt. "Mycorrhizal Fungi Mediation of Terrestrial Ecosystem Responses to Global Change: Mini-Review." *Fungal Ecology* 10 (2014): 3–19. https://doi.org/10.1016/j.funeco.2014.01.005

Murray, J. D., R. R. D. Muni, I. Torres-Jerez, Y. Tang, S. Allen, M. Andriankaja, G. Li, et al. "Vapyrin, a Gene Essential for Intracellular Progression of Arbuscular Mycorrhizal Symbiosis, Is Also Essential for Infection by Rhizobia in the Nodule Symbiosis of *Medicago truncatula*." *The Plant Journal* 65, no. 2 (2010): 244–252. https://doi.org/10.1111/j.1365-313x.2010.04415.x

Ortas, I. "The Effect of Mycorrhizal Fungal Inoculation on Plant Yield, Nutrient Uptake and Inoculation Effectiveness under Long-Term Field Conditions." *Field Crops Research* 125 (2012): 35–48. https://doi.org/10.1016/j.fcr.2011.08.005

Pellegrin, C., Y. Daguerre, J. Ruytinx, F. Guinet, M. Kemppainen, N. F. D. Frey, V. Puech-Pagès, et al. "*Laccaria bicolor* MiSSP8 is a Small-Secreted Protein Decisive for the Establishment of the Ectomycorrhizal Symbiosis." *Environmental Microbiology* 21, no. 10 (2019): 3765–3779. https://doi.org/10.1111/1462-2920.14727

Phillips, R. P., E. Brzostek, and M. G. Midgley. "The Mycorrhizal-Associated Nutrient Economy: A New Framework for Predicting Carbon–Nutrient Couplings in Temperate Forests." *New Phytologist* 199, no. 1 (2013): 41–51. https://doi.org/10.1111/nph.12221

Pimprikar, P., S. Carbonnel, M. Paries, K. Katzer, V. Klingl, M. J. Bohmer, L. Karl, D. S. Floss, M. J. Harrison, M. Parniske, and C. Gutjahr. "A CCaMK–CYCLOPS–DELLA Complex Activates Transcription of RAM1 to Regulate Arbuscule Branching." *Current Biology* 26, no. 8 (2016): 1126–1138.https://doi.org/10.1016/j.cub.2016.01.069

Plett, J. M., Y. Daguerre, S. Wittulsky, A. Vayssières, A. Deveau, S. J. Melton, A. Kohler, J. L. Morrell-Falvey, A. Brun, C. Veneault-Fourrey, and F. Martin. "Effector MiSSP7 of the Mutualistic Fungus *Laccaria bicolor* Stabilizes the *Populus* JAZ6 Protein and Represses Jasmonic Acid (JA) Responsive Genes." *Proceedings of the National Academy of Sciences* 111, no. 22 (2014): 8299–8304. https://doi.org/10.1073/pnas.1322671111

Porcel, R., J. M. Aroca, and J. M. Ruiz-Lozano. "Salinity Stress Alleviation Using Arbuscular Mycorrhizal Fungi: A Review." *Agronomy for Sustainable Development* 26, no. 3 (2006): 227–233. https://doi.org/10.1051/agro:2006021

Ramírez-Flores, M. R., E. Bello-Bello, R. Belán-Álvarez, R. J. H. Sawers, and V. Olalde-Portugal. "Inoculation with the Mycorrhizal Fungus *Rhizophagus irregularis* Modulates the Relationship between Root Growth and Nutrient Content in Maize (*Zea mays* ssp. *mays* L.)." *Plant Direct* 3, no. 12 (2019): e00192. https://doi.org/10.1002/pld3.192

Rani, B., S. Madan, Pooja, K. D. Sharma, N. Kumar, and A. Kumar. "Mitigating the Effect of Drought Stress on Yield in Wheat (*Triticum aestivum*) Using Arbuscular Mycorrhiza Fungi (*Glomus mosseae*)." *Indian Journal of Agricultural Sciences* 88, no. 12 (2018): 1903–1908. https://doi.org/10.56093/ijas.v88i12.85444

Rasmussen, H. N. *Terrestrial Orchids: From Seed to Mycotrophic Plant*. Cambridge: Cambridge University Press, 1995.

Read, D. J., and J. Perez-Moreno. "Mycorrhizas and Nutrient Cycling in Ecosystems – A Journey towards Relevance?" *New Phytologist* 157, no. 3 (2003): 475–492. https://doi.org/10.1046/j.1469-8137.2003.00704.x

Richardson, A. E., J.-M. Barea, A. M. McNeill, and C. Prigent-Combaret. "Acquisition of Phosphorus and Nitrogen in the Rhizosphere and Plant Growth Promotion by Microorganisms." *Plant and Soil* 321, no. 1–2 (2009): 305–339. https://doi.org/10.1007/s11104-009-9895-2

Rillig, M. C. "Arbuscular Mycorrhizae and Terrestrial Ecosystem Processes." *Ecology Letters* 7, no. 8 (2004a): 740–754. https://doi.org/10.1111/j.1461-0248.2004.00620.x

Rillig, M. C. "Arbuscular Mycorrhizae, Glomalin, and Soil Aggregation." *Canadian Journal of Soil Science* 84, no. 4 (2004b): 355–63.

Rillig, M. C., C. A. Aguilar-Trigueros, J. Bergmann, E. Verbruggen, S. D. Veresoglou, and A. Lehmann. "Plant Root and Mycorrhizal Fungal Traits for Understanding Soil Aggregation." *New Phytologist* 223, no. 4 (2019a): 1757–1769. https://doi.org/10.1111/nph.15767

Rillig, M. C., S. F. Wright, K. A. Nichols, W. F. Schmidt, and M. S. Torn. "Large Contribution of Arbuscular Mycorrhizal Fungi to Soil Carbon Pools in Tropical Forest Soils." *Plant and Soil* 233, no. 2 (2001): 167–177. https://doi.org/10.1023/A:1010364221169

Rillig, M. C., S. D. W. Smith, N. P. Wurzburger, M. de Lourdes A. Silva, N. E. Bruns, and J. J. Cavender-Bares. "Mycorrhizae as a Biotic Filter for Plant Community Assembly under Climate Change." *New Phytologist* 222, no. 3 (2019b): 1227–1235. https://doi.org/10.1111/nph.15673

Roy, S., and L. Choudhury. "The Symbiotic Relationship between Fungi and Plants." In *Sustainable Utilization of Fungi in Agriculture and Industry*, edited by M. Rai, 52–74. Sharjah: Bentham Science Publishers, 2022. https://doi.org/10.2174/978981504034012 2020009

Ruiz-Lozano, J. M., R. Porcel, and R. Aroca. "Regulation by Arbuscular Mycorrhizae of the Integrated Physiological and Molecular Responses to Drought Stress in Plants." In *Mycorrhizal Mediation of Soil: Fertility, Structure, and Carbon Storage*, edited by N. S. Upadhyay, 239–256. Cambridge, MA: Academic Press, 2016.

Ruiz-Lozano, J. M., R. Porcel, J. M. Aroca, and M. A. Zamarreño. "Arbuscular Mycorrhizal Symbiosis Induces Strigolactone Biosynthesis under Drought and Improves Drought Tolerance in Lettuce and Tomato." *Plant, Cell & Environment* 35, no. 9 (2012): 1723–1738. https://doi.org/10.1111/j.1365-3040.2012.02522.x

Sathiyadash, K., T. Muthukumar, V. Karthikeyan, and K. Rajendran. "Orchid Mycorrhizal Fungi: Structure, Function, and Diversity." In *Orchid Biology: Recent Trends & Challenges*, edited by T. Muthukumar, V. Karthikeyan, and K. Rajendran, 239–280. Singapore: Springer, 2020. https://doi.org/10.1007/978-981-32-9456-1_13

Schaefer, D. A., G. Heing, P. E. Mortimer, and J. Xu. "Arbuscular Mycorrhiza and Sustainable Agriculture." *Circular Agricultural Systems* 1, no. 6 (2021). https://doi.org/10.48130/CAS-2021-0006

Sharma, R. "Ectomycorrhizal Mushrooms: Their Diversity, Ecology and Practical Applications." In *Mycorrhiza – Function, Diversity, State of the Art*, edited by A. Varma, R. Prasad, and N. Tuteja, 99–131. Cham: Springer, 2017. https://doi.org/10.1007/978-3-319-53064-2_7

Shawei, Y., S. Zhaoyong, C. Xinping, G. Jiao, and W. Xinyan. "Arbuscular Mycorrhizal Fungi Increase Crop Yields by Improving Biomass under Rainfed Condition: A Meta-Analysis." *PeerJ* 10 (2022): e12861. https://doi.org/10.7717/peerj.12861

Shivanna, K. R. "Climate Change and Its Impact on Biodiversity and Human Welfare." *Journal of the Indian Institute of Science* 88, no. 2 (2022): 160–171. https://doi.org/10.1007/s43538-022-00073-6

Simard, S. W., D. A. Perry, M. D. Jones, D. D. Myrold, D. M. Durall, and R. Molina. "Carbon Transfer between Ectomycorrhizal Tree Species." *Nature* 388, no. 6642 (1997): 579–582.

Simard, S. W., D. A. Perry, M. D. Jones, D. D. Myrold, D. M. Durall, and R. Molina. "Mycorrhizal Networks: Mechanisms, Ecology and Modelling." *Fungal Biology Reviews* 26, no. 1 (2012): 39–60. https://doi.org/10.1016/j.fbr.2012.01.001

Singh, A. K., P. P. Singh, V. Tripathi, H. Verma, S. K. Singh, A. K. Srivastava, and A. Kumar. Distribution of Cyanobacteria and Their Interactions with Pesticides in Paddy Field: A Comprehensive Review. *Journal of Environmental Management* 224 (2018): 361–375. https://doi.org/10.1016/j.jenvman.2018.07.039

Singh, M., K. A. Qureshi, M. Jaremko, M. Rajput, S. K. Singh, Kaushalendra, Pandey, K. D., Ferreira, L.F.R. and Kumar, A. Bioprospects of Endophytic Bacteria in Plant Growth Promotion and Ag-Nanoparticle Biosynthesis. *Plants* 11, no. 14 (2022): 1787. https://doi.org/10.3390/plants11141787

Smith, S. E., and D. J. Read. *Mycorrhizal Symbiosis*. 3rd ed. London: Academic Press, 2010.

Smith, S. E., and F. A. Smith. "Roles of Arbuscular Mycorrhizas in Plant Nutrition and Growth: New Paradigms from Cellular to Ecosystem Scales." *Annual Review of Plant Biology* 62 (2011): 227–250. https://doi.org/10.1146/annurev-arplant-042110-103846

Tarafdar, J. C., and H. Marschner. "Phosphatase Activity in the Rhizosphere and Hyphosphere of VA Mycorrhizal Wheat Supplied with Inorganic and Organic Phosphorus." *Soil Biology and Biochemistry* 26, no. 3 (1994): 387–395. https://doi.org/10.1016/0038-0717(94)90288-7

Tedersoo, L., M. Bahram, and M. Zobel. "How Mycorrhizal Associations Drive Plant Population and Community Biology." *Science* 367, no. 6480 (2020): eaba1223. https://doi.org/10.1126/science.aba1223

Tedersoo, L., T. W. May, and M. E. Smith. "Ectomycorrhizal Lifestyle in Fungi: Global Diversity, Distribution, and Evolution of Phylogenetic Lineages." *Mycorrhiza* 20 (2010): 217–263. https://pubmed.ncbi.nlm.nih.gov/20191371/

Treseder, K. K., and M. F. Allen. "Mycorrhizal Fungi Have a Potential Role in Soil Carbon Storage under Elevated CO_2 and Nitrogen Deposition." *New Phytologist* 147, no. 1 (2000): 189–200. https://doi.org/10.1046/j.1469-8137.2000.00692.x

Ugilles, N., P. Loján, N. Aguirre, H. Blaschke, S. Günter, B. Stimm, and I. Kottke. "Application of Mycorrhizal Roots Improves Growth of Tropical Seedlings in the Nursery: A Step towards Reforestation with Native Species in the Andes of Ecuador." *New Forests* 38, no. 3 (2009): 229–239. https://doi.org/10.1007/s11056-009-9143-x

Van der Heijden, M. G. A., and T. R. Horton. "Socialism in Soil? The Importance of Mycorrhizal Fungal Networks for Facilitation in Natural Ecosystems." *Journal of Ecology* 97, no. 6 (2009): 1139–1150. https://doi.org/10.1111/j.1365-2745.2009.01570.x

Van der Heijden, M. G. A., J. N. Klironomos, M. Ursic, P. Moutoglis, R. Streitwolf-Engel, T. Boller, A. Wiemken, and I. R. Sanders. "Mycorrhizal Fungal Diversity Determines Plant Biodiversity, Ecosystem Variability and Productivity." *Nature* 396, no. 6706 (1998): 69–72. https://doi.org/10.1038/23932

Van der Heijden, M. G. A., F. M. Martin, M.-A. Selosse, and I. R. Sanders. "Mycorrhizal Ecology and Evolution: The Past, the Present, and the Future." *New Phytologist* 205, no. 4 (2015): 1406–1423. https://doi.org/10.1111/nph.13288

Wagg, C., J. Jansa, M. Stadler, B. Schmid, and M. G. A. van der Heijden. "Mycorrhizal Fungal Identity and Diversity Relaxes Plant-Plant Competition." *Ecology* 92, no. 6 (2011): 1303–1313. https://doi.org/10.1890/10-1915.1

Wang, E., S. Schornack, J. F. Marsh, E. Gobbato, B. Schwessinger, P. Eastmond, M. Schultze, S. Kamoun, and G. E. D. Oldroyd. "A Common Signaling Process that Promotes Mycorrhizal and Oomycete Colonization of Plants." *Current Biology* 22, no. 23 (2012): 2242–2246. https://doi.org/10.1016/j.cub.2012.09.043

Wei, X., W. Zhang, F. Zulfiqar, C. Zhang, and J. Chen. "Ericoid Mycorrhizal Fungi as Biostimulants for Improving Propagation and Production of Ericaceous Plants." *Frontiers in Plant Science* 13 (2022): 1027390. https://doi.org/10.3389/fpls.2022.1027390

Wilson, G. W. T., C. W. Rice, M. C. Rillig, A. Springer, and D. C. Hartnett. "Soil Aggregation and Carbon Sequestration Are Tightly Correlated with the Abundance of Arbuscular Mycorrhizal Fungi in Prairies." *Ecology Letters* 12, no. 5 (2009): 452–461. https://doi.org/10.1111/j.1461-0248.2009.01303.x

Wilson, H., B. R. Johnson, B. Bohannan, L. Pfeifer-Meister, R. Mueller, and S. D. Bridgham. "Experimental Warming Decreases Arbuscular Mycorrhizal Fungal Colonization in Prairie Plants along a Mediterranean Climate Gradient." *PeerJ* 4 (2016): e2083. https://doi.org/10.7717/peerj.2083

Wright, S. F., and A. Upadhyaya. "A Survey of Soils for Aggregate Stability and Glomalin, a Glycoprotein Produced by Hyphae of Arbuscular Mycorrhizal Fungi." *Plant and Soil* 198, no. 1 (1998): 97–107. https://doi.org/10.1023/A:1004347701584

Wu, Q.-S., and Y.-N. Xia. "Arbuscular Mycorrhizal Fungi and Drought Stress Tolerance of Plants." In *Arbuscular Mycorrhizas and Stress Tolerance of Plants*, edited by Q.-S. Wu, 55–76. New York: Nova Science Publishers, 2009.

Yang, J., X. Jia, H. Ma, X. Chen, J. Liu, Z. Shangguan, and W. Yan. "Effects of Warming and Precipitation Changes on Soil GHG Fluxes: A Meta-Analysis." *The Science of the Total Environment* 827 (2022): 154351. https://doi.org/10.1016/j.scitotenv.2022.154351

Zhang, F., L.-Y. Zou, M.-Y. Wang, and W.-F. Tan. "Arbuscular Mycorrhizal Fungi Alleviate Drought Stress in Plants by Regulating the Synthesis of Plant Hormones and the Expression of Drought-Related Genes." *Applied Soil Ecology* 124 (2018): 305–313. https://doi.org/10.1016/j.apsoil.2017.11.015

Zhou, L., X. Zhou, Y. He, Y. Fu, Z. Du, M. Lu, X. Sun, et al. "Global Systematic Review with Meta-Analysis Shows That Warming Effects on Terrestrial Plant Biomass Allocation Are Influenced by Precipitation and Mycorrhizal Association." *Nature Communications* 13, no. 1 (2022): 5074. https://doi.org/10.1038/s41467-022-32671-9

11 Plant Microbiome for Sustainable Agriculture

Charu Saklani, Aabir Chakraborty, and Anushka Pragati Lakra

11.1 INTRODUCTION

The projected population increase to over 9 billion in 2050 represents an unprecedented challenge to the world's agriculture: to meet the rising demand for food production without expanding arable land area or causing more environmental degradation (FAO, 2017). The industrial models of agriculture generated a significant increase in food production globally, but they have also triggered unintended negative consequences, including soil erosion, loss of agro-biodiversity, water scarcity, and environmental pollution (Tilman et al., 2002). All these challenges are the strong reasons for the growing trends toward embracing sustainable intensification of agriculture, an approach that strives to reach maximum crop productivity in the already employed agricultural areas with minimal environmental degradation (Pretty & Bharucha, 2014). In comparison to conventional farming relying on the excessive use of chemicals, sustainable intensification emphasizes on the effectiveness of resource use, ecological inter-balance, and long-term system stability (shown in Figure 11.1).

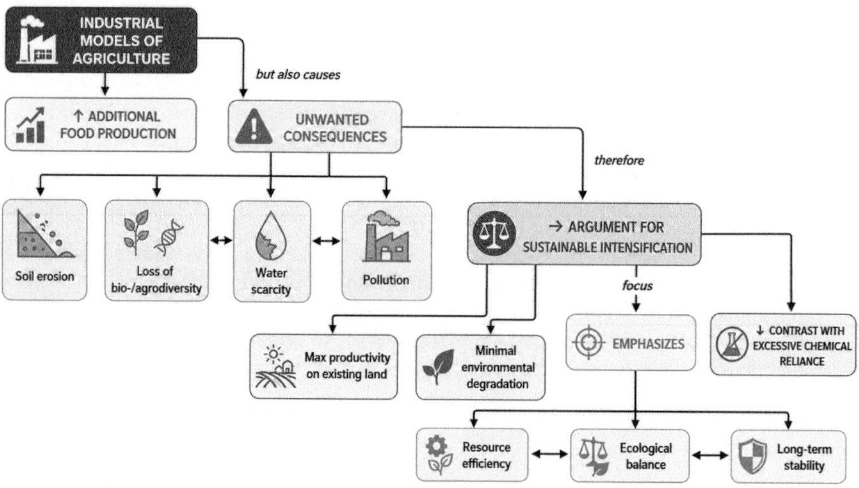

FIGURE 11.1 A cause–effect overview of industrial agriculture to sustainable intensification.

DOI: 10.1201/9781003664451-11

The plant microbiome, which is a complex assemblage of bacteria, fungi, archaea, and microorganisms that are found within the rhizosphere, endosphere, and phyllosphere, is extremely valuable for enhancing sustainable agricultural intensification (Turner et al., 2013; Kumar et al., 2021). Microbial assemblages promote nutrient uptake, and tolerance/resistance to abiotic stresses, and provide hormonal stimulation of plant growth and resistance to pathogen and insect/pest attacks (Vessey, 2003, Berendsen et al., 2012). By 2050, the global human population is projected to exceed 9 billion, and in that context, there will be an increasing pressure on the global food system, due to growing population, urbanization, and climate change, at the same time, agricultural intensification, reliant on chemical fertilizers, irrigation, and machinery, is approaching ecological limits, resulting in soil quality degradation, ecosystem biodiversity loss, and freshwater ecosystem pollution (Godfray et al., 2010; Foley et al., 2011; Pingali, 2012; FAO, 2017). Sustainable intensification seeks to increase productivity on already cleared/used farmland without damaging existing natural resources and ecological systems and based on management practices such as crop rotation, conservation tillage, and microbial biofertilizers (Garnett, 2013; Pretty & Bharucha, 2014).

Soil organic matter is significant within the sustainability framework in relation to both enrichment of soil health and cycling of nutrients, while reducing reliance on chemical inputs. In the rhizosphere, microbial communities that fix nitrogen serve as plant growth–promoting rhizobacteria (PGPR), solubilizing phosphates and stimulating ferric ion uptake using siderophores. Within plants, endophytic microbes enhance systemic resistance and tolerance to stresses, while phyllosphere microbes can minimize the risk of leaf pathogen infections (Lindow & Brandl, 2003; Compant et al., 2010; Mendes et al., 2013). These various microorganisms can boost production resilience through competition, antibiosis, and induced systemic resistance (ISR), while also being an alternative to chemical inputs (Berendsen et al., 2012; Compant et al., 2019). Establishing systems using these microorganisms is fundamentally important for generating low-input and climate-resilient production crops that not only yield, but also improve soil and ecosystem function (Singh et al., 2022; Singh et al., 2018).

This chapter summarizes biological function and applications of plant–microbe interactions, recent advances in microbiome engineering and constructing artificial microbial consortia, and integration of microbial interventions into agriculture. It briefly considers the environmental variability, host–microbe interactions, and environmental regulation that control field-scale implementation. By collating the most recent supporting evidence and examples of practice, this chapter shows that leveraging the plant microbiome can change the face of today's agriculture to become more sustainable, resilient, and high-performing.

11.2 THE PLANT MICROBIOME: AN OVERVIEW

The plant microbiome includes all the organisms associated with plants, including bacteria, fungi, archaea, viruses, and protozoa, as well as their genes and functions (Berg et al., 2014). These microbes are not simply hitchhikers, but contributors to, and members of, processes whereby plants grow better, have greater resilience to some stresses, and help promote stability of ecosystems.

11.2.1 The Plant Microbiome

The plant microbiome has spatial dimensions and its members occupy discrete ecological niches. The rhizosphere is a narrow section of soil surrounding plant roots and is distinguished from surrounding soil by being enriched by root exudates which promote and maintain dense populations of microbes (shown in Figure 11.2). This zone of influence is a dynamic site of some nutrient transformations important to plant development, such as nitrogen fixation and phosphate solubilization, and for the suppression of soil-borne plant pathogens (Philippot et al., 2013).

The phyllosphere, which encompasses all plant parts above ground, such as leaves, stems, and flowers, represents a distinct microbial habitat exposed to fluctuating environmental conditions such as temperature extremes, ultraviolet radiation, limited resources, and drying (Vorholt, 2012). Despite these stresses, diverse microbial phyla colonize the phyllosphere and form relationships that alter biological and ecological outcomes. Phyllosphere microbes are vital in foliar pathogen suppression, neutralizing biological and abiotic factors, nutrient deposition/uptake from the atmosphere, and improving the host plant's capacity to acclimate under environmental changes (Vorholt, 2012). Phyllospheres, particularly bacteria (e.g., *Pseudomonas*, *Methylobacterium*, and *Sphingomonas*), can produce antimicrobial bioactive metabolites and plant growth–derived phytohormones that provide resistance to diseases and promote growth. For example, the plant's endosphere refers to root, stem, and leaf internal tissues that comprise endophytic microorganisms that thrive in intercellular spaces, which means that they do not cause damage to the host (Hardoim et al., 2015). Endophytes improve the plant fitness by promoting better nutrient uptake, producing some plant growth hormones (e.g., auxins, gibberellins, cytokinins) and enhancing abiotic stress tolerance (e.g., drought, salinity, and temperature variability)

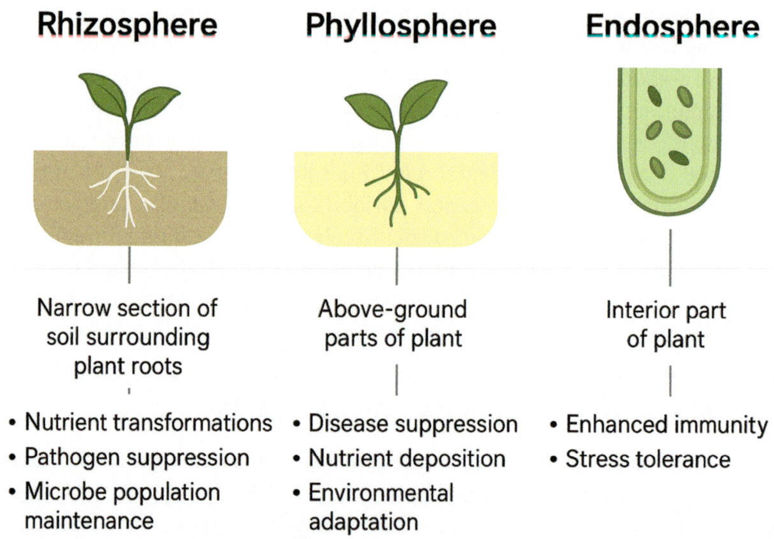

FIGURE 11.2 Types of plant microbiome and their roles.

(Compant et al., 2010). Additionally, many endophyte species promote ISR, hence preparing plants to protect from pathogens (Hardoim et al., 2015; Verma et al., 2021). Collectively, the phyllosphere and the endosphere are key units of the plant holobiont, which provide plant health, production, as well as sustainability under varied environmental conditions.

11.2.2 Prominent Microbial Players

Most studies of the microbiome have focused on bacteria, which are the dominant phyla of the microbiome such as Proteobacteria, Actinobacteria, and Firmicutes that are known for their important roles in nutrient cycling, hormones synthesis, and the prevention of pathogens (Mendes et al., 2011). Fungi, such as arbuscular mycorrhizal fungi (AMF), gain phosphorus for the plant and improve soil aggregation, while endophytic fungi offer some degree of stress tolerance. Archaeal organisms have not been studied as much, but there is a current surge of interest in their contributions to nitrogen transformations via planktonic or sediment microbes in oceans, soils, or unique environments (Bates et al., 2011). Some viruses (including bacteriophages) can serve as another type of organisms that indirectly, in certain cases, can promote plant health via the regulation of certain bacterial populations, while other viruses are responsible for causing plant disease (Suttle, 2007).

11.2.2.1 Mutualism, Commensalism, and Parasitism within Plant–Microbe Relationships

Interactions between plants and microbes can be classified based on a range of productive outcomes. A mutualistic interaction can be illustrated with rhizobium and legume symbiotic interactions: the plant takes up nitrogen and the rhizobium then gains carbon-based resources (Geisseler et al., 2024). Commensalism represents interactions that benefit microbes without functional consequences on the host such as certain epiphytic bacteria that exist on the leaf surfaces. Parasitism denotes the diseased state of plants caused by certain pathogenic microbes—the classic wilt disease caused by certain fungi (shown in Figure 11.3). It is also important to recognize that certain microbes are opportunistic and can move between mutualism and parasitism, depending on environmental stress and the host condition.

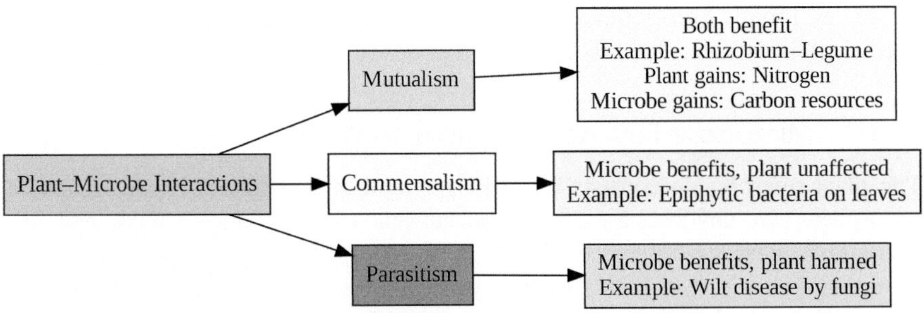

FIGURE 11.3 Plant–microbe interaction types and their outcomes (Source: Author).

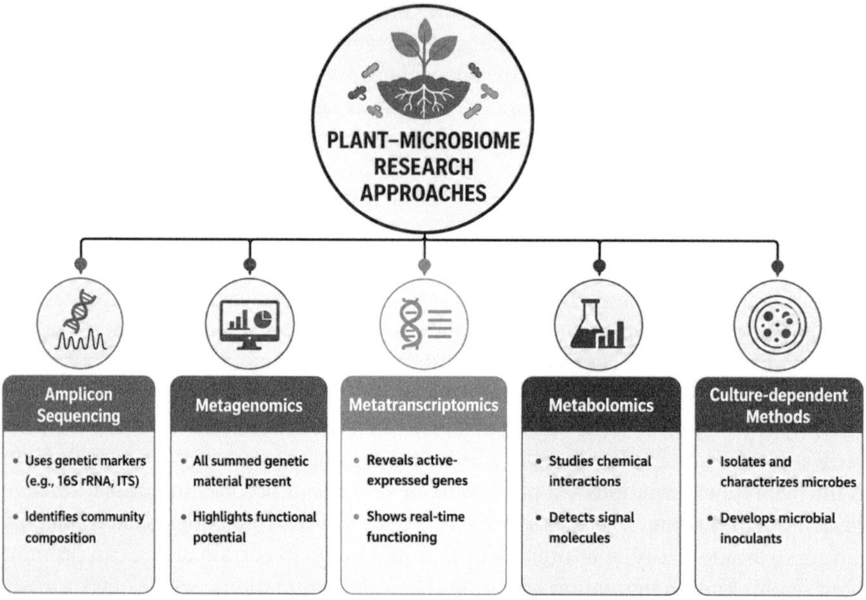

FIGURE 11.4 Integrated approaches in the analysis of microbial communities present in the microbiome of plants.

11.2.2.2 Methods for Analyzing the Microbial Communities Present in the Microbiome of Plants

Discoveries in molecular biology have changed the dynamics of the fundamental approaches to plant microbiome research (shown in Figure 11.4). Research using amplicon sequencing asks specific genetic markers, like 16S rRNA gene (bacteria and archaea), or ITS regions (fungi), to identify community composition. Metagenomics provides a wide-reaching perspective of all summed genetic material present, but more importantly, it highlights the functional potential of the microbiome of organisms (Bulgarelli et al., 2013). Metatranscriptomics examines the microbiome of plants to reveal active expressed genes at the time of sampling, which augments our understanding of the actual real-time functioning of the microbiome. Metabolomics interrogates the chemical interactions between plants and microbes including signal molecules. Finally, conventional methods using culture-dependent protocols are still intrinsic to isolating and characterizing microbes to develop inoculants.

11.2.3 Microbiome Functions Beneficial to Plants

The plant microbiome consists of various microbes that inhabit the rhizosphere, endosphere, and phyllosphere that deliver nutrient value, enhance growth, and support stress tolerance (Turner et al., 2013). Microorganisms participate in a mutualistic relationship with plants to supply nutrients via rhizobia, *Azospirillum*, and *Frankia* biological nitrogen fixation and solubilization of phosphorus by species such as *Pseudomonas*, *Bacillus*, and *Aspergillus*. By supplying nutrients to the plant, microorganisms can provide an alternative to using chemical fertilizers (Rodríguez & Fraga, 1999; Vessey, 2003). Additionally, plant-associated microbes generate phytohormones, auxins, gibberellins, and cytokinins, which stimulate root and shoot elongation. Microbes also help induce proper systemic resistance, and provide an array of antimicrobial metabolites that kill pathogens (Patten & Glick, 2002; Berendsen et al., 2012). Numerous taxa of the microbiota promote abiotic stress resistance for plants growing under conditions of drought, salinity, and heat stress, and promote growth in resiliency in these ecosystems (Compant et al., 2010; Ngumbi & Kloepper, 2016). In conclusion, the roles and contributions of these microbial communities are meant to provide a natural and valuable partner in sustainable agriculture through nutrient efficiency, tolerating stress, ecological equity, and considerations for chemical agro-input (Mendes et al., 2013; Compant et al., 2019).

11.3 Nutrient Acquisition and Cycling (Nitrogen Fixation, Phosphate Solubilization)

Efficient nutrient capture and recycling are critical in crop production for food security and maximum plant productivity. A supplementary role of the major nutrient, most importantly nitrogen and phosphorus, is the role of the plant microbiome. Despite the critical role of these nutrients in plant developmental activities, they exist in forms that are not biologically available for utilization in the plant. The rhizosphere bacteria are significant in the microbial conversion of nutrients to forms that are bioavailable, thus reducing the need for synthetic fertilizers in addition to increasing soil fertility.

As shown in Table 11.1, nitrogen fixation is a key microbial process of the plant microbiome that has been of substantial research interest. Biological nitrogen fixation is a process, to a very large extent, carried out by diazotrophic bacteria such as

TABLE 11.1
Important Microbial Processes, Mechanisms, and Benefits in Sustainable Agriculture

Microbial Process	Key Microorganisms	Mechanism	Example/ Application	Benefits
Nitrogen fixation	*Rhizobium*, *Azospirillum*, *Frankia* (diazotrophic bacteria)	Reduction of atmospheric N_2 to NH_3 via nitrogenase enzyme	Symbiosis in legumes (*Rhizobium* in root nodules); free-living and endophytic diazotrophs for nonlegume crops	Improves plant nitrogen availability, reduces chemical fertilizer use
Phosphate solubilization	*Pseudomonas*, *Bacillus*, *Aspergillus* spp.	Production of organic acids and phosphatases that solubilize insoluble phosphate	Biofertilizer application in various crops	Enhances phosphorus uptake, improves soil fertility, promotes sustainable agriculture

Rhizobium, *Azospirillum*, and *Frankia*, in that they facilitate the reduction of atmospheric nitrogen (N_2) to ammonia (NH_3) to make the nitrogen accessible to the plant (Vessey, 2003). An example is legume, in that symbionts within root nodules are species of the genus *Rhizobium* and as a result directly supply the host plant with nitrogen. Nitrogen-fixing free-living bacteria and endophytes are also able to autonomously fix nitrogen without being associated with the plant and consequently posing potential agronomic value to nonlegume crops (Lloyd et al., 2018).

Furthermore, phosphate solubilization is also a most significant microbial activity, since the greater part of the soil's phosphorus exists in the insoluble form. A few species of bacteria and fungi, like *Pseudomonas*, *Bacillus*, and *Aspergillus* sp., excrete organic acids and phosphatases that dissolve the phosphate bound and convert the same into plant-available forms (Rodríguez & Fraga, 1999). Biological solubilization enhances not only the nutrient intake of the plant, but also the soil fertility in the long run. Using these microbial activities in the form of inoculants and biofertilizers is an affordable method of attaining sustainable nutrient use. It reduces environmental pollution related to the use of chemical fertilizers while retaining good health and efficient agricultural systems.

11.3.1 Stress Tolerance

Plants are subjected to different environmental stresses, such as salt, drought, and heavy metal toxicity, that lower their productivity, growth, and survival considerably (shown in Figure 11.5). The abiotic stresses bring in the induction of damage that is oxidative, deterioration of metabolic processes, and limitation of primary nutrient use. The plant microbiome, however, is significantly effective in the enhancement of the plant's resistance in the pathogenic environment. Some of the useful microorganisms, and, most importantly, the ones living in the endosphere and rhizosphere,

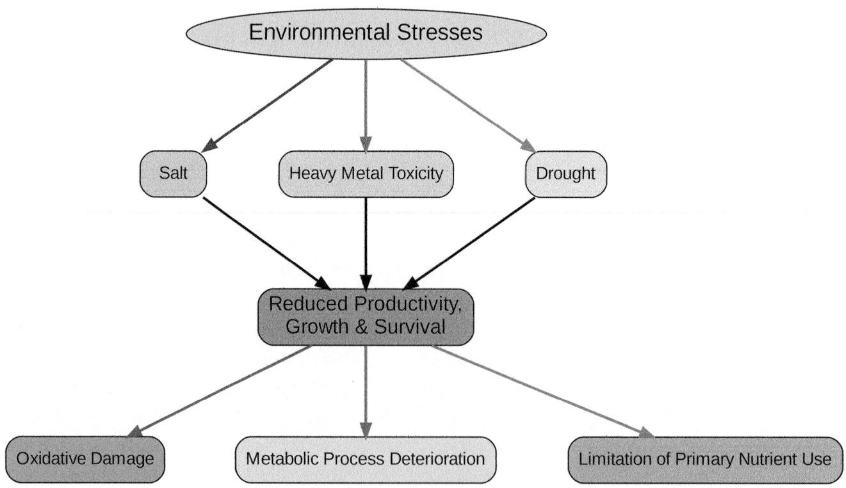

FIGURE 11.5 Effects of environmental stresses on plant physiology and productivity.

TABLE 11.2
PGPR in Enhancing Plant Growth under Water Stress

PGPR Type	Examples	Mechanism under Water Stress	Effect on Plant
Plant growth–promoting rhizobacteria	*Pseudomonas*, *Bacillus*, *Azospirillum*	• Synthesis of exopolysaccharides → improves soil water retention and aggregation • Production of abscisic acid → controls stomatal closure to conserve water	Stimulates root growth and enhances water uptake

facilitate the plant to better fit and resist the stress of the environment through different mechanisms that are both biochemical and physiological.

As shown in Table 11.2, the PGPR like *Pseudomonas*, *Bacillus*, and *Azospirillum*, when subjected to water stress conditions, will stimulate the root growth and water intake. The PGPR does this through the synthesis of exopolysaccharides that will improve the soil's water retention and clumping, and through the synthesis of elevated amounts of abscisic acid that governs the stomatal closure and water conservation (Ngumbi & Kloepper, 2016).

Microorganisms support the maintenance of plants under salinity stress through the preservation of ionic balance, osmoprotectants synthesis including proline and glycine betaine, and high activities of antioxidant enzymes (shown in Figure 11.6). Halotolerant bacteria such as *Halomonas* and certain species of the genus *Bacillus* control the uptake of potassium and sodium, thus restricting ionic toxicity and

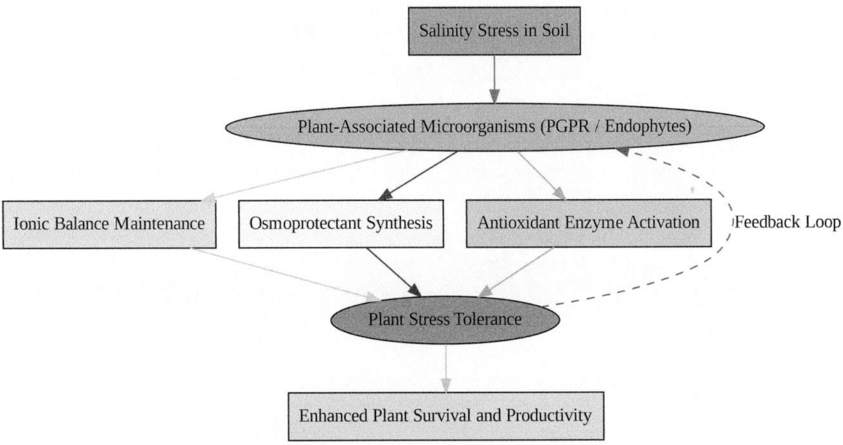

FIGURE 11.6 Microbial-mediated mechanisms for plant survival under salinity stress.

promoting plant growth (Yang et al., 2009). There are microbial strains that are involved in the alleviation of stress upon the presence of heavy metals like cadmium, lead, and arsenic through mechanisms of biosorption, bioaccumulation, and chelation. *Pseudomonas putida* and *Bacillus subtilis*, among others, have the ability to immobilize undesirable metals or convert them to less undesirable species and thus avoid injury to the plant tissue (Gadd, 2000). These microbial transformations not only render the plants tolerant to the stressful environment but also introduce the evolution of more tolerant cultivation systems. Utilization of stress-tolerant microbes in the form of bioinoculants is also a sustainable crop production approach in poor or climate-stressed environments.

11.3.2 Disease Suppression and Immune Modulation

The plant microbiome is equally instrumental in the defense of the plant against various pathogens through disease suppression and immunomodulatory mechanisms. These functions are done through the activities of the beneficial microorganisms, particularly those in the rhizosphere and endosphere, that form a barrier for pathogenic organisms, in that they fill some ecological niches in the immediate environments of plant roots; therefore, the microorganisms limit the colonization and infectability of pathogens onto and into plant tissues, a process described as competitive exclusion (shown in Figure 11.7) (Berendsen et al., 2012).

Another major disease suppression mechanism is the synthesis of antimicrobial metabolites such as antibiotics, hydrogen cyanide, and lytic enzymes that are produced by the suppressive microorganisms (shown Figure 11.8). *Pseudomonas* and *Bacillus* species, for example, are known to generate secondary metabolites that suppress the pathogens such as the phytopathogens like *Fusarium*, *Rhizoctonia*, and

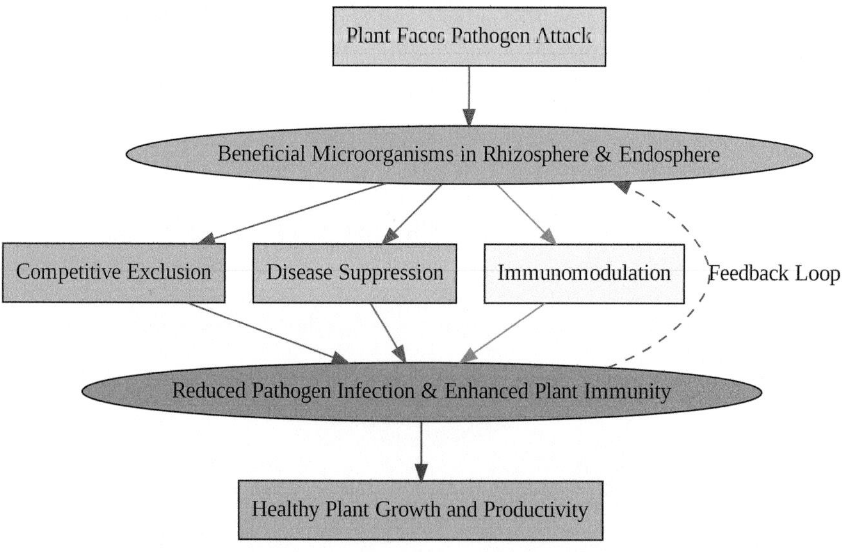

FIGURE 11.7 Microbial-mediated disease suppression and immune modulation in plants.

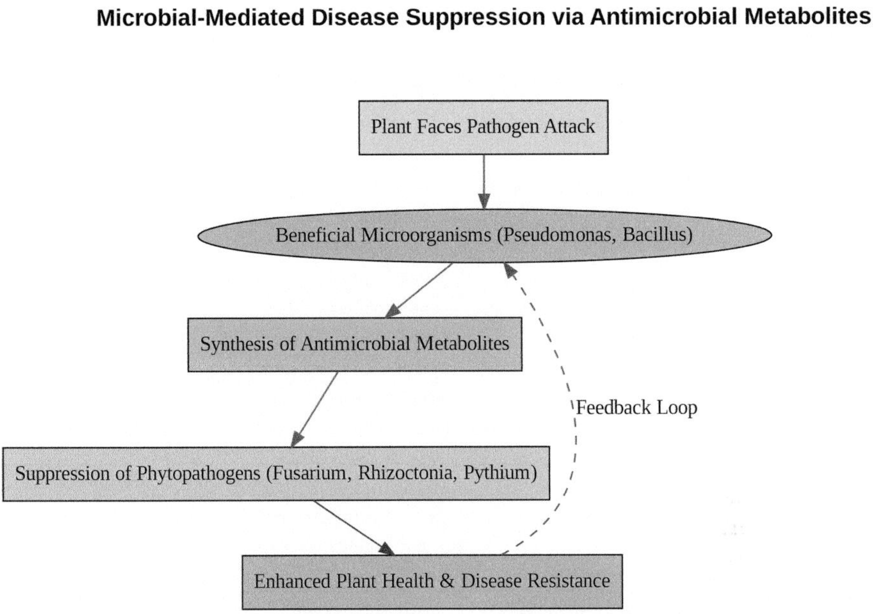

FIGURE 11.8 Microbial-mediated disease suppression via antimicrobial metabolites.

Pythium (Haas & Défago, 2005). Other than these, some microbes also generate siderophores that are most effective in capturing iron in the rhizosphere, so that the competing pathogens cannot access it and are thus suppressed.

In addition to their function of directly suppressing the pathogens, the majority of the plant-associated bacteria greatly augment the plant's defense mechanisms through a phenomenon referred to as ISR. The ISR is a priming response that prepares the plants to respond significantly and more quickly to a challenge from a pathogen. Unlike the systemic acquired resistance, initiated through an earlier infective pathogen, the ISR is usually initiated through nonpathogenic microorganisms such as certain strains of *Trichoderma*, *Pseudomonas*, and *Azospirillum* (Pieterse et al., 2014). The following microbial products reduce the use of chemical pesticides and support environmentally friendly practices of pest control. Utilizing the natural disease suppression of the plant microbiome, agricultural experts are able to generate effective and sustainable integrated pest management systems.

11.3.3 Plant Growth Promotion

The plant growth–promoting microorganisms, or bacteria and fungi that are found in the rhizosphere and endosphere, are directly responsible for the promotion of plant growth in response to the synthesis of siderophores and the secretion of phytohormones. Plant growth-promoting microorganisms (PGPMs) are the in-built plant growth promoters that regulate the physiology of plants in such a way that the root architecture is maximized, nutrient capture is optimized, and the entire biomass yield is increased (shown in Table 11.3).

TABLE 11.3
Key Mechanisms of Plant Growth Promotion by PGPMs and Their Effects

PGPM Mechanism	Examples/Microbes	Action/Function	Effect on Plant
Phytohormone production	*Azospirillum*, *Pseudomonas*, *Bacillus*	Auxins indole-3-acetic acid (IAA) → root elongation and lateral root outgrowth Cytokinins → cell division and shoot production Gibberellins → seed germination and stem elongation	Enhanced root architecture, nutrient/water uptake, shoot growth, biomass yield
Siderophore production	PGPM in rhizosphere and endosphere	High-affinity iron-sequestering molecules → make iron available to plant and limit pathogen access	Improved metabolic activities, increased plant growth, and disease suppression
Overall PGPM function	Bacteria and fungi in rhizosphere/endosphere	Regulate plant physiology via hormones and siderophores	Sustainable crop productivity, optimized nutrient capture, higher biomass, reduced need for chemical inputs

As shown in Table 11.3, among the most significant mechanisms through which microorganisms bring about plant growth is the production of phytohormones. Many biocontrol bacteria, such as bacteria of genera *Azospirillum*, *Pseudomonas*, and *Bacillus*, are able to produce plant hormones such as auxins (e.g., indole-3-acetic acid), cytokinins, and gibberellins, and these hormones are responsible for the control of the plant growth and developmental processes. For example, the root elongation and the extension of the outgrowth of lateral roots are brought about through the production of auxins, thus enhancing the contact of the root for nutrient and water absorption (Patten & Glick, 2002). Cell divisions and shoot production are processes in which cytokinins are significant, while the germination of seeds and the extension of the stem are processes in which the gibberellins are significant (Geisseler et al., 2024). In addition to the synthesis of hormones, the PGPMs are also responsible for the synthesis of siderophores, low-molecular weight, high-affinity, iron-sequestering molecules. Iron is an important part of the metabolic activities of plants but does not occur normally in the soil in soluble state. Siderophores, through their ability to dissolve and complex the iron and make this nutrient available to the plant, support increased plant growth while limiting the availability of the resultant iron to pathogens, assisting in the control of diseases (Kloepper et al., 1980). Siderophore and phytohormone syntheses in effective microbes greatly augment the effectiveness and yield of the crop. These microbial characteristics are of particular importance in sustainable low-input crop production strategies in which the use of the supplementary chemicals and fertilizers is low. These microbial activities offer a sustainable and effective method of increasing agricultural productivity.

11.3.4 Microbiome Engineering for Agriculture

Microbiome engineering is an important aspect of agricultural biotechnology that targets the intentional manipulation of plant-associated microbial communities to improve agricultural productivity, resilience, and sustainability. The intention of microbiome engineering is to use the advantageous properties of microorganisms to promote less reliance on chemical inputs, enhance nutrient cycling, and the environmental tolerance of plants (Mitter et al., 2019). The following subsections will outline how specific microbiome engineering methods related to precision rhizosphere management have been applied, their current developments, and limitations.

11.3.4.1 Inoculants and Biofertilizers

Microbial inoculants and biofertilizers are defined among the oldest microbiome-based products expected to have utility from a microbiome engineering perspective in agriculture. These products can contain beneficial microorganisms from different taxa, including *Rhizobium*, *Azospirillum*, *Pseudomonas*, and AMF, and generally stem from the intention of enhancing nutrient uptake, promoting plant growth, or suppressing plant diseases (Vessey, 2003).

More recently, advances in microbial inoculant or biofertilizer development have led to the formulation of multi-strain formulations designed to "stack" complementary microbial functions (e.g., pairing nitrogen-fixing bacteria with phosphate-solubilizing bacteria). In addition, various carrier materials have been developed, such as biochar, polymer coatings, and peat, that are intended to enhance microorganism survival during transport and storage (Bashan et al., 2014).

Despite these advances, the field performance of microbial inoculants or biofertilizers continues to exhibit moderate success, which can be attributed to environmental variability or competition with native microbiota, as well as their poor colonizing efficiency (Timmusk et al., 2014). Future refinement of standardizing their production methods, selecting specific microbial strains from specific environments, and developing delivery systems are critical aspects to improve their usage.

11.3.4.2 Engineering Plant Microbiomes by SynComs, Genome Editing, and CRISPR/Cas System

Synthetic microbial communities (SynComs) are deliberately constructed compositions of microbial strains with known and complementary functionalities, and they offer a disciplined system for promoting plant development and plant health. Compared to traditional single-strain inoculants, SynComs achieve more consistent plant responses through the combination of microorganisms that achieve specific roles, for example, nitrogen fixation, production of plant hormones, and suppression of plant disease for particular crops (e.g., maize, wheat) (Niu et al., 2017). Studies suggest that SynComs often surpass the effectiveness of single-strain inoculants in

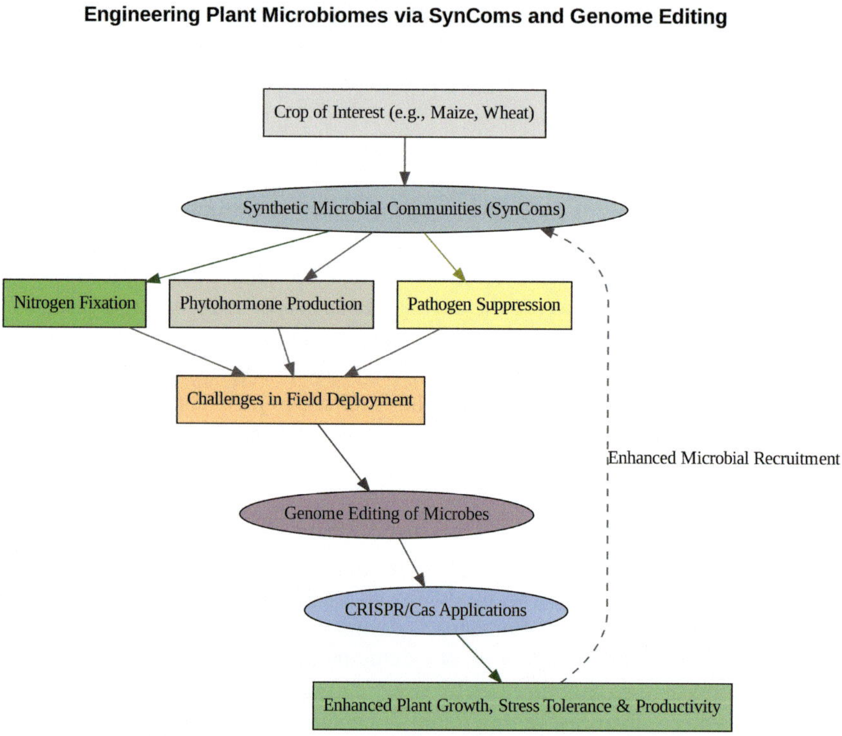

FIGURE 11.9 Harnessing synthetic microbial communities and genome editing for optimized plant performance.

controlled greenhouses because of the synergistic relationships between the microbial components. However, the difficulty of translating these success stories into field contexts remains, as native soil microbiota, abiotic stress components, and biotic relationships can compromise the consistency of SynComs and their effectiveness (Niu et al., 2017; Compant et al., 2019) (Figure 11.9).

New genome editing technologies, particularly the CRISPR/Cas system, are revolutionizing microbiome research and development of new generations of bioinoculants. By enabling precise modification of genes related to root colonization, nutrient uptake, or production of antimicrobial metabolites, researchers design microbial strains that are optimized for specific agroecosystems (Bikard & Barrangou, 2017; Johnston & Mejía, 2020). More generally, CRISPR technologies expand the understanding of plant–microbe relationships, and complement crop breeding programs seeking to discover varieties that better attract beneficial microbes and increase nutrient use efficiency. Additionally, new community-wide CRISPR strategies open up in situ microbiome editing that directly alters microbial consortia in plant habitats, in the absence of culturing individual microbes, thus opening new avenues of engineered microbiomes generating increased productivity, resilience, and sustainability in production systems (Bikard & Barrangou, 2017).

11.3.5 Role of Microbiome in Sustainable Agricultural Intensification

The concept of sustainable intensification of agriculture, to the degree possible, aims to produce more food from existing farmland while minimizing negative environmental impacts and conserving natural resources (Pretty & Bharucha, 2014). The plant microbiome is also recognized as an essential biological component that has the potential to alter current agricultural practices. Microbes found in the rhizosphere, endosphere, and phyllosphere form ecological relationships of altruistic, symbiotic, or mutual beneficial interactions that contribute to soil health, nutrient cycling, disease resistance, and stress tolerance in the plant, ultimately translating to a reduction in synthetic input (Mendes et al., 2013; Bender et al., 2016). Microbial activities, such as biological nitrogen (N) fixation by *Rhizobium* and *Azotobacter*, involve transformation of atmospheric N_2 into plant-accessible form, while phosphate-solubilizing bacteria belonging to the genera *Pseudomonas* and *Bacillus*, which solubilize insoluble phosphates into plant-accessible form, thereby enhancing nutrient uptake and reducing fertilizer inputs (Rodríguez & Fraga, 1999; Vessey, 2003).

11.3.5.1 Enhancing Yield without Excessive Input Use

Historically, existing agricultural systems have predominantly focused on maximizing agricultural production mainly through amassing and utilizing externally sourced inputs (i.e., synthetic fertilization, chemical pesticides, irrigation systems) as a strategy. Although this strategy has proven effective for raising crop yields, it has resulted in severe environmental consequences through soil erosion, water pollution, the detrimental loss of biological diversity, and accelerated emission of greenhouse gases (Foley et al., 2011). One purpose of sustainable intensification is to increase crop yields and decrease reliance on nonrenewable and environmentally detrimental inputs.

Plant-associated microbiome represents a green enhancement in crop yields through the employment of natural biological processes. Beneficial microorganisms, especially those colonizing the rhizosphere, increase plant nutrient mobilization by enhancing nutrient levels through the increase of nitrogen fixation, phosphate solubilization, and siderophore production, which enable plants to obtain nutrients while using less artificial fertilizers (Rodríguez & Fraga, 1999; Vessey, 2003). For example, nitrogen-fixing bacteria (*Rhizobium* and *Azospirillum*) can provide most if not all of the nitrogen most plants require, leading to reductions in synthetic nitrogen input.

Correspondingly, microbial inoculants that include PGPR and mycorrhizal fungi are responsible for water use efficiency, stress tolerance, hormone biosynthesis, and overall plant growth under a low-input growing environment (Bhattacharyya & Jha, 2012). Microorganisms can stimulate root growth and uptake and increase nutrient use efficiency so that the plant can completely utilize all soil resources available to it. Microbiome-based approaches in agricultural systems can enhance sustainability and innovation within agricultural production, utilizing sustainable/durable and cheaper methods as opposed to the devastating environmental impact associated with agricultural practices, while still aligning with call for sustainable intensification and efficiency and cost-effectiveness for food production systems.

11.3.5.2 Reducing Reliance on Chemical Fertilizers and Pesticides

The conventional agricultural procedures of the past decades had a high level of crop yield compared to prior years and has been accompanied by excessive amounts of chemical fertilizer and pesticide activity. There are serious ramifications, however, of chemically based approaches including environmental impacts that include soil degradation, contamination of water resources, reduction of biodiversity, and adverse effects on animal and human health (Tilman et al., 2002; Carvalho, 2017). The pursuit of sustainable agricultural intensification requires new mechanisms to increase productivity without compounding issues at the environmental and health level.

The plant microbiome offers immense value in comparison to synthetic agrochemicals, in addition to being multifaceted, by improving plant nutrition and plant defense mechanisms. Several microorganisms, such as nitrogen-fixing bacteria (e.g., *Rhizobium* and *Azospirillum*) and solubilizing bacteria (e.g., *Pseudomonas* and *Bacillus*), enhance availability and uptake of essential nutrients, thus offering a sustainable alternative to chemical fertilizers (Rodríguez & Fraga, 1999; Vessey, 2003). For instance, mycorrhizal fungi are key organisms that establish symbiotic relationships with root plants, facilitate enhanced uptake of phosphorus and other minerals, and also support soil structural properties and soil fertility (Smith & Read, 2008).

Microorganisms play an important role in plant defense against pests and diseases by using natural processes to control them. Numerous microbial strains produce substances, such as antibiotics, siderophores, or enzymes, that suppress soil-borne diseases, and a range of microbial activities lead to ISR processes which stimulate the plant immune system and can lead to greater resistance to subsequent infections (Berendsen et al., 2012; Pieterse et al., 2014). Microbial activity will not only reduce our dependence on chemical pesticides, but it may also help to promote the long-term ecological sustainability of the soil system. The use of microbial inoculants and bio-source inputs to the agroecosystem can assist farmers in being less reliant on synthetic fertilizers and pesticides as well as the production costs along with environmental impacts. These benefits present an opportunity to promote a cleaner, sustainable or "greener" agricultural system that we desperately need to sustain food supplies in the future.

11.3.6 Improving Soil Health and Biodiversity

Soil health is of critical importance to sustainable agriculture and has a strong impact on nutrient cycling, water retention, plant productivity, and disease control. However, decades of systematic crop production, including monoculture, overcultivation, and chemistries, have contributed to soil erosion, lower microbial diversity, and ecological degradation (Lal, 2015). The plant microbiome is instrumental in the recovery and sustainment of soil health through the enhancement of soil structure and functional attributes. Soil microorganisms in the rhizosphere, such as PGPR and mycorrhizal fungi, enhance soil aggregation through the production of extracellular polysaccharides and lead to improved nutrient mobilization and organic matter decomposition, all of which contribute to soil porosity and moisture-holding capacity (van der Heijden et al., 2008). The support of diverse microbial communities is necessary for soil stability, nutrient cycling, and the suppression of soil-borne pathogens (Garbeva et al., 2004). Certain farming practices, such as reduced tillage, cover cropping, and organic soil amendments, explicitly enhance microbial diversity and have been shown to build biological soil health (Schmidt et al., 2019). Ultimately, by utilizing a microbe-based soil management strategy, farmers can rehabilitate depleted soils, reinstate lost ecosystem services, and increase resilience and sustainability of food and crop systems. The microbial approach also provides productive, ecologically balanced systems that continue to allow for sustainable soil fertility management and food production. This microbial-centric approach promotes productive, ecologically balanced systems that form the foundation for long-term soil fertility and of sustainable food production.

11.4 INTEGRATION WITH ORGANIC AND PRECISION FARMING

Combining plant microbiome-based strategies with organic and precision agriculture presents a synergetic way toward climate-smart agricultural intensification, balancing ecological conservation with crop production. Organic and precision agriculture endeavor to reduce environmental degradation sustaining yield, which harmoniously aligns with the plant microbiome's service in nutrient cycling, pathogen suppression, and stress tolerance (Bhattacharyya & Jha, 2012). In organics, with reduced use of synthetic agrochemicals, PGPR and mycorrhizal fungi increase soil fertility and availability, especially of nitrogen and phosphorus, along with soil structural integrity, biodiversity, and disease resistance (Van Bruggen et al., 2018; Lori et al., 2017). Application of biofertilizers, microbial inoculums, and compost-based amendments further increases the ecological and functional integrity of these agroecosystems. In precision agriculture, the convergence of microbiome data comprising soil microbial cartography, health diagnosis, and site-specific inoculations facilitates site-specific management that proves optimum input use, enhanced uptake, and curtails fertilizer overuse (Gebbers & Adamchuk, 2010). Precision-tailored microbial intervention can thus supplement the principles of organics, with resource efficiency and environmental safeguard. In unison, the integration of microbiome science, organics, and precision technology presents a climate-smart, resource-effective framework toward next-generation farming systems, increasing crop output, sustaining ecosystem services, and sustaining the long-term (Bhattacharyya & Jha, 2012; Compant et al., 2019).

11.5 PLANT–MICROBIOME–ENVIRONMENT INTERACTIONS

The plant microbiome and the local environment interact within a constantly changing and complex triad. Moreover, these interactions are often regulated by complex contextual factors such as soil chemistry, climate, plant genotype, and agricultural practices (Berg et al., 2014). The microbiome mediates plant–environment interactions—mainly regulating nutrient uptake and stress and disease tolerance in response to variations in abiotic conditions.

Plants exercise a great deal of control over their host-associated microbiome. For example, plants exert control through host-associated compounds—called root

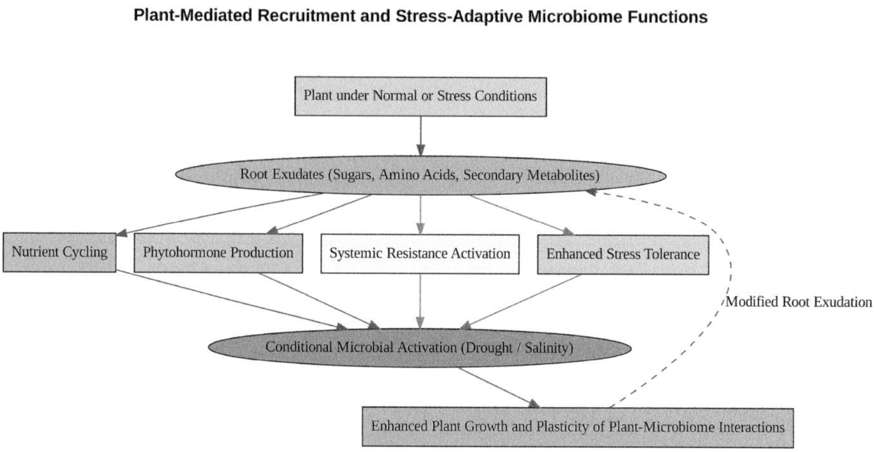

FIGURE 11.10 Plant-mediated recruitment and stress-adaptive microbiome functions.

exudates, composed of sugars, amino acids, and secondary metabolites. Root exudates are well known to help recruit beneficial microorganisms in the rhizosphere (Sasse et al., 2018). These microorganisms perform functions in nutrient cycling, create phytohormones, and initiate systemic resistance processes and enhance the plant's ability to tolerate biotic and abiotic stress (shown in Figure 11.10) (Berendsen et al., 2012). For example, during drought or salinity stress, specific microbial taxa are conditionally induced, that may serve to increase water use or osmotic adjustment, which highlights the plasticity of plant–microbiome interaction under environmental stress (Naylor & Coleman-Derr, 2018).

Soil water content, pH, and temperature are major environmental factors that directly affect microbial community composition and function. Environmental factors can cause shifts in microbial diversity and predict availability of functional attributes that improve plant health (Fierer, 2017). Natural systems are also influenced by human science and society; disturbance by agricultural practices, including intensive tillage, pesticides, and global climatic change, may negatively affect beneficial microbial communities or networks influencing plant performance. Understanding these relationships is crucial for microbiome-led farming practices. This understanding of the most critical plant–microbiome–environment interactions can drive the development of cropping systems that are adaptive, sustainable, and productive while infusing minimal ecological disturbance across changing field conditions.

11.5.1 Influence of Environmental Factors on Microbiome Composition

The ecological factors exert significant influence on the plant-associated microbiome composition and functioning, thereby shaping microbial diversity, richness, and interaction of microbes with plants (Fierer & Jackson, 2006). The type of soil, pH, temperature, humidity, and nutrient availability contribute significantly to the formation of microbial assemblages in the rhizosphere, endosphere, and phyllosphere, frequently with well-balanced gradients of microbial density and activity in space.

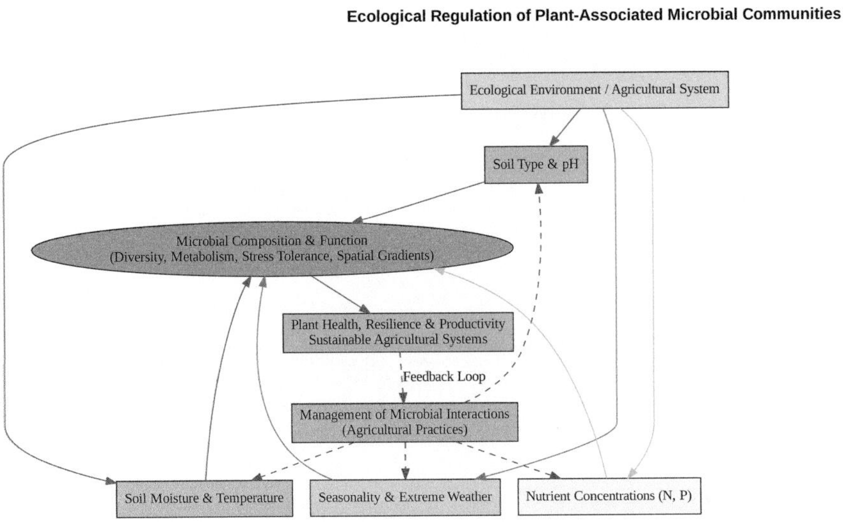

FIGURE 11.11 Ecological regulation of plant-associated microbial communities.

Soil characteristics, especially pH, also serve as important ecological factors of microbial community composition, given that some of the taxa harbor inbuilt preference for existence in acidic or alkaline soil types (Lauber et al., 2009). Soil moisture content and temperature also influence microbial populations and metabolic functions, for example, in arid conditions, where drought-tolerant taxa of *Actinobacteria* predominate, consequently permitting plant tolerance to stress (Figure 11.11).

11.5.2 Climate Change Implications

As shown in Figure 11.12, despite these impediments, the plant microbiome presents the opportunity for adaptation to climate. Some microbes have been shown to improve a plant's tolerance to heat, salinity, and drought by producing phytohormones, amassing osmolytes, and inducing genes to become stress-responsive (Ngumbi & Kloepper, 2016). The utilization of positive traits through microbial inoculants and microbiome engineering could alleviate the challenges provoked by climate change and will help establish climate-resilient and sustainable production systems that maintain productivity under stress.

11.5.3 Soil Type, Crop Rotation, and Farming Practices

Crop rotation, agricultural methods, and soil properties are significant drivers of plant microbiome structure and composition and function. They determine microbial diversity, community structure, and the frequency of benefit microbes that have important roles in improving plant health, disease resistance, and nutrient cycling (Hartmann et al., 2015).

As shown in Table 11.4, soil characteristics, including pH, mineral composition, organic matter content, and texture, remain significant predictors of microbial

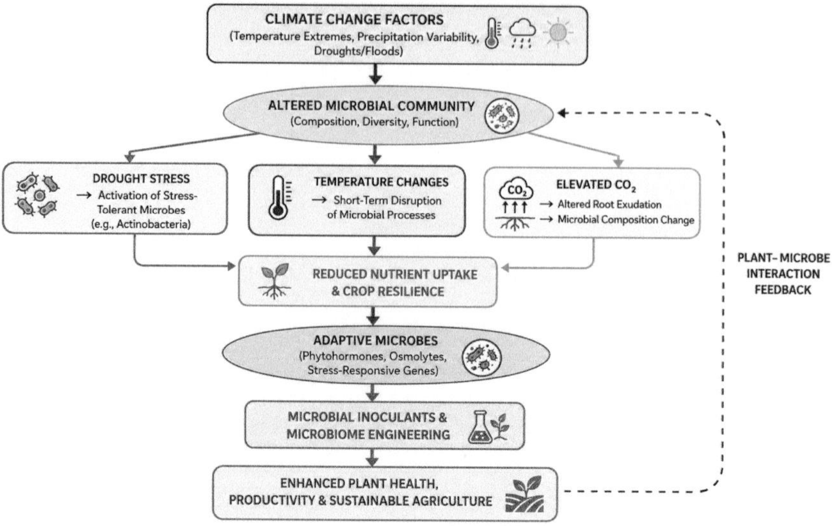

FIGURE 11.12 Climate change impacts on plants and adaptive mitigation.

TABLE 11.4
Impact of Soil Properties, Crop Management, and Farming Practices on Plant Microbiome and Agricultural Productivity

Factor	Examples/Practices	Mechanism/Impact on Microbiome	Effect on Plant/Agriculture
Soil properties	pH, mineral content, organic matter, soil texture	Determines microbial diversity, metabolic activity, and community composition	Influences plant health, nutrient cycling, and disease resistance
Soil type	Sandy soil	Low water-holding capacity → lower microbial biomass	Reduced microbial-mediated functions
	Loamy/clay soil	Higher metabolic diversity and active microbial communities	Enhanced nutrient cycling and plant support
Crop rotation	Different crops in sequence	Varying root exudates attract diverse microbial communities	Reduces crop-specific diseases, enhances nutrient cycling, supports beneficial microbes (e.g., nitrogen fixers, mycorrhizal fungi)
Crop diversity	Multiple crop rotation	Increases functional redundancy in microbial community	Greater resilience to environmental disturbances

(Continued)

TABLE 11.4 (*Continued*)
Impact of Soil Properties, Crop Management, and Farming Practices on Plant Microbiome and Agricultural Productivity

Factor	Examples/Practices	Mechanism/Impact on Microbiome	Effect on Plant/Agriculture
Farming system	Tillage intensity, organic/synthetic inputs, organic amendments	No-till, cover cropping, compost → promotes microbial stability; intensive tillage → disrupts microbial interactions	Improves soil health, microbial functions, and sustainability
	Organic farming	Higher microbial abundance and enzymatic activity	Enhanced soil fertility and crop productivity

diversity and functioning in agroecosystems. Sandy soils, for example, contain relatively higher microbial stocks, but with low biomass, for low moisture-holding properties, compared with loamy and clayey soils with higher metabolically active and diverse microbial assemblages (Lauber et al., 2009). Among all the soil attributes, pH emerges as exceptional, serving as a significant predictor of composition of bacterial communities for large spatial scales, with unique assemblages being revealed in acidic compared with alkaline soil systems (Lauber et al., 2009). Crop rotation, being an ancient but highly effective agronomic management, improves soil integrity and microbial diversity by introducing diverse root exudates with attendant particular communities of the rhizosphere. This temporal shift, in addition to minimizing crop-specialized pathogens, also enhances dynamics of nutrient cycles and positive microorganisms such as bacteria that fix nitrogen and mycorrhizal fungi.

Moreover, management of farming systems, such as tillage intensity, application of organic amendments, and use of inputs, directly impacts soil microbial community composition. Soil structure and microbial connections are destroyed by intensive tillage, while no-till, cover crops, and application of compost build microbial richness and stability (Schmidt et al., 2019). Otherwise, organic farming systems contain greater microbial abundance, enzymatic and functional diversity compared with conventional farming systems (Lori et al., 2017). Combination of these management systems with microbiome-led interventions strengthens soil health, encourages beneficial functions of microbes, and builds the ecological resilience of crop production systems, which are the main pillars of sustainably produced crops.

11.5.4 Feedback Loops between Plant Genotype and Microbial Recruitment

Plant genotype–associated microbiome interaction is significant in explaining plant–microbe interaction because plants interact selectively with particular microbial communities from the soil through mechanisms of biochemical and genetic interactions

(Lebeis et al., 2015; Pérez-Jaramillo et al., 2016). The variations in composition of the root exudates, such as sugars, amino acids, phenolics, and secondary metabolites, serve as chemical signals, as well as nutrient inputs influencing the microbiome of the rhizosphere, establishing diverse microbial assemblages across cultivars within a species (Sasse et al., 2018). Established microbial communities will interact with the plant's gene regulation, nutrient uptake, and stress responses, such as promotion of plant growth by rhizobacteria and AMF that induce nutrient uptake and abiotic stress tolerance–related genes (Vandenkoornhuyse, et al. 2015). This two-way feedback system affects the mode of subsequent exudation, solidifying positive interactions, and potentially improving the fitness of the plant over generations. Characterizing these microbiological relationships that are mediated by the plant genotype is significant for the attainment of sustainable agriculture because plant breeding in the future should integrate microbiome-aware strategies for breeding crop types that sustain positive microbial relationships that remain stable in diverse environmental conditions, thus minimizing reliance on external inputs as well as enhancing resilience.

11.5.5 Challenges and Limitations

Although the plant-associated microbiome has a potentially significant role to play in achieving sustainable agriculture, the scalability of this biosystem is fraught with multiple challenges and limitations. The first mentioned is the complex and variable nature of microbial communities in natural environments. The composition and function of the plant microbiome are influenced by multiple interacting factors including soil type, climate, crop genotype, and management practices, leading to unpredictable effects being documented in the field (Toju et al., 2018; Compant et al., 2019) (Table 11.5).

TABLE 11.5
Challenges, Limitations, and Their Potential Solutions

S. No.	Challenges and Limitations	Potential Solutions
1	Variability in field performance due to soil type, climate, crop genotype, and competition with native microbiota	Develop region- and crop-specific microbiome profiles; use resilient microbial formulations (encapsulation, cocultures, preconditioned seeds/soil); combine with organic amendments or companion crops to enhance persistence.
2	Reduced stability and persistence of introduced microbes across seasons	Engineer self-sustaining microbial populations; create supportive microhabitats; monitor microbial survival with field trials; adjust inoculant timing and placement using precision agriculture tools.
3	Technical and scientific constraints in cultivation, functional screening, and linking microbes to plant outcomes	Use high-throughput functional screening, metagenomics, bioinformatics, and AI-driven real-time monitoring; portable sequencing for adaptive management.

(*Continued*)

TABLE 11.5 (*Continued*)
Challenges, Limitations, and Their Potential Solutions

S. No.	Challenges and Limitations	Potential Solutions
4	Regulatory frameworks and biosafety concerns	Establish standardized quality control, globally harmonized biosafety protocols; conduct risk assessments for persistence, gene transfer, and effects on nontarget organisms; foster transparency and stakeholder engagement.
5	Intellectual property rights and bioprospecting issues	Develop transparent licensing frameworks; ensure fair access and benefit-sharing; balance patents to encourage innovation while maintaining open science and ethical bioprospecting.
6	Public acceptance and farmer adoption	Strengthen scientific outreach, capacity building, and extension services; conduct demonstration trials and participatory research; ensure affordable, accessible, and quality-assured microbial products; provide government incentives and training programs.

11.5.6 Variability in Field Performance of Microbial Products

The widespread use of agricultural microbiome-based practices may be difficult because of the variable effectiveness of microbial inoculants in field settings. Some microbial inoculants have some known effectiveness when tested under controlled laboratory or greenhouse conditions, but effectiveness in real field conditions tends to be much less certain (Kaminsky et al., 2019). These variations may be due to many biotic and abiotic factors which are dynamic in nature, including soil pH, soil texture, soil water-holding capacity, soil nutrient availability, and potential interactions with indigenous soil microbial communities, any of which may influence the survival, colonization, and functional efficacy of microbial inoculants (Bashan et al., 2014). Also, the performance of microbial inoculants can be complex when considering environmental variation and host specificity, as inoculant effectiveness may vary, potentially, within an inoculant itself dependent on crop type and geographic location of origin (Toju et al., 2018). Further, microbial viability may succumb to loss when subjected to adverse storage, transport, or handling conditions, which limits the confidence and shelf life of the same microbial inoculant. Current research and development is focused on a few factors that may offer some promise to address these issues with respect to microbial inoculant field performance or persistence, including evaluating multimicrobial strain consortia, region-specific formulations, and/or long-lasting inert carrier materials. Integrating microbial technologies with precision agriculture technologies may also improve the effectiveness of inoculants by refining time, location, and compatibility with environmental conditions, ultimately promoting wider consistency and scalability of microbiome-led interventions in sustainable crop production.

11.5.7 Regulatory Frameworks and Safety Concerns

As microbial products gain greater importance in the realm of sustainable agriculture, they also gain more importance in terms of regulation and biosafety assessment. The deliberate release of live microorganisms into the agroecosystems gives rise to concerns about ecological perturbation, horizontal gene flow, and potential pathogenicity, thus highlighting the need for elaborate global regulation (Mitter et al., 2019). Regulatory mechanisms for microbial inoculants are highly diverse among countries and generally lag behind chemical inputs. In many regions, microbial products fall into the profile of biofertilizers or biopesticides based on their conceptual applications, with more elaborate systems in practice by entities such as the U.S. Environmental Protection Agency and European Food Safety Authority (Backer et al., 2018). However, most countries lack standardized protocols for the identification of microbes, efficacious assessment, and risk assessment, thereby generating barriers to international trade and the safe commercialization of these products.

Major biosafety issues comprise the potential for microorganisms that have been introduced to compete with indigenous microbiota, disrupting ecological balance, or supporting the mobilization of antibiotic resistance genes. Moreover, poorly defined strains can demonstrate opportunistic pathogenicity, and consequently, pose risk to plants, animals, and humans. Overcoming such issues involves in-depth risk assessments considering persistence, gene mobilization, and nontarget effects, as well as creating standardized global protocols (Chaudhry et al., 2021). It is imperative for developing competencies of a regulatory kind, in association with open stakeholder dialogue, in order to balance biosafety with innovation, so that the increased use of plant microbiome technologies facilitates the safe and efficacious development of sustainability in agriculture.

11.5.7.1 Intellectual Property and Bioprospecting Issues

The rapid advancement of microbial technologies in agriculture has posed numerous questions of intellectual property rights and bioprospection, most notably concerning ownership, access to and benefits from microbial genetic resources. Microbial stocks, once thought to be the common heritage of mankind, are now subject to commercial interests, raising questions on their ethical, legal, and economic implications. The Convention on Biological Diversity's Nagoya Protocol attempts to provide equitable benefit-sharing, yet its application to microorganisms has raised contentious issues largely because microorganisms are ubiquitous, their origins are not identifiable, and there is a lack of national attribution, thereby complicating their enforcement and further complicating biopiracy, especially in developing countries with high levels of biodiversity (Tvedt & Young, 2007). From a legal and IP standpoint, intellectual property protection for microbial strains, genes, metabolites, and other products can reward commercialization, thereby limiting access and the ability to collaborate for researchers and small-holder producers (Van Overwalle, 2005). To facilitate microbiome innovation that is both ethical and sustainable, open licensing arrangements and global frameworks would need to be developed, ensuring the open science movement, the equitable sharing of benefits, and legal protections for those biodiversity areas will strike the right balance. Moreover, and critically, it would be necessary to develop ABS arrangements that are fair and equitable while still being ethical and promote increased ethical IP-associated work to promote the use of microbiome products and services as a basis for sustainable agriculture, inclusion, and any other socioeconomic or environmental advances needed, globally.

11.5.8 Public Acceptance and Farmer Adoption

Although there is increasing scientific proof for plant microbiome–based technologies' advantages, their acceptability and uptake continue as main impediments for widespread application in farms. Acceptance by farmers influenced by agronomic performance, socioeconomic situations, and cultural beliefs is essential for the successful application of microbial products as biofertilizers, biopesticides, and microbial consortia (Chandler et al., 2011). The limited knowledge and awareness of microbial benefits in the farming community, especially in rural and developing countries where chemical fertilizers and pesticides provide faster and more visible results, is widely seen as an issue (Alori et al., 2017). Other issues include storage and application complexity, compatibility with existing agrochemicals, and variability in product consistency and performance (Timmusk et al., 2014). Trust among the public is also affected by legal insecurity, inadequate scientific outreach, and limited extension service, with the consequence of impairing better-informed decision-making and faith in these technologies. To foster enhanced uptake, capacitation, policy, and participatory research with the involvement of the farmer community is essential. Demonstrative trails, training initiatives, and incentives supported by governments can improve user faith, such that microbiome-based remedies become key components of reliable and society-acceptable farming systems.

11.6 FUTURE DIRECTIONS AND RESEARCH PRIORITIES

The growing knowledge of plant microbiome has provided multiple opportunities for changing agriculture toward sustainable practices. To capitalize on the opportunities that the plant microbiome provides, will require strategic research, development of technology, and incorporation with policy. Potential future directions include microbiome-based breeding of crops, aspects of artificial intelligence (AI) and big data, incorporation of microbiomes in agricultural policy, and collaboration across disciplines.

11.6.1 Microbiome-Based Breeding of Crops

Traditionally, crop breeding has relied upon the selection of common plant traits (i.e., yield, disease resistance, stress tolerance); however, new evidence has shown that plant genotype can have a direct impact on the composition and function of the microbiome associated to the plant (Berg et al., 2014). Well-designed breeding programs may encompass microbiome-optimized genotypes where cultivars can better recruit beneficial microbes to provide plants with nutrients, suppress disease, or reduce abiotic stresses. Unique maize cultivars have been identified to have selective enriched rhizosphere communities of nitrogen-fixing bacteria (Walters et al., 2018), allowing for a genetic gain to decrease fertilizer use. The focus of a plant breeding program may shift to dual selection by selecting for both the plant genotype and its microbiome.

11.6.2 AI and Big Data in Microbiome Research

A plant–microbe system is often multifaceted and heterogeneous; therefore, we require analytical frameworks that incorporate high-throughput sequencing, environmental attributes, and plant traits in an unobtrusive manner. Building upon increases to the use of AI, algorithms, and machine learning approaches to predicting microbial functions, identifying keystone taxa, and modifying plant–microbe–environmental interactions in these meta-systems (Trivedi et al., 2020), are big data platforms which offer flexibility for the researcher to query microbiome databases and navigate datasets from a global to a local scale for identifying cross-comparative meta-analyses informing global trends and local adaptations. Considering plant–microbe systems from this perspective provides novel opportunities for generating standardized AI-assisted data pipelines linking omics technologies, metagenomics, metatranscriptomics, and metabolomics, prior to developing targeted and elaborate microbiome interventions.

11.6.3 Integrating Microbiome Strategies into Agri-Policy

Utilization of microbiome-based strategies will require supportive policy instruments. Governments could support/invest in microbial technology programs via sustainable intensification programs, subsidy programs, and carbon credits. Policy instruments could strengthen advisory standards for microbial quality, thus addressing poorly produced or ineffective microbes (Sessitsch et al., 2019). An additional policy is also required in national agriculture policy to promote locally relevant products and build capacity for farmers to utilize microbial amendments which effectively impresses long-term benefits. In addition, internationally, agreements on chemicals such as the Nagoya Protocol need to broaden their approach to microbial genetic resources, to formulate standards for fair and equitable benefit-sharing across "communities" and enforce ethical guidelines.

11.7 CONCLUSION

The plant microbiome is a revolutionary strategy for sustainable intensification of agriculture, with green solutions for nutrient acquisition, stress tolerance, disease suppression, and crop promotion (Berg et al., 2014). By deploying plant–microbe relationships, researchers can boost productivity while rebuilding soil health and biodiversity to counter long-term decline from input-intense, chemical-based, and land-depleting conventional farming. To realize it, there is a need for a mechanistic description of plant–microbe–environment relationships, combining omics and ecological studies to distinguish keystone taxa and design optimal microbial consortia for specific soil, climate, and genotype applications (Busby et al., 2017). To practitioners, microbial inputs must be contemplated as components of integrated crop management systems, optimally used in combination with crop rotation, organic amendments, conservation tillage, and precision management (Backer et al., 2018). Training schemes and local adaptation remain critical to achieving consistent performance in the fields and safe application on a long-term perspective. Policymakers need to build enabling environments through quality assurance standards, site-specific incentives, and global cooperation in line with the Nagoya Protocol for fair benefit-sharing and biosafety (Sessitsch et al., 2019). Scaling of microbiome applications depends on sectoral cooperation, predictive tools based on AI, and design-informed by the farmer for expanding uptake and malleability. If appropriately integrated, plant microbiomes can, at the same time, expand productivity, reduce climate change, preserve biodiversity, and build global food security that represent a watershed move in developing a resilient and sustainable food system.

REFERENCES

Alori, E. T., M. O. Dare, and O. O. Babalola. 2017. "Microbial Inoculants for Soil Quality and Plant Health." In *Sustainable Agriculture Reviews*, edited by E. Lichtfouse, 281–307. Cham: Springer International Publishing. https://doi.org/10.1007/978-3-319-48006-0_9

Backer, R., J. S. Rokem, G. Ilangumaran, J. Lamont, D. Prabhu, and P. J. Smith. 2018. "Plant Growth-Promoting Rhizobacteria: Context, Mechanisms of Action, and Roadmap to Commercialization of Biostimulants for Sustainable Agriculture." *Frontiers in Plant Science* 9:1473. https://doi.org/10.3389/fpls.2018.01473

Bashan, Y., L. E. de-Bashan, S. R. Prabhu, and J.-P. Hernández. 2014. "Advances in Plant Growth-Promoting Bacterial Inoculant Technology: Formulations and Practical Perspectives (1998–2013)." *Plant and Soil* 378 (1): 1–33. https://doi.org/10.1007/s11104-013-1956-x

Bates, S. T., D. Berg-Lyons, J. G. Caporaso, W. A. Walters, R. Knight, and N. Fierer. 2011. "Examining the Global Distribution of Dominant Archaeal Populations in Soil." *ISME Journal* 5 (5): 908–917. https://doi.org/10.1038/ismej.2010.171

Bender, S. F., C. Wagg, and M. G. A. van der Heijden. 2016. "An Underground Revolution: Biodiversity and Soil Ecological Engineering for Agricultural Sustainability." *Trends in Ecology & Evolution* 31 (6): 440–452. https://doi.org/10.1016/j.tree.2016.02.016

Berendsen, R. L., C. M. J. Pieterse, and P. A. H. M. Bakker. 2012. "The Rhizosphere Microbiome and Plant Health." *Trends in Plant Science* 17 (8): 478–486. https://doi.org/10.1016/j.tplants.2012.04.001

Berg, G., M. Grube, M. Schloter, and K. Smalla. 2014. "The Plant Microbiome and Its Importance for Plant and Human Health." *Frontiers in Microbiology* 5: 491. https://doi.org/10.3389/fmicb.2014.00491

Bhattacharyya, P. N., and D. K. Jha. 2012. "Plant Growth-Promoting Rhizobacteria (PGPR): Emergence in Agriculture." *World Journal of Microbiology & Biotechnology* 28 (4): 1327–1350. https://doi.org/10.1007/s11274-011-0979-9

Bikard, D., and R. Barrangou. 2017. "Using CRISPR–Cas Systems for Genome Engineering." *Current Opinion in Microbiology* 37: 155–161. https://doi.org/10.1016/j.mib.2017.05.010

Bulgarelli, D., K. Schlaeppi, S. Spaepen, E. V. L. van Themaat, and P. Schulze-Lefert. 2013. "Structure and Functions of the Bacterial Microbiota of Plants." *Annual Review of Plant Biology* 64: 807–838. https://doi.org/10.1146/annurev-arplant-050312-120106

Busby, P. E., C. Soman, M. R. Wagner, M. L. Friesen, J. Kremer, et al. 2017. "Research Priorities for Harnessing Plant Microbiomes in Sustainable Agriculture." *PLoS Biology* 15 (3): e2001793. https://doi.org/10.1371/journal.pbio.2001793

Carvalho, F. P. 2017. "Pesticides, Environment, and Food Safety." *Food and Energy Security* 6 (2): 48–60. https://doi.org/10.1002/fes3.108

Chandler, D., A. S. Bailey, G. M. Tatchell, G. Davidson, E. Greaves, and J. Grant. 2011. "The Development, Regulation and Use of Biopesticides for Integrated Pest Management." *Crop Protection* 30 (3): 143–152. https://doi.org/10.1016/j.cropro.2010.08.003

Chandler, D., A. S. Bailey, G. M. Tatchell, G. Davidson, J. Greaves, and W. Grant. 2011. "The Development, Regulation and Use of Biopesticides for Integrated Pest Management." *Philosophical Transactions of the Royal Society B: Biological Sciences* 366 (1573): 1987–1998. https://doi.org/10.1098/rstb.2010.0390

Chaudhry, V., P. Runge, P. Sengupta, G. Doehlemann, J. E. Parker, and E. Kemen. 2021. "Shaping the Leaf Microbiota: Plant–Microbe–Microbe Interactions." *Journal of Experimental Botany* 72 (1): 36–56. https://doi.org/10.1093/jxb/eraa417

Classen, A. T., M. K. Sundqvist, J. A. Henning, G. S. Newman, J. A. M. Moore, M. A. Cregger, et al. 2015. "Direct and Indirect Effects of Climate Change on Soil Microbial and Soil Microbial–Plant Interactions: What Lies Ahead?" *Ecosphere* 6 (8): 130. https://doi.org/10.1890/ES15-00217.1

Compant, S., C. Clément, and A. Sessitsch. 2010. "Plant Growth-Promoting Bacteria in the Rhizo- and Endosphere of Plants: Their Role, Interactions, and Biotechnological Applications."*FEMS Microbiology Ecology* 74 (1): 1–13. https://doi.org/10.1111/j.1574-6941.2010.00906.x

Compant, S., A. Samad, H. Faist, and A. Sessitsch. 2019. "A Review on the Plant Microbiome: Ecology, Functions, and Emerging Trends in Microbial Application." *Journal of Advanced Research* 19: 29–37. https://doi.org/10.1016/j.jare.2019.03.004

Fierer, N. 2017. "Embracing the Unknown: Disentangling the Complexities of the Soil Microbiome." *Nature Reviews Microbiology* 15: 579–590. https://doi.org/10.1038/nrmicro.2017.87

Fierer, N., and R. B. Jackson. 2006. "The Diversity and Biogeography of Soil Bacterial Communities." *Proceedings of the National Academy of Sciences* 103 (3): 626–631. https://doi.org/10.1073/pnas.0507535103

Foley, J. A., N. Ramankutty, K. A. Brauman, E. S. Cassidy, J. S. Gerber, M. Johnston, N. D. Mueller, C. O'Connell, D. K. Ray, P. C. West, C. Balzer, E. M. Bennett, S. R. Carpenter, J. Hill, and C. Monfreda. 2011. "Solutions for a Cultivated Planet." *Nature* 478 (7369): 337–42. https://doi.org/10.1038/nature10452

Food and Agriculture Organization (FAO). 2017. *The Future of Food and Agriculture — Trends and Challenges*. Rome: FAO.

Gadd, G. M. 2000. "Bioremedial Potential of Microbial Mechanisms of Metal Mobilization and Immobilization." *Current Opinion in Biotechnology* 11 (3): 271–279. https://doi.org/10.1016/S0958-1669(00)00095-1

Garbeva, P., J. A. van Veen, and J. D. van Elsas. 2004. "Microbial Diversity in Soil: Selection of Microbial Populations by Plant and Soil Type and Implications for Disease Suppressiveness." *Annual Review of Phytopathology* 42: 243–270. https://doi.org/10.1146/annurev.phyto.42.012604.135455

Garnett, T. 2013. "Food Sustainability: Problems, Perspectives and Solutions." *Proceedings of the Nutrition Society* 72 (1): 29–39. https://doi.org/10.1017/S0029665112002947

Gebbers, R., and V. I. Adamchuk. 2010. "Precision Agriculture and Food Security." *Science* 327 (5967): 828–31. https://doi.org/10.1126/science.1183899

Geisseler, D., and K. M. Scow. 2014. "Long-Term Effects of Mineral Fertilizers on Soil Microorganisms — A Review." *Soil Biology & Biochemistry* 75: 54–63. https://doi.org/10.1016/j.soilbio.2014.03.023

Godfray, H. C. J., J. R. Beddington, I. R. Crute, L. Haddad, D. Lawrence, J. F. Muir, J. Pretty, S. Robinson, S. M. Thomas, and C. Toulmin. 2010. "Food Security: The Challenge of Feeding 9 Billion People." *Science* 327 (5967): 812–818. https://doi.org/10.1126/science.1185383

Haas, D., and G. Défago. 2005. "Biological Control of Soil-Borne Pathogens by Fluorescent Pseudomonads." *Nature Reviews Microbiology* 3 (4): 307–319. https://doi.org/10.1038/nrmicro1129

Hardoim, P. R., L. S. van Overbeek, and J. D. van Elsas. 2015. "The Hidden World within Plants: Ecological and Evolutionary Considerations for Defining Functioning of Microbial Endophytes." *Microbiology and Molecular Biology Reviews* 79 (3): 293–320. https://doi.org/10.1128/MMBR.00050-14

Hartmann, M., B. Frey, J. Mayer, P. Mäder, and F. Widmer. 2015. "Distinct Soil Microbial Diversity under Long-Term Organic and Conventional Farming." *The ISME Journal* 9 (5): 1177–94. doi:10.1038/ismej.2014.210

Johnston -Monje, D., and J. L. Mejía. 2020. "Botanical Microbiomes on the Cheap: Inexpensive Molecular Fingerprinting Methods to Study Plant-Associated Communities of Bacteria and Fungi." *Applications in Plant Sciences* 8 (8): e11334. https://doi.org/10.1002/aps3.11334

Kaminsky, L. M., R. V. Trexler, R. J. Malik, K. L. Hockett, and T. H. Bell. 2019. "The Inherent Conflicts in Developing Soil Microbial Inoculants." *Trends in Biotechnology* 37 (2): 140–151. https://doi.org/10.1016/j.tibtech.2018.11.011

Kloepper, J. W., M. N. Schroth, and T. D. Miller. 1980. "Effects of Rhizosphere Colonization by Plant Growth-Promoting Rhizobacteria on Potato Plant Development and Yield." *Phytopathology* 70: 1078–1082.

Kumar, A., S. K. Singh, C. Kant, H. Verma, D. Kumar, P. P. Singh, A. Modi, S. Droby, M. S. Kesawat, H. Alavilli, and S. K. Bhatia. 2021. "Microbial Biosurfactant: A New Frontier for Sustainable Agriculture and Pharmaceutical Industries." *Antioxidants* 10 (9): 1472. https://doi.org/10.3390/antiox10091472

Lal, R. 2015. "Sustainable Intensification for Adaptation and Mitigation of Climate Change and Advancement of Food Security in Africa." In *Sustainable Intensification to Advance Food Security and Enhance Climate Resilience in Africa*, edited by R. Lal, B. R. Singh, D. L. Mwaseba, D. Kraybill, D. O. Hansen, and L. O. Eik, 3–17. Cham: Springer. https://doi.org/10.1007/978-3-319-09360-4_1

Lauber, C. L., M. Hamady, R. Knight, and N. Fierer. 2009. "Pyrosequencing-Based Assessment of Soil pH as a Predictor of Soil Bacterial Community Structure at the Continental Scale." *Applied and Environmental Microbiology* 75 (15): 5111–5120. https://doi.org/10.1128/AEM.00335-09

Lebeis, S. L., et al. 2015. "Salicylic Acid Modulates Colonization of the Root Microbiome by Specific Bacterial Taxa." *Science* 349 (6250): 860–864. https://doi.org/10.1126/science.aaa8764

Lindow, S. E., and M. T. Brandl. 2003. "Microbiology of the Phyllosphere." *Applied and Environmental Microbiology* 69 (4): 1875–1883. https://doi.org/10.1128/AEM.69.4.1875-1883.2003

Lloyd, K. G., A. D. Steen, J. Ladau, J. Yin, and L. Crosby. 2018. "Phylogenetically Novel Uncultured Microbial Cells Dominate Earth Microbiomes." *m*Systems 3 (5): e00055–18. https://doi.org/10.1128/mSystems.00055-18

Lori, M., S. Symnaczik, P. Mäder, G. De Deyn, and A. Gattinger. 2017. "Organic Farming Enhances Soil Microbial Abundance and Activity — A Meta-analysis and Meta-Regression." *PLoS One* 12 (7): e0180442. https://doi.org/10.1371/journal.pone.0180442

Mendes, R., P. Garbeva, and J. M. Raaijmakers. 2013. "The Rhizosphere Microbiome: Significance of Plant Beneficial, Plant Pathogenic, and Human Pathogenic Microorganisms." *FEMS Microbiology Reviews* 37 (5): 634–663. https://doi.org/10.1111/1574-6976.12028

Mendes, R., M. Kruijt, I. de Bruijn, E. Dekkers, M. van der Voort, J. H. M. Schneider, Y. M. Piceno, T. Z. DeSantis, et al. 2011. "Deciphering the Rhizosphere Microbiome for Disease-Suppressive Bacteria." *Science* 332 (6033): 1097–1100. https://doi.org/10.1126/science.1203980

Mitter, B., G. Brader, N. Pfaffenbichler, and A. Sessitsch. 2019. "Next Generation Microbiome Applications for Crop Production — Limitations and the Need of Knowledge-Based Solutions." *Current Opinion in Microbiology* 49: 59–65. https://doi.org/10.1016/j.mib.2019.10.006

Naylor, D., and D. Coleman-Derr. 2018. "Drought Stress and Root-Associated Bacterial Communities." *Frontiers in Plant Science* 8: 2223. https://doi.org/10.3389/fpls.2017.02223

Ngumbi, E., and J. W. Kloepper. 2016. "Bacterial-Mediated Drought Tolerance: Current and Future Prospects." *Applied Soil Ecology* 105: 109–125. https://doi.org/10.1016/j.apsoil.2016.04.009

Niu, B., and R. Kolter. 2017a. "Complete Genome Sequences of Seven Strains Composing a Model Bacterial Community of Maize Roots." *Genome Announcements* 5 (36): e00997–17. https://doi.org/10.1128/genomeA.00997-17

Niu, B., J. Paulson, X. Zheng, and R. Kolter. 2017b. "Simplified and Representative Bacterial Community of Maize Roots." *Proceedings of the National Academy of Sciences of the United States of America* 114 (12): E2450–E2459. https://doi.org/10.1073/pnas.1616148114

Patten, C. L., and B. R. Glick. 2002. "Role of *Pseudomonas putida* Indoleacetic Acid in the Development of the Host Plant Root System." *Applied and Environmental Microbiology* 68 (8): 3795–3801. https://doi.org/10.1128/AEM.68.8.3795-3801.2002

Pérez -Jaramillo, J. E., R. Mendes, and J. M. Raaijmakers. 2016. "Impact of Plant Domestication on Rhizosphere Microbiome Assembly and Functions." *Plant Molecular Biology* 90 (6): 635–644. https://doi.org/10.1007/s11103-015-0337-7

Philippot, L., J. M. Raaijmakers, P. Lemanceau, and W. H. van der Putten. 2013. "Going Back to the Roots: The Microbial Ecology of the Rhizosphere." *Nature Reviews Microbiology* 11: 789–799. https://doi.org/10.1038/nrmicro3109

Pieterse, C. M. J., C. Zamioudis, R. L. Berendsen, D. M. Weller, and S. C. M. van Wees. 2014. "Induced Systemic Resistance by Beneficial Microbes." *Annual Review of Phytopathology* 52: 347–375. https://doi.org/10.1146/annurev-phyto-082712-102340

Pingali, P. L. 2012. "Green Revolution: Impacts, Limits, and the Path Ahead." *Proceedings of the National Academy of Sciences* 109 (31): 12302–12308. https://doi.org/10.1073/pnas.0912953109

Pretty, J., and Z. P. Bharucha. 2014. "Sustainable Intensification in Agricultural Systems." *Annals of Botany* 114 (8): 1571–1596. https://doi.org/10.1093/aob/mcu205

Rodríguez, H., and R. Fraga. 1999. "Phosphate Solubilizing Bacteria and Their Role in Plant Growth Promotion." *Biotechnology Advances* 17 (4–5): 319–339. https://doi.org/10.1016/S0734-9750(99)00014-2

Sasse, J., E. Martinoia, and T. Northen. 2018. "Feed Your Friends: Do Plant Exudates Shape the Root Microbiome?" *Trends in Plant Science* 23 (1): 25–41. https://doi.org/10.1016/j.tplants.2017.09.003

Schmidt, J. E., A. D. Kent, A. C. M. Gaudin, et al. 2019. "Agricultural Management and Plant Selection Interactively Affect Rhizosphere Microbial Community Structure and Nitrogen Cycling." *Microbiome* 7:146. https://doi.org/10.1186/s40168-019-0756-9

Sessitsch, A., N. Pfaffenbichler, and B. Mitter. 2019. "Microbiome Applications from Lab to Field: Facing Complexity." *Trends in Plant Science* 24 (3): 194–198.

Singh, A. K., P. P. Singh, V. Tripathi, H. Verma, S. K. Singh, A. K. Srivastava, and A. Kumar. 2018. "Distribution of Cyanobacteria and Their Interactions with Pesticides in Paddy Field: A Comprehensive Review." *Journal of Environmental Management* 224: 361–375. https://doi.org/10.1016/j.jenvman.2018.07.039

Singh, M., K. A. Qureshi, M. Jaremko, M. Rajput, S. K. Singh, Kaushalendra, K. D. Pandey, L. F. R. Ferreira, and A. Kumar. 2022. "Bioprospects of Endophytic Bacteria in Plant Growth Promotion and Ag-Nanoparticle Biosynthesis." *Plants* 11 (14): 1787. https://doi.org/10.3390/plants11141787

Smith, S. E., and D. J. Read. 2008. *Mycorrhizal Symbiosis*. 3rd ed. London: Academic Press.

Suttle, C. A. 2007. "Marine Viruses — Major Players in the Global Ecosystem." *Nature Reviews Microbiology* 5 (10): 801–812. https://doi.org/10.1038/nrmicro1750

Tilman, D., K. G. Cassman, P. A. Matson, R. Naylor, and S. Polasky. 2002. "Agricultural Sustainability and Intensive Production Practices." *Nature* 418 (6898): 671–677. https://doi.org/10.1038/nature01014

Timmusk, S., I. A. Abd El-Daim, L. Copolovici, T. Tanilas, A. Kännaste, L. Behers, E. Nevo, G. A. Seisenbaeva, E. Stenström, and Ü. Niinemets. 2014. "Drought-Tolerance of Wheat Improved by Rhizosphere Bacteria from Harsh Environments: Enhanced Biomass Production and Reduced Emissions of Stress Volatiles." *PLoS One* 9 (5): e96086. https://doi.org/10.1371/journal.pone.0096086

Toju, H., K. G. Peay, M. Yamamichi, K. Narisawa, K. Hiruma, K. Naito, S. Fukuda, et al. 2018. "Core Microbiomes for Sustainable Agroecosystems." *Nature Plants* 4 (5): 247–257. https://doi.org/10.1038/s41477-018-0139-4

Trivedi, P., J. E. Leach, S. G. Tringe, T. Sa, and B. K. Singh. 2020. "Plant–Microbiome Interactions: From Community Assembly to Plant Health." *Nature Reviews Microbiology* 18: 607–621. https://doi.org/10.1038/s41579-020-0412-1

Turner, T. R., E. K. James, and P. S. Poole. 2013. "The Plant Microbiome." *Genome Biology* 14: 209. https://doi.org/10.1186/gb-2013-14-6-209

Tvedt, M. W., and T. Young. 2007. Beyond Access: Exploring Implementation of the Fair and Equitable Sharing Commitment in the CBD. Gland: IUCN (IUCN Environmental Policy and Law Paper No. 67/2).

Van Bruggen, A. H. C., M. M. He, K. Shin, V. Mai, K. C. Jeong, M. R. Finckh, and others. 2018. "Environmental and Health Effects of the Herbicide Glyphosate." *Science of the Total Environment* 616–617: 255–268. https://doi.org/10.1016/j.scitotenv.2017.10.309

van der Heijden, M. G. A., et al. 2008. "The Unseen Majority: Soil Microbes as Drivers of Plant Diversity and Productivity in Terrestrial Ecosystems." *Ecology Letters* 11 (3): 296–310. https://doi.org/10.1111/j.1461-0248.2007.01139.x

Van Overwalle, G. 2005. "Protecting and Sharing Biodiversity and Traditional Knowledge: Holder and User Tools." *Ecological Economics* 53 (4): 585–607. https://doi.org/10.1016/j.ecolecon.2004.10.014

Vandenkoornhuyse, P., A. Quaiser, M. Duhamel, A. Le Van, and A. Dufresne. 2015. "The Importance of the Microbiome of the Plant Holobiont." *New Phytologist* 206: 1196–1206. https://doi.org/10.1111/nph.13312

Verma, H., D. Kumar, V. Kumar, M. Kumari, S. K. Singh, V. K. Sharma, S. Droby, G. Santoyo, J. F. White, and A. Kumar. 2021. "The Potential Application of Endophytes in Management of Stress from Drought and Salinity in Crop Plants." *Microorganisms* 9(8): 1729. https://doi.org/10.3390/microorganisms9081729

Vessey, J. K. 2003. "Plant Growth-Promoting Rhizobacteria as Biofertilizers." *Plant and Soil* 255 (2): 571–586. https://doi.org/10.1023/A:1026037216893

Vorholt, J. A. 2012. "Microbial Life in the Phyllosphere." *Nature Reviews Microbiology* 10 (12): 828–840. https://doi.org/10.1038/nrmicro2910

Walters, W. A., Z. Jin, N. Youngblut, et al. 2018. "Large-Scale Replicated Field Study of Maize Rhizosphere Identifies Heritable Microbes." *Proceedings of the National Academy of Sciences* 115 (28): 7368–7373. https://doi.org/10.1073/pnas.1800918115

Yang, J., J. W. Kloepper, and C.-M. Ryu. 2009. "Rhizosphere Bacteria Help Plants Tolerate Abiotic Stress." *Trends in Plant Science* 14 (1): 1–4. https://doi.org/10.1016/j.tplants.2008.10.004

12 Microbial Solutions to Drought Agriculture

Nisha Yadav, Neha Singh, and Shaili Srivastava

12.1 INTRODUCTION

A major global concern that threatens crop growth and productivity is drought; this occurs annually and has a severe influence on crop yields. A meta-analysis indicated that global crop losses from drought stress likely outnumbered those from all other abiotic stresses combined, with drought-stressed crops seeing a 48% fall in yield. Based on the causes drought can be divided into several categories, such as meteorological drought, which is defined by a sharp decline in precipitation; agricultural drought, which is defined as a lack of moisture that affects crop productivity; hydrological drought, which is defined as a decrease in water levels of river, lakes, etc.; and socioeconomic drought, which is defined as a negative impact on communities and economies as a result of inadequate water resources (Cohen et al., 2021).

Drought occurrences have become more frequent and severe in recent decades, a trend that is closely associated with human-caused climate change. In many areas, the probability of protracted and severe droughts is increased by rising global temperatures, which also worsen evaporation rates and change precipitation patterns. Drought-prone regions are expected to keep expanding, according to climate projections, which will have a big impact on ecosystem resilience, agricultural sustainability, and water resource management. To effectively reduce the negative consequences of drought and improve the resilience of both natural and human systems, it is imperative to comprehend these tendencies (Patil et al., 2023).

For agricultural production and the survival of natural ecosystems, drought tolerance is an essential quality. In the field of agriculture, food security is directly impacted by crops' resilience to dry spells. The need for resilient agricultural systems to guarantee a steady supply of food and reduce the risks of crop failure grows as the world's population rises and climate variability increases. In areas prone to frequent droughts, increasing crop resilience to drought can result in more consistent yields, less reliance on irrigation, and better financial results for farmers (Ilyas et al., 2021).

Beyond agriculture, the well-being and functional efficiency of natural ecosystems depend heavily on their inherent drought tolerance. Sufficient water availability is essential to the health of a variety of biological communities and ecosystem functions in natural environments, such as wetlands, grasslands, and forests. Stress from drought can cause plants to grow less, become more vulnerable to pests and diseases, and even cause important species to die off in large numbers. These modifications not only reduce biodiversity but also interfere with ecological services

DOI: 10.1201/9781003664451-12

such as soil stabilization, carbon sequestration, and water purification. Natural ecosystems must possess robust drought resilience to preserve ecological balance and sustain the critical services that underpin human health and environmental integrity (Oszust et al., 2023).

Drought is one of these factors that harms plant productivity and growth. Numerous physiological and biochemical activities of plants, including photosynthesis, respiration, transpiration, metabolism of carbohydrates, and nutrient absorption, translocation, and assimilation, are negatively impacted by this stress (Khan et al., 2018). Effective and sustainable management strategies are essential to mitigate drought-induced yield losses and maintain agricultural productivity. Traditional breeding techniques and crop genetic engineering for drought resistance are the two main strategies that have been used in previous years to address drought stress in plants.

Plant growth–promoting microorganisms (PGPMs) are gaining popularity as an attractive and affordable alternative strategy for mitigating drought stress. Actinomycetes, mycorrhiza, and plant growth–promoting rhizobacteria (PGPRs) have been shown in numerous studies to impact plant growth and drought stress tolerance (Gontia-Mishra et al., 2016). Current knowledge regarding drought stress, plants' physiological and molecular reactions, is compiled in this chapter. The chapter's latter part emphasizes the critical PGPM processes for reducing plant drought stress.

12.2 IMPACT OF DROUGHT STRESS ON CROP GROWTH

Crops that are under water stress usually show reduced leaf area, higher root-to-shoot ratios, and poor growth. The formation of reactive oxygen species (ROS), photosynthesis, protein synthesis, food intake and assimilation, and many other plant cellular processes are all negatively impacted by prolonged dryness.

12.2.1 Germination

For seedling establishment and growth, successful seed germination is essential. Many crops' most vulnerable phases to water scarcity are seed germination and early seedling growth, with drought stress adversely affecting metrics including seed vigor index and seed germination index for different crop species. Putting a desiccated seed in water causes it to germinate in three different stages for nondormant seeds. Water absorption and imbibition are part of phase I of the germination process. In drought stress conditions, water potential gradient between the seed's interior and exterior decreases gradually, that hinders water entry into the seed coat and subsequent water absorption. This leads to decreased seedling vigor and prevents germination (Batool et al., 2023). Phase II, sometimes referred to as the lag phase, is when reserve mobilization starts and water intake stops. Drought stress lengthens the time needed for carbohydrate hydrolysis and results in unsuccessful seed germination by reducing the activity of hydrolytic enzymes such amylase, which catalyzes the breakdown of starch into simpler sugars (Khan et al., 2023). Plant hormones including gibberellic acid (GA) and abscisic acid (ABA) are essential for seed germination; ABA has the opposite effect of GA, which activates hydrolytic enzymes. Lack of water causes ABA levels to rise while GA levels fall, which prevents the production of hydrolytic enzymes and reserves from being mobilized (Hassan et al., 2021). In addition to hydrolytic enzyme activity, water scarcity also negatively impacts the activities of other enzymes, including proteases, α-amylase, and β-amylase, which further impedes the mobilization of seed reserves (Hameed et al., 2021). Radicles emerging from the seed coat indicate phase III (Khan et al., 2023). When drought stress is only applied during Phase II, it has no effect on radicle emergence in common beans, but when drought stress is applied after Phase II, it delays radicle emergence (Stephan & Laval-Martin, 2000).

Due to decreased enzymatic activity, the antioxidant defense system is unable to scavenge the increased ROS levels under drought stress, causing ROS generation to exceed its capacity. The lack of seed germination is thus the result of decreased respiration and ATP generation, and impaired metabolic activity (Wei et al., 2023).

12.2.2 Photosynthesis

A key factor in influencing photosynthetic capability and, in turn, crop output is chlorophyll, the major pigment involved in photosynthetic processes. Drought exposure is frequently linked to a lower chlorophyll content in crops including canola, soybean, quinoa, and wheat. Crop yields are consistently greater in genotypes with a higher chlorophyll content because they are better at photosynthetic processes (Zhang et al., 2015). The reason for a lower chlorophyll content is the decreased availability of vital elements like N, Fe, and Mg, which are necessary for the synthesis of chlorophyll (Six, 2011). Chlorophyll precursors are broken down and chlorophyllase activity is elevated when ROS buildup during drought stress causes chloroplasts to degrade. Moreover, photoreaction rates are negatively impacted by an imbalance between light absorption and utilization brought on by variations in the ratio of chlorophyll to carotenoid (Hussain et al., 2018). Numerous studies have documented the varying reactions of carotenoids to water scarcity. Some have found that plants like sunflower, sorghum, and wheat have reduced levels of carotenoid content (Manivannan et al., 2014), while others have found higher levels (Gajanayake & Reddy, 2016). These conflicting results most likely result from variations in species, duration, and severity of drought stress. By reducing oxidative damage brought on by drought stress, carotenoid concentrations are expected to rise and fall under moderate and severe drought stress, respectively. In these situations, they are essential to the plant's antioxidant defense mechanism.

Water stress reduces the buildup of photosynthetic products by influencing both the light and dark processes. Numerous studies have examined the decreased stomatal conductance under water stress, which is caused by an increased ABA content in guard cells (Elferjani & Soolanayakanahally, 2018). The main way that plants react to drought stress is through ABA-mediated stomatal closure, which limits CO_2 intake and reduces water loss through transpiration, maintaining hydraulic function but eventually lowering photosynthetic rates.

Additionally, mesophyll conductance limits crop photosynthesis. Under conditions of water stress, non-stomatal limitations of photosynthesis, such as ribulose-1, 5-bisphosphate carboxylase/oxygenase (Rubisco) carboxylation capacity or mesophyll conductivity, are significantly more noticeable than stomatal limitations (Li et al., 2022). Leaf anatomical features including leaf and cell wall thickness and chloroplast morphology are the main determinants of mesophyll conductivity. Crop species, the length and severity of stress, and the resulting decreased mesophyll conductance play a significant role in lowering photosynthetic rates. Reduced CO_2 diffusion within the mesophyll during drought stress is influenced by decreased carbonic anhydrase and aquaporin activities. Reduced photosynthetic electron transport components build up as a result of low chloroplastic CO_2 concentrations during water shortage, which may have an impact on molecular oxygen generation and produce ROS (Basu et al., 2016). The oxidation of chloroplast proteins, DNA, and lipids (Batool et al., 2023) and the decrease of enzymes necessary for carbon uptake and utilization are caused by ROS overabundance. For instance, under water scarcity, Rubisco activity and abundance drastically decrease, which has an adverse effect on photosynthesis (Hassan et al., 2020).

12.2.3 Biomass and Yield

The main cause of decreased yields and food shortages is drought (Al-Qubati et al., 2023). Inadequate watering during the early vegetative and reproductive phases significantly reduces crop yield and biomass. Cell growth is hindered by low photosynthetic rates, limited uptake of water and nutrients, and reduced leaf area, internode elongation, and plant height (Zia et al., 2021). Additionally, the final yield is primarily determined by vegetative growth during stem elongation, and poorer flower and grain development results from less carbon assimilation (Sinclair & Jamieson, 2006). The negative effects of drought are particularly noticeable during the reproductive stage, when a lack of water affects pollen development, flowering, fertilization, and grain growth in addition to impeding flower initiation, ovary and embryo development (Barnabas et al., 2008). Water scarcity has a substantial and adverse effect on important agricultural production components, including rice, wheat, cotton, potatoes, maize, soybeans, and sugar beet (Alavilli et al., 2023). Zhang et al. (2018) conducted a meta-analysis and found that drought stress reduced rice and wheat yields by 25.4% and 27.5%, respectively.

12.3 PLANT–MICROBE INTERACTION UNDER DROUGHT STRESS

The responses of microbial communities are governed by changes in crop morphology and metabolism under drought conditions. A prior study found that the abundance of 163 metabolites changed significantly during water stress in soybean roots (Tripathi et al., 2016). This was due to the inhibition of photosynthetic activity due to stomatal closure, which causes an increase in root depth and density and, consequently, an increase in the root/shoot ratio, which in turn affects the composition and distribution of microbiomes. Root exudates are the ultimate designers of the rhizosphere microbiome, which is regulated by plants to enhance plant adaptation in a particular ecosystem and encourage the growth of microorganisms. The level of drought stress generally causes an increase in root exudates, which include soluble sugars, amino acids, and organic acids. Microbial activities are disrupted, and root-associated microbial populations are shaped by altered carbon exudation through rhizodeposition. Preece and

Peñuelas (2016) concluded that while rhizodeposition is very species-dependent, so are the consequences on soil communities. Glycerol-3-phosphate, a glycolysis intermediate and a precursor for peptidoglycan biosynthesis and cell wall construction in bacteria, is one of the several groups of chemicals that increases during water deficiency stress in sorghum.

12.4 BENEFICIAL MICROBES FOR DROUGHT TOLERANCE

12.4.1 Plant Growth–Promoting Rhizobacteria

PGPRs are a broad group of bacteria that invade plant roots and provide their host plants with a variety of advantages, especially when they are stressed by environmental factors like drought. Genera like *Azospirillum*, *Bacillus*, *Rhizobium*, and *Pseudomonas* are notable instances of PGPR. Their capacity to flourish in the rhizosphere—the small area of soil that is immediately impacted by root secretions—and form advantageous relationships with plants is what distinguishes these bacteria. PGPR promotes plant growth in a number of ways, both directly and indirectly. Indole-3-acetic acid (IAA) and other phytohormones are synthesized through direct processes. Furthermore, by removing nitrogen from the atmosphere, dissolving phosphates, and releasing other vital minerals, PGPR can improve soil fertility and plant health by facilitating nutrient acquisition (Santoyo et al., 2021).

An indirect way that PGPR provides drought tolerance is through the synthesis of exopolysaccharides, which enhance soil structure and water retention and lessen the effects of water stress. Furthermore, by promoting the expression of genes that respond to stress, PGPR can help plants develop systemic resistance, strengthening their natural defenses against oxidative stress brought on by drought. By lowering ethylene levels through the synthesis of the enzyme 1-aminocyclopropane-1-carboxylate (ACC) deaminase, PGPR also shields the plant from stress. PGPR is essential to creating sustainable farming methods that increase crop drought resistance because of these complex relationships (Basu et al., 2021).

12.4.2 Mycorrhizal Fungi

Most terrestrial plants have symbiotic relationships with mycorrhizal fungi in their roots, which are essential for improving plant water intake and drought resistance in general. Both ectomycorrhizae and arbuscular mycorrhizae are the two main forms of mycorrhizae, and they vary in their functional contributions and structural relationships with plant roots. Plant roots' cortical cells are penetrated by arbuscular mycorrhizal fungi (AMF), which are members of the phylum Glomeromycota. They create highly branching structures called arbuscules. By facilitating the flow of nutrients, these structures allow the fungi to help the plant absorb water and minerals, especially phosphorus, in return for the carbohydrates that get produced during photosynthesis. Ectomycorrhizal fungi, which are mostly found in woody plants and trees, increase the surface area available for absorbing water and nutrients by forming a sheath around the roots and extending their hyphae into the surrounding soil (Branco et al., 2022).

In drought situations, mycorrhizal connections have numerous advantages. In times of scarcity, mycorrhizal fungi efficiently increase the plant's capacity to absorb water by improving its ability to get water from a larger volume of soil. Additionally, they increase soil porosity and decrease evaporation and water runoff by linking soil particles into aggregates. Furthermore, by altering hormonal pathways and strengthening the plant's antioxidant defenses, mycorrhizal fungi help plants tolerate stress and lessen the negative consequences of oxidative stress brought on by drought. Mycorrhizal fungi are essential for sustainable agriculture and ecosystem resilience since they have been demonstrated to increase plant survival rates, biomass accumulation, and overall growth performance in drought-prone environments (Mikiciuk et al., 2019).

12.4.3 Endophytes

Endophytic microbes, often known as endophytes, live inside plant tissues and don't seem to be harmful. These microorganisms, which include fungus and bacteria, have close relationships with the plants they inhabit and provide many advantages that improve plant growth and resistance to stress. Endophytes have the ability to colonize a variety of plant components, such as roots, leaves, and even seeds. They establish a symbiotic and protective association that is especially beneficial during dry spells. Endophytes can directly affect a plant's physiological functions by living inside its tissues, which increases the plant's resistance to water deficit stress (Salvi et al., 2022).

Endophytes play a variety of roles in improving drought resilience. They can generate phytohormones, including cytokinins and gibberellins, which control plant growth and development and promote improved root structure and more effective water absorption. By producing osmoprotectants like proline and trehalose, which support cellular water balance and shield cellular structures from dehydration, endophytes also aid in osmotic adjustment. Additionally, by strengthening the plant's antioxidant defense mechanisms, endophytic microorganisms might lessen the buildup of ROS, which are normally increased during drought stress. The plant's overall resistance to drought is improved by this antioxidant defense, which reduces cellular damage and preserves cellular integrity. Endophytes' potential as natural agents for enhancing plant tolerance to drought is further highlighted by their capacity to alter gene expression linked to stress responses (Koopman et al., 2018).

12.4.4 Other Beneficial Microbes

In addition to PGPRs, mycorrhizal fungi, and endophytes, a number of different microorganism groups make distinct contributions to plants' ability to withstand drought. Known for their filamentous growth, actinomycetes are a class of Gram-positive bacteria that are well known for producing a variety of bioactive substances, such as enzymes that can improve plant stress tolerance and antibiotics. By breaking down organic debris, these microbes increase soil fertility and structure, which improves water retention and plant availability. Furthermore, like PGPR, actinomycetes can develop symbiotic partnerships with plants to support nutrient intake and root growth (Oyedoh et al., 2023).

Blue-green algae, also beneficial and play a crucial role in drought tolerance. Because of their ability to fix atmospheric nitrogen, these photosynthetic bacteria can increase soil fertility and lessen the need for chemical fertilizers. Because they secrete extracellular polysaccharides that aid in the production of soil aggregates, cyanobacteria help to stabilize soil and retain water. In dry and semiarid areas that are vulnerable to drought, they are useful partners due to their capacity to increase soil moisture content.

Additionally, some bacterial and fungal species have special skills, such as the ability to produce volatile organic compounds (VOCs) that might alter plant development and stress reactions. Plants can develop systemic tolerance as a result of these VOCs, making them more resilient to drought. Furthermore, certain microbes have the capacity to break down soil contaminants, enhancing soil quality and fostering an environment that is more favorable for plant growth under stressful situations (Chieb & Gachomo, 2023).

The variety of helpful microorganisms goes beyond the well-researched endophytes, mycorrhizal fungi, and PGPRs. Through different processes, cyanobacteria, actinomycetes, and other specialized microorganisms each contribute differently to drought tolerance and increase plant resilience. In the face of growing water scarcity, the incorporation of these varied microbial populations into ecological and agricultural processes holds enormous promise for creating effective solutions to counteract drought stress and guarantee sustained plant productivity.

12.5 MECHANISM OF DROUGHT TOLERANCE BY SOIL MICROBIOME

Global food security is severely threatened by drought stress, which is exacerbated by climate change. It can affect arable land worldwide and result in significant crop losses. Sustainable agricultural techniques are needed to address this issue, and PGPMs are showing promise to improve crop drought resilience and help achieve the International Union for Conservation of Nature (IUCN)'s Sustainable Development Goal 2 (SDG 2), Zero Hunger. To deal with drought, plants use a variety of strategies, including as morphological, physiological, and molecular adjustments. An economical and environmentally responsible method of sustainable agriculture is provided by microbial metabolites and related genes, which are essential in reducing drought stress. Reduced photosynthesis, stomatal closure, and increased photorespiration are some of the negative consequences of drought on crops that result in salt buildup and nutrient deficits. These elements put the world's food security at risk by reducing crop growth and yield. Physiological adaptations like osmotic regulation, morphological changes like decreased leaf area and altered root systems, and the synthesis of antioxidants to fight oxidative stress are some of the ways that plants react to drought. Beneficial soil bacteria use a variety of strategies to lessen the effects of drought stress. This comprises the creation of phytohormones, the synthesis of volatile compounds by soil bacteria, changes in the root's morphology, the action of ACC deaminase, the accumulation of osmolytes, the performance of exopolysaccharides, antioxidant resistance, and molecular techniques.

12.5.1 Alteration of Phytohormones Production in Plants

Growth and development of plants depend on phytohormones such as ethylene, GA, cytokinins, and ABA. Plants can escape or survive stressful situations thanks in large part to phytohormones (Yadav 2021). PGPRs, including *Bacillus thuringiensis*, *Azospirillum lipoferum*, and *Phyllobacterium brassicacearum*, can produce phytohormones (cytokinin, ABA, gibberellins, and IAA) that promote cell division and growth, increase the uptake of water and nutrients, and trigger the stress transduction signal pathways that reduce leaf transpiration in a variety of crops (Arzanesh et al., 2011). Some PGPR isolates generated IAA, which helped the seedlings withstand drought stress by increasing their root volume. Hussain et al. (2014) reported that under drought stress, *Bacillus* sp. (12D6) and *Enterobacter* sp. (16i) metabolized salicylic acid (SA) and IAA, which markedly changed root morphology and stimulated growth in wheat (*Triticum aestivum*) and maize (*Zea mays*) crops.

12.5.2 Functions of Volatile Compounds

Under different stressors, plants accumulate volatile compounds. These volatile substances trigger signals for various metabolic processes within and between plants.

Pseudomonas chlororaphis produced 2R, 3R-butanediol, and volatile metabolites that stop water loss by causing stomatal closure in the plant through an SA-dependent mechanism when it was present in the rhizosphere of the *Arabidopsis* plant. In contrast, no tolerance mechanism has been demonstrated when bacteria are absent. According to Garbeva and Weisskopf (2020), rhizobacteria release a variety of volatile organic chemicals that also boost resistance to abiotic stress.

12.5.3 Modification in Root Morphology

One of the earliest steps toward increased resilience to drought stress was the change of the flexibility of the root cell membrane by bacteria. Rhizobacteria that promote plant growth greatly improve the integrity of cell membranes by inducing the antioxidant defense system and mediating drought tolerance (Gusain et al., 2015).

12.5.3.1 ACC Deaminase Production

As a precursor to ethylene, 1-aminocyclopropane-1-carboxylate synthase transforms s-adenosylmethionine (S-AdoMet) into ACC in ethylene metabolic pathways. During drought, ethylene regulates plant growth and development to preserve plant homeostasis endogenously (Gontia-Mishra et al., 2020). In order to give bacteria nitrogen and energy, bacterial ACC deaminase depleted the sequestered plant ACC. Under drought stress, a lower ACC concentration in the plant promotes plant development and mitigates the harmful effects of ethylene. ACC-producing bacteria eliminated the impact of drought stress on pea yield and yield characteristics during ripening in both pot and field experiments. Similarly, in order to mitigate drought stress, *Bacillus licheniformis* K11-infected pepper secreted ACC deaminase and six distinct stress proteins, which were identified by two-dimensional polyacrylamide gel electrophoresis. In tomato plants, Rhizo SF 48, which produces ACC deaminase, increases resistance to oxidative damage in conditions of moisture stress and stimulates plant development.

12.5.3.2 Osmolytes Production

Through metabolic alterations, plant drought tolerance is associated with the accumulation of several ideal electrolytes/osmolytes, such as choline, betaines, polyamines, carbohydrates, proline, quaternary ammonium compounds, polyhydric alcohols, other amino acids, and water stress proteins like dehydrin. PGPR secretes osmolytes in response to drought stress, which complement those produced by plants. Paul et al. (2008) claim that it promotes the growth and development of plants. In plants under moisture stress, *Pseudomonas putida* GAP-P45 inoculation enhanced proline accumulation, leading to improvements in dry matter weight (DMW), relative water (RWC) content of the leaf, and leaf water efficiency. Trehalose metabolism by rhizobacteria is crucial for determining plant growth, productivity, and resistance to biotic and abiotic stressors, much like it is for proline. Furthermore, crops are significantly impacted agronomically by trehalose management. *Bacillus subtilis* GB03, *Klebsiella variicola* F2 (*Arabidopsis*), and *Pseudomonas fluorescens* YX2 (maize) all produce choline as a precursor to glycine betaine metabolism, which raises RWC and DMW and mediates the dry tolerance function. Moreno-Galván et al. (2020) discovered that proline accumulation and glutathione reductase activity stimulation were the primary causes of drought stress in Guinea grass.

12.5.3.3 Production of Exopolysaccharides

Rhizobacteria are protected from harsh environments and are able to flourish because of the extracellular polymeric substances (EPS) they produce (Vives-Peris et al., 2020). By creating cation bridges, H-bonding, Waals forces, and adsorption mechanisms, rhizobacteria release EPS into the soil as capsular or slime materials that adhere to soil colloidal surfaces and increase the stability of soil aggregates. In order to protect microorganisms and roots from drought stress, EPS provide a micro-environment that retains available water and keeps the environment from drying out. Under water-deficit stress, plants inoculated with EPS-producing bacteria have demonstrated higher organic osmolyte accumulation.

12.6 AGRICULTURAL INNOVATIONS AND APPLICATIONS

A variety of novel techniques focused on improving crop resilience and productivity in drought-prone environments have been sparked by the use of beneficial bacteria in agriculture. Precision technologies for monitoring, the creation of synergistic microbial communities, sophisticated microbial delivery methods, and the application of successful case studies that illustrate the usefulness of microbe-based drought mitigation are all examples of these agricultural innovations. This section delves further into these innovations, showcasing the technology and approaches used, their effectiveness, adoption rates, and the observable results attained in diverse agricultural contexts (Ault, 2020).

12.6.1 Beneficial Microbes for Seed Coating and Soil Amendments

Incorporating helpful microorganisms into agricultural systems through seed coating and soil amendments is essential for promoting improved plant growth and drought resistance from the very beginning of plant development. By directly applying microbial inoculants to seeds prior to planting, seed coating guarantees the presence of advantageous microbes from the very beginning of germination. This technique makes use of a range of materials and technologies to produce coatings that effectively shield the microorganisms and guarantee their survival and effectiveness. Common seed coating materials include biopolymers like chitosan and alginate, which create protective matrices that enclose the microorganisms and offer a stable environment for handling and storing seeds. To improve the adherence and dispersion of microorganisms on seed surfaces, transporters such as humic materials, clay minerals, and inert fillers are also used (Etesami, 2021).

However, the process of adding microbial inoculants to soil in several forms, such as granular formulations, liquid suspensions, and carriers based on biochar, is known as soil amendment. By increasing the diversity and activity of soil microbes, these additions aim to increase the soil's fertility, structure, and ability to retain water. Technologies including encapsulation methods and slow-release formulations are used to extend the effectiveness and persistence of introduced microbes in the soil environment (Zhang et al., 2022).

The effectiveness of soil amendment and seed coating procedures has been proven in several research studies, showing notable improvements in drought tolerance, yield stability, and plant growth indices. The application of PGPR-coated seeds, for instance, has been demonstrated to enhance biomass accumulation, root architecture development, and water-use efficiency, thereby facilitating plant survival during periods of water scarcity. Similarly, the integration of AMF as soil amendments improves nutrient bioavailability, mitigates soil erosion, and enhances moisture retention, collectively bolstering plant resilience to drought stress (Etesami, 2021).

Several factors, including cost, accessibility, and farmer awareness, influence the adoption rates of these technologies across various agricultural sectors and geographical areas. The growing understanding of microbial benefits in sustainable agriculture has driven their wider application, particularly within areas vulnerable to frequent water shortages and drought conditions. The creation of farmer support networks, including extension services and training courses, as well as the development of user-friendly formulations, has further facilitated the widespread adoption of soil amendment and seed coating techniques (Sudheer et al., 2020).

In general, seed coating and soil amendments containing advantageous microorganisms are successful agricultural technologies that offer a basis for improving crop drought resistance. Promoting microbial symbiosis from the earliest stages of plant growth establishes robust, sustainable systems that maintain productivity despite the challenges of arid and semiarid conditions.

12.6.2 Precision Agriculture and Microbe Monitoring

Various environmental and soil characteristics that affect microbial activity and plant health are monitored using sensors and Internet of Things (IoT) technology. By providing constant information on soil conditions, temperature probes, pH sensors, and soil moisture sensors help farmers make well-informed decisions on the use of microbial inoculants, fertilizer, and irrigation. Microbial sensors, like biosensors that identify particular metabolites or genetic markers, can also reveal information about the existence and activity of helpful microorganisms in the soil–plant system. The integration of real-time monitoring facilitates the spatiotemporal optimization of microbial inoculant delivery, enabling precision application that mitigates drought stress by targeting interventions only where and when physiological thresholds are met (Haddi et al., 2020).

In precision agriculture, data-driven methods such as predictive analytics and machine learning algorithms are essential for maximizing microbiological applications. These methodologies are capable of detecting patterns and connections among environmental factors, microbial activity, and plant performance by examining huge information produced by sensors and IoT devices. Predictive models that predict crop results, microbial reactions, and drought episodes can be created using this data, allowing for proactive and flexible management techniques. Machine learning–based predictive models utilize high-resolution environmental data, including temperature fluctuations, phenological growth stages, and soil water status, to determine the optimal timing and dosage for microbial inoculation. This spatiotemporal optimization maximizes the efficacy of bio-stimulants while significantly minimizing input wastage and environmental leakage (Agrawal et al., 2024).

Additionally, precision agriculture makes it easier to adapt microbial treatments to particular crop needs and field conditions. By using site-specific management techniques that customize microbial inoculants to the particular requirements of various regions within a field, spatial heterogeneity in soil characteristics, moisture content, and microbial populations can be handled. By preventing overapplication and making sure that microorganisms are used effectively, this focused strategy not only increases the efficacy of microbial therapies but also lessens their negative effects on the environment (Dehnavi & Ebrahimipour, 2022).

In a variety of agricultural contexts, the combination of precision agriculture and microbial treatments has shown notable advantages. Precision irrigation systems with real-time microbiological monitoring have improved crop yields, improved plant health, and reduced water consumption in drought-prone areas of Australia. Similar to this, the precise application of mycorrhizal fungi in southeast Asian rice paddies has been made possible by the use of IoT-enabled sensors to monitor soil moisture and microbial activity. This has improved water retention, decreased the need for irrigation, and raised rice productivity in water-limited environments. Ultimately, the adoption of advanced tools like spatiotemporal monitoring and precision delivery is paramount for optimizing bio-inoculant performance, thereby fostering sustainable agricultural systems capable of mitigating drought-induced yield losses. Through the utilization of cutting-edge technologies and data-driven methodologies, these techniques provide the accurate control of advantageous microorganisms, maximizing their contributions to crop performance and drought resistance. Precision agriculture technologies have enormous potential to revolutionize farming methods, guarantee food security, and encourage sustainable farming in a time of growing water scarcity and climatic variability if they are developed and integrated further.

12.7 MICROBIAL DROUGHT TOLERANCE AND ZERO HUNGER

In order to achieve zero hunger (SDG 2), microbial drought tolerance is essential because it improves crop resilience, stabilizes yields, and permits sustainable agriculture in areas with limited water resources. Particularly in marginal soils, drought stress significantly restricts plant growth by lowering biomass accumulation, photosynthetic efficiency, and germination rates (Kaushal & Wani, 2016). By modifying phytohormones, generating osmolytes, and enhancing root shape, beneficial soil microbes such as endophytes, mycorrhizal fungi, and PGPRs help lessen these impacts (Nadeem et al., 2014). Under drought conditions, crops can survive and produce more because these microbial interactions improve water uptake and nutrient acquisition. This has a special effect on smallholder farmers in dry and semiarid areas, where climate unpredictability makes food insecurity worse (FAO, 2023). In line with the tenets of climate-resilient agriculture, microbial inoculants and biofertilizers provide affordable, scalable solutions that lessen reliance on chemical inputs and restore soil health (UNDP, 2022).

Moreover, microbial techniques support sustainable intensification, which increases output without enlarging agricultural area or harming ecosystems. These bacteria directly assist SDG 2 targets on food availability, nutrition, and agricultural sustainability by enhancing crop performance under stress, hence reducing yield gaps and ensuring a steady supply of food (UN, 2023). Ultimately, leveraging microbial mechanisms for drought resilience offers a crucial nature-based solution for ecosystem restoration, reinforcing global food security, and providing socioeconomic empowerment to marginalized populations.

12.8 FUTURE PERSPECTIVES AND RESEARCH DIRECTIONS

12.8.1 Innovations in Microbiome Analysis and Genomics

Rapid advances in microbiome analysis and genetics have transformed our understanding of and capacity to work with microbial populations. High-throughput sequencing technologies, such as third-generation sequencing platforms and next-generation sequencing, have made it possible to thoroughly profile the microbiomes associated with plants and soil. In example, metagenomics makes it possible to investigate the genetic variety seen in microbial communities without the need to cultivate individual species. This method offers priceless insights into the ecological functions and functional possibilities of microorganisms engaged in drought tolerance (Sharma et al., 2022).

In general, the ongoing development of technology for microbiome analysis and genetics holds great potential for revealing the intricate processes via which helpful microorganisms improve drought resistance. Researchers can find new microbial interactions and traits more quickly by using these methods, opening the door to creative uses in sustainable agriculture.

12.8.2 The Fields of Microbe Engineering and Synthetic Biology

Innovative possibilities for developing and altering microorganisms with improved capacities to promote plant drought tolerance are presented by synthetic biology. Through the modification of microbial genomes, researchers can engineer bespoke strains optimized for specific pedo-climatic conditions and agricultural objectives. This precision engineering approach utilizing techniques such as CRISPR-Cas systems or synthetic biology facilitates the introduction of novel traits, including the biosynthesis of osmoprotectants, enhanced nutrient sequestration, and superior rhizosphere colonization (Sanjuán et al., 2023; Meier et al., 2021).

In this context, developing bacteria to overproduce phytohormones like gibberellins and IAA, which encourage root growth and water uptake, is one of the main uses of synthetic biology. Furthermore, microorganisms can be modified to generate greater amounts of osmolytes, like trehalose and proline, which support cellular defense and osmotic adjustment in drought-stressed environments. By adding synthetic gene circuits, microorganisms may also detect and react to environmental stimuli, guaranteeing that advantageous features are expressed at the right time (Palazzolo et al., 2023; Li et al., 2023).

Additionally, synthetic biology makes it easier to produce artificial microbial communities with distinct interactions and redundant functions. Researchers can improve the overall stability and robustness of microbial therapies by creating consortia with complementary characteristics. For example, a synthetic consortium may consist of drought-tolerant PGPRs, phosphate-solubilizing fungi, and nitrogen-fixing bacteria, each of which performs unique tasks that together improve plant resilience (Tripathi et al., 2020).

But there are drawbacks to using synthetic biology in agriculture as well, such as opposition from the public, regulatory barriers, and the possibility of unforeseen ecological effects. Strict safety evaluations and ethical considerations are necessary to guarantee that modified bacteria do not endanger human health or disturb natural ecosystems. The potential advantages of synthetic biology in creating tailored microbial solutions for drought tolerance are enormous, providing a route to more robust and sustainable agricultural systems in spite of these obstacles.

The effective application of advantageous microbial technology in agriculture depends on cooperation and policy support. Stakeholders may jointly promote the development and adoption of microbial solutions that improve drought tolerance and support resilient and sustainable agricultural systems by encouraging multidisciplinary research, establishing partnerships, and advocating for supporting policies.

12.9 CONCLUSION

The complex interrelationships between plants and the microbiomes that accompany them present a viable path toward improving agricultural resistance to the growing severity of drought. This chapter describes how drought stress significantly affects a number of vital physiological functions in crops, such as photosynthesis, nitrogen metabolism, germination, and, eventually, biomass and yield. However, these negative impacts can be considerably lessened by advantageous plant–microbe interactions, especially those involving endophytes, mycorrhizal fungi, and PGPRs. A wide range of strategies are used by these microbial companions to provide drought resistance. Plant development and stress responses can be influenced by their ability to adjust the production of phytohormones (plant hormones). Bacterial VOCs and modifications to root architecture significantly improve plant communication and resource uptake. Through the synthesis of ACC deaminase, osmolytes, and antioxidant enzymes, microbes directly mitigate drought stress by suppressing ethylene, maintaining cellular turgor, and scavenging ROS. Integrating these microbial solutions into agriculture is essential for developing sustainable, resilient food systems and achieving zero hunger. Continued research into the plant–microbe–environment interaction remains critical to unlocking the full potential of these biological tools for global food security.

REFERENCES

Agrawal, A.; Maharana, J.K.; Patel, A.K. "Microbial inoculants and their potential application in bioremediation." In A. Kumar, L. Shukla, J. Singh, and L. F. R. Ferreira (Eds.), *Microbes Based Approaches for the Management of Hazardous Contaminants*. Wiley: Hoboken, NJ, USA, (2024): 321–344.

Alavilli, H.; Yolcu, S.; Skorupa, M.; Aciksoz, S.B.; Asif, M. "Salt and drought stress-mitigating approaches in sugar beet (Beta vulgaris L.) to improve its performance and yield." *Planta* (2023): 258, 30.

Al-Qubati, A.; Zhang, L.; Pyarali, K. "Climatic drought impacts on key ecosystem services of a low mountain region in Germany." *Environ. Monit. Assess.* (2023): 195, 800.

Arzanesh, M.H.; Alikhani, H.A.; Khavazi, K.; Rahimian, H.A.; Miransari, M. "Wheat (*Triticum aestivum* L.) growth enhancement by *Azospirillum* sp. under drought stress." *World J. Microbiol. Biotechnol.* (2011): 27,197–205.

Ault, T.R. "On the essentials of drought in a changing climate." *Science* (2020): 368, 256–260.

Barnabas, B.; Jager, K.; Feher, A. "The effect of drought and heat stress on reproductive processes in cereals." *Plant Cell Environ.* (2008): 31, 11–38.

Basu, A.; Prasad, P.; Das, S.N.; Kalam, S.; Sayyed, R.Z.; Reddy, M.S.; El Enshasy, H. "Plant Growth Promoting Rhizobacteria (PGPR) as green bioinoculants: Recent developments, constraints, and prospects." *Sustainability* (2021): 13, 1140.

Basu, S.; Ramegowda, V.; Kumar, A.; Pereira, A. "Plant adaptation to drought stress." *F1000Research* (2016): 5, 1554.

Batool, M.; El-Badri, A.M.; Hassan, M.U.; Yang, H.; Wang, C.; Yan, Z.; Kuai, J.; Wang, B.; Zhou, G. "Drought stress in Brassica napus: Effects, tolerance mechanisms, and management strategies." *J. Plant Growth Regul.* (2023): 42, 21–45.

Branco, S.; Schauster, A.; Liao, H.; Ruytinx, J. "Mechanisms of stress tolerance and their effects on the ecology and evolution of mycorrhizal fungi." *New Phytol.* (2022): 235, 2158–2175.

Chieb, M.; Gachomo, E.W. "The role of plant growth promoting rhizobacteria in plant drought stress responses." *BMC Plant Biol.* (2023): 23, 407.

Cohen, I.; Zandalinas, S.I.; Huck, C.; Fritschi, F.B.; Mittler, R. "Meta-analysis of drought and heat stress combination impact on crop yield and yield components." *Physiol. Plant.* (2021): 171, 66–76.

Dehnavi, S.M.; Ebrahimipour, G. "Comparative remediation rate of biostimulation, bioaugmentation, and phytoremediation in hydrocarbon contaminants." *Int. J. Environ. Sci. Technol.* (2022): 19, 11561–11586.

Elferjani, R.; Soolanayakanahally, R. "Canola responses to drought, heat, and combined stress: Shared and specific effects on carbon assimilation, seed yield, and oil composition." *Front. Plant Sci.* (2018): 9, 1224.

Etesami, H. "Potential advantage of rhizosheath microbiome, in contrast to rhizosphere microbiome, to improve drought tolerance in crops." *Rhizosphere* (2021): 20, 100439.

FAO. 2023. *The State of Food Security and Nutrition in the World.* Rome: Food and Agriculture Organization of the United Nations. https://www.fao.org

Gajanayake, B.; Reddy, K.R. "Sweet potato responses to mid- and late-season soil moisture deficits." *Crop Sci.* (2016): 56, 1865–1877.

Garbeva P, Weisskopf L. "Airborne medicine: Bacterial volatiles and their influence on plant health." *New Phytol.* (2020): 226(1), 32–43.

Gontia-Mishra, I.; Sapre, S.; Deshmukh, R.; Sikdar, S.; Tiwari, S. "Microbe-mediated drought tolerance in plants: Current developments and future challenges." In A. Yadav, J. Singh, A. Rastegari, and N. Yadav (Eds.), *Plant Microbiomes for Sustainable Agriculture.* Cham: Springer, (2020): 351–379.

Gontia-Mishra I, Sapre S, Sharma A, Tiwari S. "Amelioration of drought tolerance in wheat by the interaction of plant growth-promoting rhizobacteria." *Plant Biol.* (2016): 18, 992–1000.

Gusain, Y.S.; Singh, U.S.; Sharma, A.K. "Bacterial mediated amelioration of drought stress in drought tolerant and susceptible cultivars of rice (Oryza sativa L.)." *African J. Biotechnol.* (2015): 14, 764–773.

Haddi, K.; Turchen, L.M.; Viteri Jumbo, L.O.; Guedes, R.N.; Pereira, E.J.; Aguiar, R.W.; Oliveira, E.E. "Rethinking biorational insecticides for pest management: Unintended effects and consequences." *Pest. Manag. Sci.* (2020): 76, 2286–2293.

Hameed, A.; Farooq, T.; Hameed, A.; Sheikh, M.A. "Sodium nitroprusside mediated priming memory invokes water-deficit stress acclimation in wheat plants through physio-biochemical alterations." *Plant Physiol. Biochem.* (2021): 160, 329–340.

Hassan, M.J.; Geng, W.; Zeng, W.; Raza, M.A.; Khan, I.; Iqbal, M.Z.; Peng, Y.; Zhu, Y.; Li, Z. "Diethyl aminoethyl hexanoate priming ameliorates seed germination via involvement in hormonal changes, osmotic adjustment, and dehydrins accumulation in white clover under drought stress." *Front. Plant Sci.* (2021): 12, 709187.

Hassan, M.U.; Aamer, M.; Chattha, M.U.; Haiying, T.; Shahzad, B.; Barbanti, L.; Nawaz, M.; Rasheed, A.; Afzal, A.; Liu, Y.; et al. "The critical role of zinc in plants facing the drought stress." *Agriculture* (2020): 10, 396.

Hussain, H.A.; Hussain, S.; Khaliq, A.; Ashraf, U.; Anjum, S.A.; Men, S.; Wang, L. "Chilling and drought stresses in crop plants: Implications, cross talk, and potential management opportunities." *Front. Plant Sci.* (2018): 9, 393.

Hussain, M.B.; Zahir, Z.A.; Asghar, H.N.; Asghar, M. "Exopolysaccharides producing rhizobia ameliorate drought stress in wheat." *Int. J. Agric. Biol.* (2014): 16, 3–13.

Ilyas, M.; Nisar, M.; Khan, N.; Hazrat, A.; Khan, A.H.; Hayat, K.; Fahad, S.; Khan, A.; Ullah, A. "Drought tolerance strategies in plants: A mechanistic approach." *J. Plant Growth Regul.* (2021): 40, 926–944.

Kaushal, M.; Wani, S.P. "Plant-growth-promoting rhizobacteria: Drought stress alleviators to ameliorate crop production in drylands." *Ann. Microbiol.* (2016): 66 (1), 35–42. https://doi.org/10.1007/s13213-015-1112-3

Khan, M.N.; Fu, C.; Li, J.; Tao, Y.; Li, Y.; Hu, J.; Chen, L.; Khan, Z.; Wu, H.; Li, Z. "Seed nanopriming: How do nanomaterials improve seed tolerance to salinity and drought?." *Chemosphere* (2023): 310, 136911.

Khan, N.; Zandi, P.; Ali, S.; Mehmood, A.; Adnan Shahid, M. "Impact of salicylic acid and PGPR on the drought tolerance and phytoremediation potential of *Helianthus annus*." *Front. Microbiol.* (2018): 9, 2507.

Koopman, J.; Dajdok, Z.; Więcław, H.; Marinetto, E.; Grulich, V.; Řepka, R.; Jimenez-Mejas, P. "Global distribution of *Carex buekii* (Cyperaceae) reappraised." *Phytotaxa* (2018): 358, 139–161.

Li, G.; Wang, K.; Qin, Q.; Li, Q.; Mo, F.; Nangia, V.; Liu, Y. "Integrated microbiome and metabolomic analysis reveal responses of rhizosphere bacterial communities and root exudate composition to drought and genotype in rice (Oryza sativa L.)." *Rice* (2023): 16, 19.

Li, S.-L.; Tan, T.-T.; Fan, Y.-F.; Raza, M.A.; Wang, Z.-L.; Wang, B.-B.; Zhang, J.-W.; Tan, X.-M.; Chen, P.; Shafiq, I.; et al. "Responses of leaf stomatal and mesophyll conductance to abiotic stress factors." *J. Integr. Agric.* (2022): 21, 2787–2804.

Manivannan, P.; Rabert, G.A.; Rajasekar, M.; Somasundaram, R. "Drought stress-induced modification on growth and pigments composition in different genotypes of Helianthus annuus L." *Curr. Bot.* (2014): 5, 7–13.

Meier, M.A.; Lopez-Guerrero, M.G.; Guo, M.; Schmer, M.R.; Herr, J.R.; Schnable, J.C.; Alfano, J.R.; Yang, J. "Rhizosphere microbiomes in a historical maize-soybean rotation system respond to host species and nitrogen fertilization at the genus and subgenus levels." *Appl. Environ. Microbiol.* (2021): 87, e03132–20.

Mikiciuk, G.; Sas-Paszt, L.; Mikiciuk, M.; Derkowska, E.; Trzciński, P.; Głuszek, S.; Lisek, A.; Wera-Bryl, S.; Rudnicka, J. "Mycorrhizal frequency, physiological parameters, and yield of strawberry plants inoculated with endomycorrhizal fungi and rhizosphere bacteria." *Mycorrhiza* (2019): 29, 489–501.

Moreno-Galván, A.E.; Cortés-Patiño, S.; Romero-Perdomo, F.; Uribe-Vélez, D.; Bashan, Y.; Bonilla, R.R. "Proline accumulation and glutathione reductase activity induced by drought-tolerant rhizobacteria as potential mechanisms to alleviate drought stress in guinea grass." *Appl. Soil Ecol.* (2020): 147, 103367.

Nadeem, S.M.; Naveed, M.; Farooq, M.; Wahid, A. "Role of PGPR in improving abiotic stress tolerance in plants." In *Plant Microbes Symbiosis: Applied Facets*, edited by Naveen Kumar Arora. New Delhi: Springer, (2014): 1–36. https://doi.org/10.1007/978-81-322-2068-8_1

Oszust, K.; Szpilska, K.; Gryta, A.; Panek, J.; Pylak, M.; Lipa, T.; Frąc, M. "New biotechnological solutions in biocontrol and molecular diagnostics of *Neofabraea* spp. in Apples—A review." *Postharvest Biol. Technol.* (2023): 204, 112442.

Oyedoh, O.P.; Yang, W.; Dhanasekaran, D.; Santoyo, G.; Glick, B.R.; Babalola, O.O. "Rare rhizo-actinomycetes: A new source of agroactive metabolites." *Biotechnol. Adv.* (2023): 67, 108205.

Palazzolo, N.; Peres, D.J.; Bonaccorso, B.; Cancelliere, A. "A probabilistic analysis of drought areal extent using SPEI-based severity-area-frequency curves and reanalysis data." *Water* (2023): 15, 3141.

Patil, R.; Polisgowdar, B.S.; Rathod, S.; Bandumula, N.; Mustac, I.; Srinivasa Reddy, G.V.; Wali, V.; Satishkumar, U.; Rao, S.; Kumar, A.; et al. "Spatiotemporal characterization of drought magnitude, severity, and return period at various time scales in the Hyderabad Karnataka region of India." *Water* (2023): 15, 2483.

Paul, M.J.; Primavesi, L.F.; Jhurreea, D.; Zhang, Y. "Trehalose metabolism and signaling." *Ann. Rev. Plant. Biol.* (2008): 59, 417–441.

Preece, C.; Peñuelas, J. "Rhizodeposition under drought and consequences for soil communities and ecosystem resilience." *Plant Soil* (2016): 409, 1–17.

Salvi, P.; Mahawar, H.; Agarrwal, R.; Kajal; Gautam, V.; Deshmukh, R. "Advancement in the molecular perspective of plant-endophytic interaction to mitigate drought stress in plants." *Front. Microbiol.* (2022): 13, 981355.

Sanjuán, J.; Nápoles, M.C.; Pérez-Mendoza, D.; Lorite, M.J.; Rodríguez-Navarro, D.N. "Microbials for agriculture: Why do they call them biostimulants when they mean probiotics?" *Microorganisms* (2023): 11, 153.

Santoyo, G.; Urtis-Flores, C.A.; Loeza-Lara, P.D.; Orozco-Mosqueda, M.D.C.; Glick, B.R. "Rhizosphere colonization determinants by Plant Growth-Promoting Rhizobacteria (PGPR)." *Biology* (2021): 10, 475.

Sharma, A.; Usha; Gupta, S.; Chaubey, K.K.; Singh, S.V. "An overview of microbial-mediated alleviation of abiotic stress response in plant." In A. Vaishnav, S. Arya, and D.K. Choudhary (Eds.), *Plant Stress Mitigators*. Singapore: Springer Nature Singapore, (2022): 581–596.

Sinclair, T.R.; Jamieson, P.D. "Grain number, wheat yield, and bottling beer: An analysis." *Field Crops Res.* (2006): 98, 60–67.

Six, J. "Plant nutrition for sustainable development and global health." *Plant Soil* (2011): 339, 1–2.

Stephan, C.; Laval-Martin, D.L. "Changes in NAD+ kinase activity during germination of *Phaseolus vulgaris* and *P. acutifolius*, and effects of drought stress." *J. Plant Physiol.* (2000): 157, 65–73.

Sudheer, S.; Bai, R.G.; Usmani, Z.; Sharma, M. "Insights on engineered microbes in sustainable agriculture: Biotechnological developments and future prospects." *Curr. Genom.* (2020): 21, 321–333.

Tripathi, A.N.; Meena, B.R.; Pandey, K.K.; Singh, J. "Microbial bioagents in agriculture: Current status and prospects. In A. Rakshit, H. Singh, A. Singh, U. Singh, and L. Fraceto (Eds.), *New Frontiers in Stress Management for Durable Agriculture*." Singapore: Springer Singapore, (2020): 331–368.

Tripathi, P.; Rabara, R.C.; Reese, R.N.; Miller, M.A.; Rohila, J.S.; Subramanian, S.; Shen, Q.J.; Morandi, D.; Bucking, H.; Shulaev, V.; et al. "A toolbox of genes, proteins, metabolites and promoters for improving drought tolerance in soybean includes the metabolite coumestrol and stomatal development genes." *BMC Genom.* (2016): 17, 102.

UN. 2023. "Goal 2: Zero Hunger." United Nations Sustainable Development Goals. https://sdgs.un.org/goals/goal2

UNDP. 2022. *Microbial Solutions for Climate-Resilient Agriculture*. New York: United Nations Development Programme. https://www.undp.org

Vives-Peris, V.; de Ollas, C.; Gómez-Cadenas, A.; Pérez-Clemente, R.M. "Root exudates: From plant to rhizosphere and beyond." *Plant Cell Rep.* (2020): 39(1), 3–17.

Wei, J.; Xu, L.; Shi, Y.; Cheng, T.; Tan, W.; Zhao, Y.; Li, C.; Yang, X.; Ouyang, L.; Wei, M.; et al. "Transcriptome profile analysis of Indian mustard (Brassica juncea L.) during seed germination reveals the drought stress-induced genes associated with energy, hormone, and phenylpropanoid pathways." *Plant Physiol. Biochem.* (2023): 200, 107750.

Yadav, A.N. "Beneficial plant-microbe interactions for agricultural sustainability." *J. Appl. Biol. Biotechnol.* (2021): 9, 1–4.

Zhang, J.; Mason, A.S.; Wu, J.; Liu, S.; Zhang, X.; Luo, T.; Redden, R.; Batley, J.; Hu, L.; Yan, G. "Identification of putative candidate genes for water stress tolerance in canola (Brassica napus)." *Front. Plant Sci.* (2015): 6, 1058.

Zhang, K.; Khan, Z.; Yu, Q.; Qu, Z.; Liu, J.; Luo, T.; Zhu, K.; Bi, J.; Hu, L.; Luo, L. "Biochar coating is a sustainable and economical approach to promote seed coating technology, seed germination, plant performance, and soil health." *Plants* (2022): 11, 2864.

Zia, R.; Nawaz, M.S.; Siddique, M.J.; Hakim, S.; Imran, A. "Plant survival under drought stress: Implications, adaptive responses, and integrated rhizosphere management strategy for stress mitigation." *Microbiol. Res.* (2021): 242, 126626.

Index

Note: **Bold** page numbers refer to tables; *italic* page numbers refer to figures.

A

ABA. *See* Abscisic acid (ABA)
Abiotic factors, 49, 56, 103, 143–144, 159, 248, 267
Abiotic stress, 1, 4, 119, 151, 152
  boost resistance to, 284
  endophytic bacteria in mitigating, 112–114
  factor, 223
  improving tolerance to, 41
  plant resilience to, 117
  protection against, 49–52
  resistance for plants growing, 251
  tolerance, 110, 248
Abiotic stressors, 57, 103, 104, *109,* 150, 157, 285
Abscisic acid (ABA), 11, 45, 88, 110, 224, 225, 278
Acyl-homoserine lactones (AHLs), 45, 48, 165
Agricultural innovations and applications, 285–286
Agricultural intensification, role of microbiome in sustainable, 259–260
Agricultural microbiome-based practices, 267
Agricultural soils, 6, 219
Agriculture
  industrial models of, 246
  microbial consortia in, 55–56
  microbiome engineering for, 257–258
Agri-policy, integrating microbiome strategies into, 270
Agri-productivity
  ecological strategies on, 137
  sustainable, importance of, 125–126
Agroecosystem, 258, 260, 261, 265, 268
  arid and semiarid, 145
  nutrient-rich, 137
  resilience of, 16, 72, 76, 138
  rice, 154
AHLs. *See* Acyl-homoserine lactones (AHLs)
Air pollution, mitigation of, 76–77
AM. *See* Arbuscular mycorrhiza (AM)
AMF. *See* Arbuscular mycorrhizal fungi (AMF)
1-Aminocyclopropane-1-carboxylate (ACC) deaminase, **3,** 51, 151, 157, 161, 281, 284–285
Ammonia-oxidizing bacteria (AOB), 83, 84
Anammox, 82, 84
Anammoxosome, 84
Antagonism, 12, 26, 193, 200
Antibiosis, 26, 115, 193, 247
Antibiotic production, 191
Anticipated remedial tools, 114–115
Antimicrobial compounds, 26, 46–47, 52
Antioxidant enzymes, 8, 11, 13, 51, 88, 112, 119, 147, 152, 158, 200
Antioxidative activity, 152
Antioxidative enzyme production, 113
AOB. *See* Ammonia-oxidizing bacteria (AOB)
Apoplast, 7, 104
Arbuscular mycorrhiza (AM), 210, 211, 217–218
Arbuscular mycorrhizal fungi (AMF), 72, 75, 76, 114, 153, 188, 192–193, 211–213, *212,* 222, 226, 249, 281
Arbuscules, 192, 193, 210, 211, 220–221, 281
Arid and saline soils, 144–145
Arid and semiarid areas, plant biofertilizers in, 157–158
Arid legume cultures, rhizobia use in, 155
Aspergillus flavus, 13, **15**
Autoinducers, 165
Auxins, 45, 114, 189, 193, 194, 222, 225, 256
Azospirillum brasilense, 83, **89,** 112, 194, 199

B

Bacillus cereus, 9, 77, 113
Bacillus megaterium, **15,** 86, 113, 194
Bacillus subtilis, **15,** 47, 48, 51, 52, 54, 55, 93, 112, 113, 194
Bacillus velezensis, **15,** 47, 55
Bacteria
  beneficial, 46–48
  methylotrophic, 117
  phosphorus-solubilizing, 43
  plant growth–promoting, 160–161
Bacterial communication, 164
Bacterial hormones, 45–46
Bacterial phytohormones, role of, 150–151
Beneficial bacteria, 46–48
Beneficial fungi, 193
Beneficial microbes, 282–283
Berchemia polyphylla var. *leioclada,* 12
BFBFs. *See* Biofilmed biofertilizers (BFBFs)
BGA. *See* Blue-green algae (BGA)
Bio-Bricks, 198
Biochar (BC), 116
Biochar–microbe synergy, 116–117

Biochemical process, 2–3
Biocontrol agents, 200
Biodiversity, soil health and, 260–261
Bioengineering (synthetic biology), 197
Biofertilizers, 14–16, 115, 176, 257
next-generation, 200
Biofilmed biofertilizers (BFBFs), 115
Biofilm formation, 192
Bioinoculants, mycorrhizal, 230, *230*
Biological nitrate elimination, 83
Biological nitrogen fixation (BNF), *5*, 5–6, 82–83, 154, 189
Biological oxygen demand (BOD), 73, 80
Biomass and yield, 280
Bioremediation, microbes-based environmental, 175
Biostimulants
cyanobacteria-based, 116
enhancing crops with, 29
microbial, 116
Biotic stress, protection against, 52–54
Blue-green algae (BGA), 154, 283
BNF. *See* Biological nitrogen fixation (BNF)
BOD. *See* Biological oxygen demand (BOD)
Breeding
microbiome-aware, 117
programs, well-designed, 269
Brevibacterium linens RS16, 15
Burkholderia sp. strain P50, 163

C

Candidatus scalindua, 84
Carbon, 82
cycle and soil organic matter, 81–82
Carboxylates, 7
Catalase (CAT), 192
Catecholates, 7
Cell divisions, 256
Chemical fertilizers
intensive use of, 16
reducing reliance on, 260
Chemical oxygen demand (COD), 73, 80
Chlorophyll precursors, 279
Chrome azurol S (CAS) medium, 8
CKs. *See* Cytokinins (CKs)
Climate change implications, 263
CLPs. *See* Cyclic lipopeptides (CLPs)
CMN. *See* Common mycorrhizal network (CMN)
Co-inoculation, 164
Commensalism, 249
Common mycorrhizal network (CMN), 227
Community engineering, microbial consortia and, 193–194
Complementary organisms, 193
Condensation (C) domain, 44
Convention on Biological Diversity's Nagoya Protocol, 268
Copper, mechanism of uptake of, 221–222
Corchorus capsularis, 8
Cover crops, 201
CRISPR/Cas system, 257–258
Crop
breeding, 269
fields, 49
growth, impact of drought stress on, 278
management, 264
microbiome-based breeding of, 269–270
Cropping system, integration into, 228–229
Crop rotation, 263
and diversification, 200–201
Crop selection and management
community-driven documentation frameworks, 138
plant conservation techniques and practices conducive to microbes, 138
preservation and documentation of local knowledge, 137
traditional knowledge in, 136–137
Crop yield and productivity, impact of interaction between Grasses and microorganisms on, 87–94, **89–92,** *94*
Curcuma longa, 129, 132–133
Curcumin, 132–133
Customized plant microbiome engineering, 117–118
Cyanobacteria (blue-green algae), 114, 154
Cyanobacteria-based biostimulants, 116
Cyclic lipopeptides (CLPs), 47, 48
Cytokinins (CKs), 46, 112–114
Cytoplasm, 7

D

Degraded soils, remediation of, 14–16
DELLA proteins, 234, 235
Denitrification, 83
Detoxification, heavy metal, 11
Dietary legumes, 158
Diffusible signal factor (DSF)-mediated signaling, 165
Disease suppression and immune modulation, 254–255
Diversification, crop rotation and, 200–201
Drought, 277, 278
exposure, 279
Drought-exposed leguminous plants, 156–157
Drought-prone regions, 277
Drought stress, 110, 143, 156–157
plant–microbe interaction under, 280–281
Drought tolerance
beneficial microbes for, 281–283
by soil microbiome, 283–284
Drumstick tree. *See Moringa oleifera*
Dryland agricultural system 163

E

Ebiotic stress mitigation, endophytic bacteria for, 114
EcM. *See* Ectomycorrhiza (EcM)
Eco-FABs. *See* Fabricated ecosystem devices (Eco-FABs)
Ecological significance and ecosystem services, 225–226
  impact on plant biodiversity and community dynamics, 227
  influence on ecosystem functions under changing water availability, 227–228
  soil health and structure improvement, 226–227
Ecology, saline soil microorganism, 145–147
Economic revenue of water, mycorrhizae enhancing, 232–233
Ecosystem
  grassland, 81, 84
  natural, 277
  role of grasses in, 74–76
  terrestrial, 86, 215
Ectoine, 149, 150
Ectomycorrhiza (EcM), 210–211, 213–214, 216, 218, 228
Edible microbiome, 28
Edible plant microbiome, 24
  factors influencing, **26**
Effective consortia, design principles for, 194
  ecological compatibility, 194–195
  functional complementarity, 194
  scalability, 195
  stability and resilience, 195
Effective microorganisms (EM), 114
Emissions, greenhouse gas, 115
Endoglucanase, 8
Endophyte-mediated mechanisms, of plant protection, **15**
Endophytes, 1–3, 6, 8, 248, 282
  applications in sustainable agriculture, 14–16, **15**
  facultative, 105
  micronutrient synthesis induced by, 3–5, *4*
  non-native, 12
  non-rhizobial, 157–158
  opportunistic, 106
  passenger, 106–107
  in plant health, 12
  role in plant health, 12
  utilization of, 14
Endophyte–soil microbe interactions, 12–13
Endophyte–soil microbiome interactions, 12–13
Endophytic *Bacillus* strains, **15**
Endophytic bacteria, 114
  for abiotic stress mitigation in plants, 114
  challenges and limitations, 118
  classification and characteristics, 104–107
  habitat diversity of, 107
  host range and specificity of, 107–108
  industrial applications, 117
  in mitigating abiotic stress, 112–114
  in nitrogen-limited soils., 6
  plant-beneficial, 16
  presence of, 9
  research on, 118–119
Endophytic communities, composition of, 107
Endophytic microbes, 282
Endophytic microorganisms, 1, 2, 16, 17
  enhancing soil health and resilience, *10*
  as soil scavengers, 10, *10*
Endophytic organisms, **109**
Endosphere, *248*
Engineered microbes, 200
Engineering plant microbiomes, by SynComs, 257–258
Enterobactin, 7
Ent-kaurene, 46
Environmental bioremediation, microbes-based, 175
Environmental Protection Agency, 268
Environmental stress, 144
Enzyme glutamine synthetase, 42
Enzyme-linked immunosorbent assay (ELISA), 54
Enzyme production, antioxidative, 113
Enzymes
  antioxidant, 147
  phytase, 43–44
  production of, 6–7
  protective, production of, 192
Epiphytes, 1
Ericoid mycorrhiza (ErM), 210–211, 214–216, 218
ErM. *See* Ericoid mycorrhiza (ErM)
Essential nutrient uptake, 161–162
Ethnobotanical plants, with nutritional and therapeutic value, 127–128
Ethnobotany, 127
  and traditional knowledge, 124
Ethylene (ET), 9, 151
European Food Safety Authority, 268
Exopolysaccharides (EPS), 13, 114
  production of, 285
  secretion of, 2
'Exposome', 24
Extracellular polymeric substances (EPS), *10*
  dula role in soil health and gene transfer, 13–14
Extracellular polysaccharides (EPS), 51, 114–115

F

Fabricated ecosystem devices (Eco-FABs), 196, 198–199
Facultative endophytes, 105
Farming practices, 263

Farming system, management of, 265
Farmlands, mycorrhizal inoculants in, 231
'Farm-to-gut' integrated vision, 31
Fengycin, 47
Fenugreek. *See Trigonella foenum-graecum* (Fenugreek)
Ferrioxamines, 7
Fertilization, rock phosphate (RP), 86
Fertilizers, intensive use of chemical, 16
Finger millet, 78
Fixation
  biological nitrogen, *5,* 5–6
  nitrogen, 42–43
Flavonoids, 52, 53
Food-associated microbes, 24
Food safety imperative, 30
Food security, global, 283
Fruits, diverse microbial world of, 25
Fungal EPS, 13
Fungal species, arbuscular mycorrhizal, 153
Fungi
  arbuscular mycorrhizal, 192–193
  beneficial, 193
  ectomycorrhizal, 216
  mycorrhizal, 36, 81, 85, 281–282
Fur boxes, 44–45

G

Gadgil effect, 214
Gene transfer, EPS role in, 13–14
Genome editing, 257–258
Germination, 278–279
Gibberellic acid (GA), 278
Gibberellins (GA), 112, 114
  biosynthesis, 15–16
Global food security, 283
Global regulators, 45
Global warming, 143
Glomalin-related soil proteins (GRSPs), 224, 226, 228
Glomeromycota, 210
Glomus mosseae, 230
Glycine betaine (GB) treatments, 15
Gram-negative bacteria, 113
Gram-positive bacteria, 282
Grasses, 72, 80
  and microorganism, impact of interaction between, 87–94, **89–92,** 94
  multifunctional roles of, 74
  resilience of, 73
  role of grasses in ecosystems, 74–76
  and soil microorganism, symbiotic relationship between, 87
Grassland ecosystem, 81, 84
Grass–microbe interactions, 93, *94*
  benefits of, 93
Great Oxidation Event, 87
Green energy technology, microbes in industries and, 178–179
Greenhouse gas emissions, 115
Green revolution, 125
GRSPs. *See* Glomalin-related soil proteins (GRSPs)
Gut bacteria, modulation of, 27

H

Habitat diversity of endophytic bacteria, 107
Halanaerobium praevalens, 147
Halophilic bacteria, 147
  adapt to saline environments, 148–150, *149*
  classification of, 147
  plant growth under salt stress, 150–152
Halophilic microbes, to reduce salt stress, 165–166
Halophilic PGPB, 165
"Hartig net", 213
Health-promoting effects, 27
Heat shock proteins (HSPs), 111
Heavy metal detoxification, 11
Heavy metal stress, 111
Herbaspirillum huttiense, **15**
High-temperature stress, 143
Himalayan hot springs, metagenomic work from, 86–87
Holy Basil. *See Ocimum sanctum* (Holy Basil)
Hormones, 147
  bacterial, 45–46
Horseradish tree. *See Moringa oleifera*
Host–microbe interactions, 177
HSPs. *See* Heat shock proteins (HSPs)
Human gut microbiota, impact on, 27
Hunger alleviation, region-specific ethnobotanical crops for, 135–136
Hydroxamates, 7

I

IAA. *See* Indole-3-acetic acid (IAA)
Immune modulation, disease suppression and, *254,* 254–255
Immune system, plant, 52
Indigenous farming systems and practices, 137
Indigenous plant species, integration of, 137
Indole-3-acetamide (IAM) pathway, 45
Indole-3-acetic acid (IAA), 8, 45, 51, 112, 158, 186, 281
Indole-3-acetonitrile pathway (IAN), 45
Indole-3-pyruvic acid pathway (IPA/IPyA), 45
Induced systemic resistance (ISR), 9, 41, 47, 49, 114, 186, 191, 255
Inoculants, microbial, 257
Inorganic phosphate, solubilization of, 220
Integrated pest management (IPM), 201

Intercropping systems, 93
Intervention strategies, targeted and untargeted, 199, *199*
In vivo studies demonstrating medicinal effects, **130–131**
Ion exclusion, 147
IPM. *See* Integrated pest management (IPM)
IPP. *See* Isopentenyl pyrophosphate (IPP)
Iron bioavailability, siderophores for, 44–45
Iron, uptake in rhizosphere, 7–8
Irrigation salinity, 144
Isopentenyl pyrophosphate (IPP), 46
ISR. *See* Induced systemic resistance (ISR)
Iturins, 47

J

Jasmonic acid (JA), 9, 48

K

Klebsiella pneumoniae, 135

L

Legumes, 75, 164
L-glutamate, 149–150
Linear lipopeptides (LLPs), 47
Lipopeptides, 46–47
LLPs. *See* Linear lipopeptides (LLPs)
Long-read sequencing technologies, 196
Lytic enzyme secretion, 191

M

Mesorhizobium ciceri (LL10), 156
Metabolic disorders, management, 133–134
Metabolism, of phytochemicals, 27
Metabolites
  secondary, 9
  therapeutic potential of, 17
Metabolomics, untargeted, 196
Metagenomics, 250
  and microbiome analysis, 196–197, *197*
  shotgun, 196
Metagenomic work, from Himalayan hot springs, 86–87
Metarhizium brunneum, 12
Metatranscriptomics, 196, 250
Methylotrophic bacteria, 117
Microbes
  beneficial, 282–283
  endophytic, 282
  engineered, 200
  food-associated, 24
  in human and veterinary wellness, 176–178, *177*
  in industries and green energy technology, 178–179
  monitoring, precision agriculture and, 287–288
  plant-associated, 251
  role in sustainable environmental management, 175–176
  thermophilic, 117
Microbes-based environmental bioremediation, 175
Microbial activities, 280–281
Microbial assemblages, 247
Microbial biomass (MB), soil, 115
Microbial biostimulants, 116
Microbial communities
  methods for analyzing, 250
  natural and synthetic, 194
Microbial consortia, 34, 35, 46, 50, *50*
  in agriculture, 55–56
  beneficial effects on crop plants, **37–40**
  and community engineering, 193–194
  contribution of molecular tools for studying, 54–55
  direct mechanisms, 35, *36*
  economic and environmental implications
    economic benefits for farmers, 202
    market size and growth, 202
    market trends and economic viability, 202
    reduction in pesticide use, 203
    yield and input savings, 202
  fabricated ecosystem devices (Eco-FABs), 198–199
  indirect mechanisms, 41
  metagenomics and microbiome analysis, 196–197, *197*
  quality control and production challenges, 195
    batch consistency, 196
    formulation stability, 195
    regulatory compliance, 195–196
    strain authentication, 195
  for soil enhancement, 193
  synthetic biology applications, 197–198
Microbial decomposition transforms, 81
Microbial drought tolerance and zero hunger, 288
Microbial functions, 186
Microbial inoculants, 257
Microbial-mediated protection against stress, 49–54
Microbial omics, diverse applications of, *176*
Microbial plant biostimulants, 116
Microbial process, 117
  in phosphorus cycle, 85
Microbial products
  regulatory frameworks and safety concerns, 268
  variability in field performance of, 267
Microbial risk assessment (MRA), 30
Microbial technologies, intellectual property and bioprospecting issues, 268
Microbiology, 174

Microbiome
  diverse industrial applications of, 179–180
  of plants, 250
  role in postharvest health, 26–27
  of soil, 185
Microbiome analysis and genomics, innovations in, 288–289
Microbiome analysis, metagenomics and, 196–197, *197*
Microbiome-aware breeding, 117
Microbiome-based approach, 177, 186
  in agricultural systems, 259
Microbiome-based breeding of crops, 269–270
Microbiome-based research, 178
Microbiome-based strategies, utilization of, 270
Microbiome–based technology, public acceptance and farmer adoption, 269
Microbiome composition, influence of environmental factors on, 262–263, *263*
Microbiome engineering, 118–119
  for agriculture, 257–258
  customized plant, 117–118
  intention of, 257
Microbiome functions, beneficial to plants, 250–251
Microbiome research, AI and big data in, 270
Microbiota, impact on human gut, 27
Micronutrient, 2, 75
  mobilization, 2–3, **3**
  secondary metabolites and absorption of, 8–9
Micronutrient synthesis, induced by endophytes, 3–5, *4*
Microorganism, 83, 143
  ecology, saline soil, 145–147
  endophytic (*See* Endophytic microorganisms)
  nitrogen-cycling, 84
  plant growth promotion by, 34, 255
  role in regulation of nutrient cycles, 80–81
Mitigation
  of soil pollution, 77–79
  of water pollution, 79–80
Mobilization, micronutrient, 2–3, **3**
Moringa oleifera, 128–129
MRA. *See* Microbial risk assessment (MRA)
Mutualism, 249
Mutualistic interaction, 249
Mycorrhiza, 210
  bioinoculants and formulation, 229–231, *230*
  hyphal extension and soil exploration, 220
  in wheat cultivation, 232
Mycorrhizae
  enhancing the economic revenue of water, 232–233
  nutrient acquisition mechanism by, 217–218
Mycorrhizal association, 218
  biotechnological innovations in, 237
  and fungal partners, arbuscular mycorrhizal fungi systems, 192–193
  genes involved in, **234**
  impact of climate change on, 237–238
  regulatory genes involve in, **236**
Mycorrhizal bioinoculants, 230, *230*
Mycorrhizal contribution to water stress resistance, 223–224
  modulation of stomatal conductance and hormonal balance, 225
  role in drought tolerance and hydration efficiency, 224–225
  water absorption and retention, 224
Mycorrhizal fungi, 36, 81, 85, 281–282
  application in sustainable agriculture, 228
    challenges in field applications, 231
    integration into cropping system, 228–229
    mycorrhiza bioinoculants and formulation, 229–231, *230*
    opportunities and advancements, 231–232
  establishment of, 233
  types, 211, *211*
Mycorrhizal inoculants, 200
  in farmlands, 231
Mycorrhizal relationships, 210
Mycorrhizal symbiosis, genetic and molecular understandings of, 233–235, **234**

N

Natural and synthetic microbial communities, 194
Natural ecosystems, 277
Net effect, 84
Next-generation biofertilizers, 200
nifH gene, 5, 42
NIS. *See* Nonribosomal-independent synthesis (NIS)
Nitrate elimination, biological, 83
Nitrogen, 155
  mechanism of uptake of, 221–222
  rhizobia fix, 43
Nitrogenase complex, 42
Nitrogenase enzyme, 5
Nitrogen cycle, 82–84
Nitrogen-cycling microorganisms, 84
Nitrogen fixation, 42–43, 113, 251–252
  biological, *5,* 5–6, 82–83
Nitrogen-fixing bacteria, 189
Nitrogen-fixing symbioses, 144
Nitrogen-limited soils, endophytic bacteria in, 6
Nitrous oxide, 84
Non-native endophytes, 12
Non-rhizobial endophytes, 157–158
Nonribosomal-independent synthesis (NIS), 44
Nonribosomal lipopeptides (NRLPs), 47
Nonribosomal peptide synthetase (NRPS), 44, 46
Novel functional foods, 29–30

NRLPs. *See* Nonribosomal lipopeptides (NRLPs)
NRPS. *See* Nonribosomal peptide synthetase (NRPS)
Nucleic acid–based methodologies, 55
Nutrient acquisition and cycling, 251–252
Nutrient acquisition mechanism, by mycorrhizae, 217–218
  enhancement of phosphorus acquisition, 219–221
  nutrient mobilization and uptake pathways, 218–219
Nutrient cycles, microorganisms role in regulation of, 80–81
Nutrient imbalance, 112
Nutrient mobilization, 113
Nutrient-rich agroecosystem, 137
Nutrients and space, competition for, 192
Nutritional and therapeutic value, ethnobotanical plants with, 127–128, *128*
Nutritional quality, improvement of, 232

O

Obligate endophytes, 105–106
Ocimum sanctum (Holy Basil), 134–135
OM. *See* Orchid mycorrhizae (OM)
Omics Era, 55
'One Health' model, 24
Opportunistic endophytes, 106
Orchid mycorrhizae (OM), 211, 214–215
Organic acids, **3**
  production and secretion of, 43
  production of, 6–7
Organic amendments, 201
Organic carbon (OC) levels, 154
Osmolyte, 147, 285

P

Paraburkholderia phytofirmans PsJN, 14–15
Parasitism, 249
Particulate matter (PM), 76
Passenger endophytes, 106–107
Pathogenesis-related (PR) proteins, 48, 53
PCR. *See* Polymerase chain reaction (PCR)
Peptidyl carrier protein (PCP) domain, 44
Peroxidases, 192
Pesticides, reducing reliance on, 260
Petrobactin, 7
PGP. *See* Plant growth promotion (PGP)
PGPMs. *See* Plant growth–promoting microorganisms (PGPMs)
PGPR. *See* Plant growth–promoting rhizobacteria (PGPR)
Phenazines, 47
Phenolic compounds, 52
Phosphatase, **3**
Phosphate solubilization, 35, 113, 151–152, 251–252
Phosphate-solubilizing microbes, 85
Phosphate-solubilizing microorganisms (PSMs), 85
Phosphorus, 74
Phosphorus acquisition, enhancement of, 219–220
Phosphorus cycle, 85–86
  microbial process in, 85
Phosphorus solubilization, 6–7, 43–44
Phosphorus-solubilizing bacteria, 43
Phosphorus uptake, plant barriers to, 220
Photoreaction rates, 279
Photosynthesis, 77, 279–280
Photosynthetic electron transport components, 280
Phyllosphere, 248, *248*
Phytases, **3,** 43–44
Phytoalexins, 53
Phytodiversity, 127
Phytohormones, 8
  bacterial, role of, 150–151
  production, 161, 189, **256**
Phytopathogens, substances for protection against, 9–10
Phytoremediation, 117
PKS. *See* Polyketide synthases (PKS)
Plant
  abiotic stress and mechanisms, 108–112
  alteration of phytohormones production in, 284
  biofertilizers in arid and semiarid areas, 157–158
  defense responses, induction, 47–49
  endosphere, 248
  genotype and microbial recruitment, feedback loops between, 265–266
  and microorganisms, interaction with, 8–9
Plant-associated microbes, 251
Plant-associated microbiome, 259
Plant-beneficial endophytic bacteria, 16
Plant–endophyte interactions, 4
Plant-friendly microbes. *See* Plant growth-promoting microorganisms (PGPMs)
Plant growth-promoting (PGP), 156
Plant growth–promoting bacteria (PGPB), 144, 160–161
  application of PGPB in salt-stressed soil, 162–163
  consortia for mitigating drought and salinity stress in legumes, 164
Plant growth–promoting microorganisms (PGPMs), 29, 187–188, 255, 278
  direct mechanisms, 189
    biological nitrogen fixation, 189
    organic acid production, 190–191
    phosphate solubilization, 189–190

Plant growth–promoting microorganisms (*cont.*)
phytohormone production, 189
potassium and zinc solubilization, 190
siderophore production, 190
diversity and mechanism, 188–189
indirect mechanism
antibiotic production, 191
biofilm formation, 192
competition for nutrients and space, 192
induced systemic resistance (ISR), 191
lytic enzyme secretion, 191
production of protective enzymes, 192
volatile organic compound production, 192
rhizosphere, 188
Plant growth–promoting rhizobacteria (PGPR), 41, 53, 87, 88, 151, 165–166, 199, 247, 281
elicitors, 54
enhancing plant growth under water stress, *253*
thrives, 27
Plant growth–promoting substances, production of, 112–113
Plant growth promotion (PGP), 2, 255–257
by microorganisms, 34
molecular pathways of, 41–42
Plant health, endophytes role in, 12
Plant hormones, 278
Plant immune system, 52
Plant interactions, 54–55
Plant–microbe interaction, 154–155, *249*
under drought stress, 280–281
effects of salinity and aridity on microbial communities and, 159–160
Plant–microbe relationships, 249
Plant–microbe system, 270
Plant microbiology, 4
Plant microbiome, 247, 249, 270
and agricultural productivity, **264–265**
Plant microbiome-based strategies, 261
Plant–microbiome–environment interactions, 261–262, *262*
Plant microbiota, 143–144
Plant probiotics, enhancing crops with, 29
Plant protection
endophyte-mediated mechanisms of, **15**
molecular pathways of, 46–49
PM. *See* Particulate matter (PM)
Polygalacturonidase, 8
Polyketides, 47
Polyketide synthases (PKS), 46
Polymerase chain reaction (PCR), 5, 7, 8
Potassium, 74
solubilization, 6–7, 35
Precision agriculture and microbe monitoring, 287–288
Precision farming, integration with organic and, 261
Priming, 41
Prominent microbial players, 249
Proso millet *(Panicum miliaceum),* 78
Protection against stress, microbial-mediated, 49–54
Protective enzymes production, 192
Proteins, pathogenesis-related (PR), 48
Pseudomonas, 47
P. chlororaphis, 284
P. savastanoi, 45
PSMs. *See* Phosphate-solubilizing microorganisms (PSMs)

Q

Quorum sensing (QS), 165
and signaling molecules, 113–114

R

Reactive oxygen species (ROS), 52, 104, 152
Relative water content (RWC), 225
Remediation of degraded soils, 14–16
Renewable energy, 117
Rhizobacteria, plant growth–promoting, 199, *199*
Rhizobia
fix nitrogen, 43
soil bacteria, 42
use in arid legume cultures, 155
Rhizobium inoculants, 164
Rhizobium legume symbiosis, 156
Rhizophagus irregularis, 221
Rhizosphere, 154–155, 188, *248*
alteration in root architecture and, 222–223
of cactus plants, 159
feedback loop, 155
iron uptake in, 7–8
Rhizosphere-derived beneficials from soil to plate, 27–28
Rhizosphere–PGPB interactions, cell-to-cell communication, 164–165
Rock phosphate (RP) fertilization, 86
Root exudates, 280
Root morphology modification, 284
ROS. *See* Reactive oxygen species (ROS)
RWC. *See* Relative water content (RWC)

S

Salicylic acid (SA), 9
Saline environments, halophilic bacteria adapt to, 148–150, *149*
Saline soil
application, 158–159
arid and, 144–145
microorganism ecology, 145–147
Salinity stress, 110–111
microbial-mediated mechanisms for plant survival under, *253*

Salt-affected soils, 159, 160
Salts in soil water, 51
Salt stress, 152
halophilic bacteria's function in promoting plant growth under, 150–152
halophilic microbes to reduce, 165–166
Salt-stressed soil, application of PGPB in, 162–163
Salt-tolerant consortia, 41
Salt-tolerant range, halophilic bacterial species with, **148**
Scavengers, endophytic microorganisms as soil, *10*
SDGs. *See* Sustainable Development Goals (SDGs)
Secondary metabolites, 9
Shotgun metagenomics, 196
Siderophore, **3,** 7, 49, 256
for iron bioavailability, 44–45
producers, 152
production, 7–8, **256**
Siderophore biosynthesis, regulation of, 44
Signaling molecules, quorum sensing (QS) and, 113–114
Sodic soils, 145
Soil
agricultural, 6, 219
arid and saline, 144–145
endophytic bacteria in nitrogen-limited, 6
microbes, 81, 82
microbiome of, 185
microorganisms, 260
properties, 264
sodic, 145
type, 263
Soil amendments, beneficial microbes for seed coating and, 286
Soil bacteria, 13
rhizobia, 42
Soil-borne diseases, 155
Soil ecosystem, health of, 185
Soil enhancement, microbial consortia for, 193
Soil exploration, mycorrhiza hyphal extension and, 220
Soil fertility, 76, 186
biochemical processes and impact on, *5,* 5–6
Soil health
and biodiversity, 260–261
EPS role in, 13–14
preserving, 185
Soil microbial biomass (MB), 115
Soil microbiome, mechanism of drought tolerance by, 283–284
Soil nutrients
dynamics, 74
regulation of, 74–76
Soil organic matter (SOM), 72, 75, 247
carbon cycle and, 81–82
Soil pollution, mitigation of, 77–79
Soil salinity and sodicity, 145
Soil scavengers, endophytic microorganisms as, 10, *10*
Soil water, salts in, 51
Solubilization
of phosphate and potassium, 35
phosphorous, 43–44
Soluble salts, 144
SOM. *See* Soil organic matter (SOM)
SRMs. *See* Sulphate-reducing microbes (SRMs)
Staphylococcus hominis, 8
Stenotrophomonas maltophilia, 165
Stress tolerance, *252,* 252–254
Strigolactones (SLs), 235
Sulphate-reducing microbes (SRMs), 86
Sulphur cycle, 86–87
Superoxide dismutase (SOD), 51, 110, 152, 153, 191, 192
Surfactin, 47
Sustainable agricultural intensification, role of microbiome in, 259–260
Sustainable agriculture
microbial processes, mechanisms, and benefits in, **251**
Sustainable agri-productivity, 124–125
importance of, 125–126, *126*
scope of ethnobotany and traditional knowledge, 126–127
Sustainable Development Goals (SDGs), 166
Sustainable environmental management, microbes role in, 175–176
Sustainable intensification, 247
cause-effect overview of industrial agriculture to, *246*
Switchgrass, 84
Synchronization, 2
SynComs. *See* Synthetic microbial communities (SynComs)
Synthetic biology, fields of microbe engineering and, 289
Synthetic microbial communities (SynComs), 29, 116, 194
engineering plant microbiomes by, 257–258

T

Temperature stress, 111
Terrestrial ecosystems, 86, 215
Thermophilic microbes, 117
Tightly bound EPS (TBEPS), 14
"Topsoil foraging" method, 222
Transcription factors (TFs), **236**
Trichoderma, 193
Trichoderma virens SB10, 15

Trigonella foenum-graecum (Fenugreek), 133–134
Tropical tree seedlings, mycorrhizal roots improve the growth of, 233
Tryptamine pathway (TAM), 45
Tryptophan side-chain oxidase (TSO) pathway, 45
Tulsi. *See Ocimum sanctum* (Holy Basil)
Turmeric. *See Curcuma longa*
Typha latifolia litter, 75
Typha species, 75

U

Untargeted ecosystem approaches, 200
 crop rotation and diversification, 200–201
 reduced or no-till farming, 201
Untargeted metabolomics, 196
UV-B radiation stress, 111–112

V

Vegetables, diverse microbial world of, 25
Veterinary wellness, microbes in human and, 176–178, *177*
Vitamin C, 132
Vitamin E, 132
Volatile organic compounds (VOCs), 49, 73, 77, 114, 192, 283, 284

W

Waterlogging and flooding, 111
Water pollution, mitigation of, 79–80
Water stress, 279
 PGPR enhancing plant growth under, **253**
 resistance, mycorrhizal contribution, 223–224
Well-designed breeding programs, 269
WGS. *See* Whole-genome sequencing (WGS)
Wheat cultivation, mycorrhiza in, 232
Whole-genome sequencing (WGS), 193

Z

Zero hunger, microbial drought tolerance and, 288
Zinc, mechanism of uptake of, 221–222